Aerobic and Anaerobic Microbial Treatment of Industrial Wastewater

Industrial wastewater can contain many toxic pollutants as well as varying concentrations of organic and inorganic matter. These pollutants can be carcinogenic, mutagenic, or hardly biodegradable, which could cause serious human health risks and also affect other aquatic and terrestrial biota as well. Biological treatment techniques for industrial wastewater, including aerobic and anaerobic digestion, are known to be environmentally friendly, clean, and generally superior to other physicochemical techniques. *Aerobic and Anaerobic Microbial Treatment of Industrial Wastewater* presents the latest information on multiple bioremediation treatment techniques; summarizes the sources, occurrence, and removal of industrial pollutants; and suggests the most appropriate treatment options for different scenarios.

- Describes the biochemistry of pollutant removal by aerobic and anaerobic digestion.
- Highlights emerging pollutants as well as resource recovery techniques from contaminated industrial wastewater.
- Emphasizes the role of both conventional and innovative novel technologies in aerobic and anaerobic microbial bioremediation of pollutants originated from industrial wastewater.

Aerobic and Anaerobic Microbial Treatment of Industrial Wastewater

Edited by
Maulin P. Shah

CRC Press is an imprint of the
Taylor & Francis Group, an **informa** business

Designed cover image: Shutterstock

First edition published 2025
by CRC Press
2385 NW Executive Center Drive, Suite 320, Boca Raton FL 33431

and by CRC Press
4 Park Square, Milton Park, Abingdon, Oxon, OX14 4RN

CRC Press is an imprint of Taylor & Francis Group, LLC

ISBN: 978-1-032-46358-2 (hbk)
ISBN: 978-1-032-46359-9 (pbk)
ISBN: 978-1-003-38132-7 (ebk)

DOI: 10.1201/9781003381327

Typeset in Times
by KnowledgeWorks Global Ltd.

Contents

Author Biography

Dr. Maulin P. Shah has been an active researcher and scientific writer in his field for over 20 years. He received a B.Sc. degree (1999) in Microbiology from Gujarat University, Godhra (Gujarat), India. He earned his Ph.D. degree (2005) in Environmental Microbiology from Sardar Patel University, Vallabh Vidyanagar (Gujarat), India. His research interests include biological wastewater treatment, environmental microbiology, biodegradation, bioremediation, and phytoremediation of environmental pollutants from industrial wastewaters. He has published more than 350 research papers in national and international journals of repute on various aspects of microbial biodegradation and bioremediation of environmental pollutants. He is the editor of 150 books of international repute. He has edited 25 special issues specifically in industrial wastewater research, microbial remediation, and biorefinery of wastewater treatment areas. He is associated as an Editorial Board Member in 25 highly reputed journals.

Contributors

Bianca Ramos Estevam
Laboratory of Optimization
Design and Advanced Control
Department of Process and Product Development
School of Chemical Engineering
University of Campinas (UNICAMP)
Campinas, SP, Brazil

Sachin Rameshrao Geed
CSIR-North East Institute of Science and Technology
Jorhat, Assam, India

Sougata Ghosh
Department of Microbiology
School of Science
RK University
Rajkot, Gujarat, India

Latonglila Jamir
Department of Environmental Science
Nagaland University
Lumami, Zunheboto, Nagaland, India

Kalpana Katiyar
Department of Biotechnology
Dr. Ambedkar Institute of Technology for Handicapped
Kanpur, Uttar Pradesh, India

C. Nagendranatha Reddy
Department of Biotechnology
Chaitanya Bharathi Institute of Technology
Gandipet, Hyderabad, Telangana, India

Rym Salah-Tazdaït
Department of Environmental Engineering
National Polytechnic School
Algiers, Algeria

Anita Rani Santal
Centre for Biotechnology
M. D. University
Rohtak, Haryana, India

Rajasri Yadavalli
Department of Biotechnology
Chaitanya Bharathi Institute of Technology
Gandipet, Hyderabad, Telangana, India

1 Aerobic and Anaerobic Treatment of Textile Wastewater Using Membrane Technology

Yashika Rani, Jyoti Gulia, Amit Lath, Sushil Kumar, Nater Pal Singh, and Anita Rani Santal

1.1 INTRODUCTION

Rapid urbanization and industrial growth, coupled with the inescapable rise in population, have exacerbated environmental deterioration in developing nations. Acids, dyes, surfactants, metals, and dispersion agents are just a few examples of the harmful compounds eliminated in industrial wastewater discharge (Alsukaibi, 2022). Many processing steps in the textile industry use a significant volume of water in diverse operations. As a result, the textile sector releases a significant amount of effluent into the environment. Sizing, de-sizing, scouring, bleaching, mercerizing, dyeing, and finishing are procedures in the textile processing technique that are linked to the discharge of various pollutants. Notably, chromophore groups such as azo, carbonyl, and nitro are used in cotton dyeing, which primarily increase the unacceptable toxicity of textile wastewater (Chequer et al., 2013). Textile wastewater is often made up of organic matter, suspended solids, dyes, and a variety of additional chemicals, including solvents, detergents, refractory substances, heavy metal ions, and polycyclic aromatic hydrocarbons. Several types of operations lead to different types of wastewater discharges produced by the textile industry (Kishore et al., 2021). Agriculture is the main source of income for developing nations, and about 70% of all water is used in this sector (De Bon et al., 2010). The main target pollutant is color, which, if not eliminated, can be disruptive for receiving water bodies and agricultural use in addition to high organic loads from textile effluents. The dyeing processing stage, which uses water as a typical medium for dye application, is primarily responsible for the colorants that are released in textile outflow (Al-Tohamy et al., 2022). The majority of colors used in dyeing processes are complex organic compounds that are resistant to degradation and released as target pollutants in water bodies (Lellis et al., 2019). Furthermore, biocides, stain repellents, sequestering, anti-creasing, sizing, softening, and wetting chemicals are present in textile wastewater (Sarayu and Sandhya, 2012). Surface and groundwater sources nearby are contaminated when textile wastewater is released into the environment (Sheng et al., 2018). In textile effluent, the majority of the contaminants are biologically resistant to degradation (Mojsov et al., 2016). The ineffectiveness of textile wastewater treatment facilities

DOI: 10.1201/9781003381327-1

is due to the complexity of textile wastewater. Aquatic life is put in peril when textile effluent is released into the ecosystem, which is a notable cause of aesthetic pollution. Also, many dyes are made from aromatic chemicals, some of which are carcinogenic (Sarayu and Sandhya, 2012). While biological techniques are rarely used, physical and chemical treatments are frequently used to clean textile effluent. Advanced oxidation is one of the suggested approaches for treating textile effluent. In advanced oxidation, highly reactive oxidants called hydroxyl radicals are produced to degrade contaminants. Advanced oxidation is an environmentally favorable approach since it produces no solid waste. Nevertheless, because advanced oxidation uses a lot of energy and chemicals, it might not be economically feasible (Asghar et al., 2015). The process of coagulation/flocculation depends on adding chemicals to textile effluent to change the contaminants' physical condition and encourage the flocs' ability to settle (Dotto et al., 2019; Shah Maulin, 2021a). This process generates a lot of sludge, which causes disposal issues. Under decreasing conditions, dye breakdown in textile effluent with azo chromophore can be accomplished in an anaerobic bioreactor (Bidu et al., 2023). Anaerobic conditions cause the –N=N– azo linkages to disintegrate, creating aromatic amines that must be mineralized under aerobic conditions. Moreover, researchers have discovered that many dyes used in textiles are degraded by oxidoreductive enzymes (Saratale et al., 2013; Shah Maulin, 2020). The azo dyes' degradation produces colorless aromatic amines. The microbial community that upholds the reducing conditions for the decolorization of textile dyes is impacted by the toxicity caused by aromatic amines. As a result, an increase in aromatic amines in textile effluents affects the effectiveness of biological color removal. The removal of color may also be inhibited by heavy metals, sulfides, and salts as contaminants (Zafar et al., 2022).

1.2 FIBERS USED IN THE TEXTILE INDUSTRY

From 5000 years ago until now, the textile industry has had the richest history. The textile industry not only fulfills our desire for attractive clothing but also provides job opportunities to many skilled and unskilled people, resulting in the economic growth of the country (Ghaly et al., 2014). Textiles have always been an essential component of human life to provide comfort and sustainability. Different types of fibers like rayon, linen, cotton, and hemp manufactured from cellulose fiber and wool, cashmere, silk, and angora manufactured from protein fiber, and polyester, nylon, spandex, and acetate manufactured from synthetic fiber are used in the textile industry. Some examples of different kinds of fibers and their manufacturing processes are given in Figure 1.1. Natural fiber obtained from natural resources is environment-friendly, but due to some cons like fading of color over time, not being that strong and durable, getting damaged by insects and moths, and being heavy, it is replaced by man-made fiber or synthetic fiber. Humans desire more durable and light garments, and with the advancement of science, many synthetic fibers like polyester, nylon, polyamide, and polyacrylonitrile have come into existence (Table 1.1). Being synthetic, they are prepared in labs and factories by chemical reactions. Generally, petrochemicals are used in the manufacturing of synthetic fibers. These are formed by the polymerization of products obtained from petrochemicals or by reacting some chemicals with cellulose.

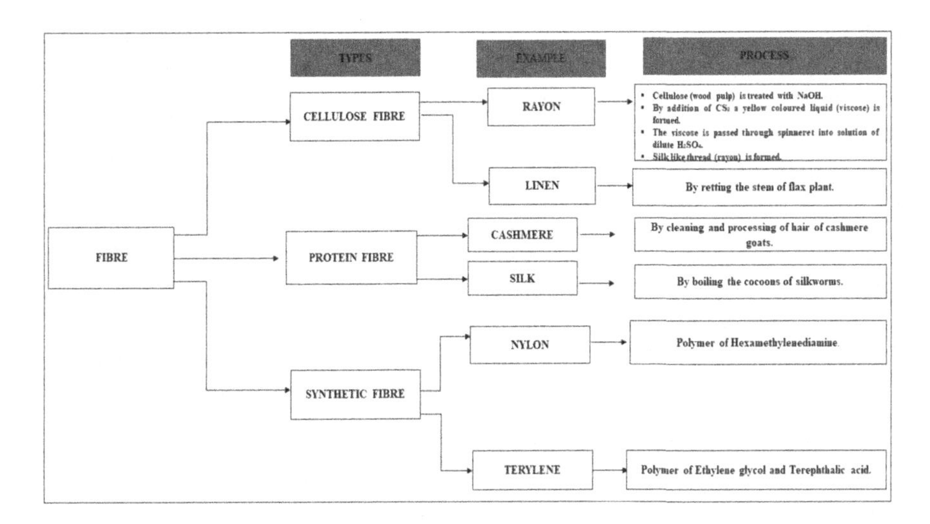

FIGURE 1.1 Types of fibers used in textile industries.

TABLE 1.1
Some Common Synthetic Fibers and Their Monomer

Sr. No.	Synthetic Fibers	Monomer
1.	Nylon-6	Caprolactam
2.	Nylon-6,6	Adipic acid + hexamethylene diamine
3.	Melamine	Formaldehyde + melamine
4.	Bakelite	Phenol + formaldehyde
5.	Teflon	Tetrafluoroethylene
6.	Perspex	Methyl methacrylate
7.	Polythene	Ethylene
8.	Buna-S rubber	Styrene + butadiene
9.	Buna-N rubber	Butadiene + acrylonitrile
10.	Neoprene	Chloroprene

Whether it is manufactured from natural fiber or synthetic fiber, a piece of fabric has to undergo various processes to be made and marketed, and for convenience, all the processes are divided into two groups (Mani et al., 2019):

- Dry process – It includes all the assembling and engineering processes that do not involve water as a raw material or part of them (weaving, willowing, carding, combing, etc.).
- Wet process – It includes those processes that involve water as the raw material, part of the process, or both (bleaching, coloring, washing, etc.)

Dry processes, as mentioned earlier, use very little water, whereas wet processes that involve chemical operations use a significant amount of water. Out of all the processes discussed, a major portion of water is used as a solvent for dying purposes, a medium to transfer dyes, rinsing, washing, bleaching, cleaning, boiling, steaming, etc. The amount of water to be used largely depends on the type of fiber used because natural fiber tends to absorb water very effectively, which leads to increased water usage. On the other hand, the preparation of synthetic fiber requires less amount of water. A portion of the water used in textile industries is discharged as effluent with various types of chemicals, trace elements, intermediate compounds, solid waste, toxic chemicals, acids, alkalis, etc. The amount of effluent and the type of pollutant discharged largely depend on the type of process. Dry processes produce comparatively less effluent. Most of the water pollution caused by textile industries is produced by wet processes. The dye concentration present in the effluent is dangerous for the environment. Dyes are organic compounds that absorb light of the wavelength in the range of visible region (Azanaw et al., 2022; Shah Maulin, 2021a). They get attached to fiber with the help of chemical and physical bonds. Dyes are made up of two parts – chromophore (responsible for imparting color) and functional group (which increases the affinity of the dye for fiber). There are various types of dyes having their unique application method, properties, source, and mechanism of action (Ghaly et al., 2014). One such classification of dyes based on their source is presented in Figure 1.2. Dyes can also be classified based on their nuclear structure – cationic dyes (basic dyes that are used to impart color to nylon, polyester, hemp, wool, cotton, and rayon), anionic dyes (acidic dyes, reactive dyes, and direct dyes that are suitable to dye protein fiber and are also highly water soluble), and nonionic dyes (dispersed dyes) (Nidheesh et al., 2013). Various types of chromophore groups like azo, cyanine, xanthene, nitro, quinone-imine, indigoid, acridine, oxazine, anthraquinone, triarylmethane, phthalein, triphenylmethane dyes, nitroso, and diarylmethane are present in synthetic dyes (Benkhaya et al., 2020). Based on their functional group, choice of color, method of application, type of fiber, and color imparting affinity, dyes are selected.

The dyeing process consumes a large quantity of water, which, after the completion of the dyeing process, is discharged as effluent. Natural dyes and the intermediate

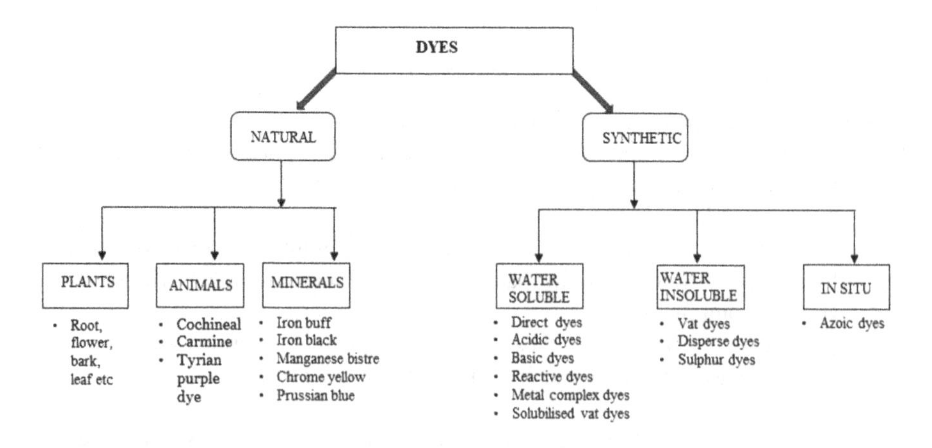

FIGURE 1.2 Classification of dyes based on the sources.

compounds produced from them are not harmful because they are biodegradable. But on the other hand, effluent produced from dyeing the fiber with synthetic dyes contains lots of toxic chemicals, harmful products, ammonium salts, solid impurities, organic acids, hydrosulfite salts, sulfuric acid, biocides, heavy metals, chlorinated organic solvents, etc. (Mullai et al., 2017). As we know, water is one of the basic requirements and the most important thing on which life depends. However, rapid industrialization and more and more reckless usage of chemicals have polluted it.

1.3 CHARACTERISTICS OF TEXTILE WASTEWATER

Textile industries are also a part of such organizations, which have a hand in increasing water pollution. There is the presence of many chemicals, like sulfur, nitrate, chromium compounds, soaps, acids, and heavy metals, in these effluents. Because of that, discharged water becomes highly damaging, and it also has a high pH and temperature (Table 1.2). It also disrupts oxygen transfer (Chowdhary et al., 2020). The water discharged without treatment properly and completely can cause many health risks like cancer, dermatitis, etc. Their discharged effluent has been proven to carry carcinogenic, teratogenic, and mutagenic properties by several researchers. Toxic chemicals present in such water may also end up getting into the food chain. Their dark color due to the presence of chemicals and dyes also resists sunlight passing, hence preventing photosynthesis, which can cause eutrophication. These discharged effluents have a negative impact not only on water but also on the soil. If this polluted

TABLE 1.2
Characteristics of Textile Wastewater

Sr. No.	Parameter	Permissible Range	References
1	Total suspended solids (TSS)	100 mg/L	Sathiyaraj et al. (2017)
2	Total dissolved solids (TDS)	2000 mg/L	Bhatia et al. (2018)
3	pH	6.5–8.5	GilPavas et al. (2020)
4	Temperature	<40°C	Bhatia et al. (2018)
5	Color	100 units	Khan and Malik (2018)
6	Chemical oxygen demand (COD)	250 mg/L	Manikandan et al. (2015)
7	Biological oxygen demand (BOD)	30 mg/L	Manikandan et al. (2015)
8	Oil and Grease	10 mg/L	Patel et al. (2013)
9	Potassium (K)	12 mg/L	David Noel and Rajan (2015)
10	Sodium (Na)	200 mg/L	Roy et al. (2018)
11	Nitrogen	50 mg/L	Patel et al. (2013)
12	Chloride	250 mg/L	Bhuvaneswari (2021)
13	Dyes	Between 10 and 200 mg/L	Doble and Kumar (2005)
14	(Heavy metals)		Mubashar et al. (2020)
	Chromium	1 mg/L	
	Copper	1 mg/L	
	Lead	0.5 mg/L	
	Cadmium	0.1 mg/L	

water reaches agricultural land, it can block the pores of the soil and restrict oxygen transfer and water penetration, leading to plant death. If these pollutants reach sewage, they can also block pipelines, cause foul smells, and generate various pathogens. If it gets reached to river, it will affect the quality of potable water. It also depletes the dissolved oxygen level of water, hence hindering the self-purification process (Khan and Malik, 2014).

1.4 VARIOUS METHODS TO TREAT TEXTILE WASTEWATER

Wastewater produced from the textile industry is of complex composition due to the usage of a large amount of several different chemicals, and treatment of this complex discharge has always been a challenge for engineers and environmentalists. But its treatment is very necessary, not only for the environment but for the sustainability of life too. There have been several existing technologies for the treatment of water. These can be classified as physical, chemical, and biological treatment methods (Figure 1.3).

1.4.1 Physical and Chemical Treatment Methods

These methods include the usage of substances and chemicals to remove pollutants from effluent. Elimination of pollutants in these treatments depends largely on the type of pollutant, type of chemical or substance used, pH of the wastewater, and type of process used (Samer, 2015). Physical and chemical treatment methods include many different types of processes that are used independently or simultaneously with each other. But these processes differ in the type of operation, the principle of mechanism, the efficiency of the substance or chemical used, etc. In coagulation, pollutants are made to clump together by using positive-charge particles like alum.

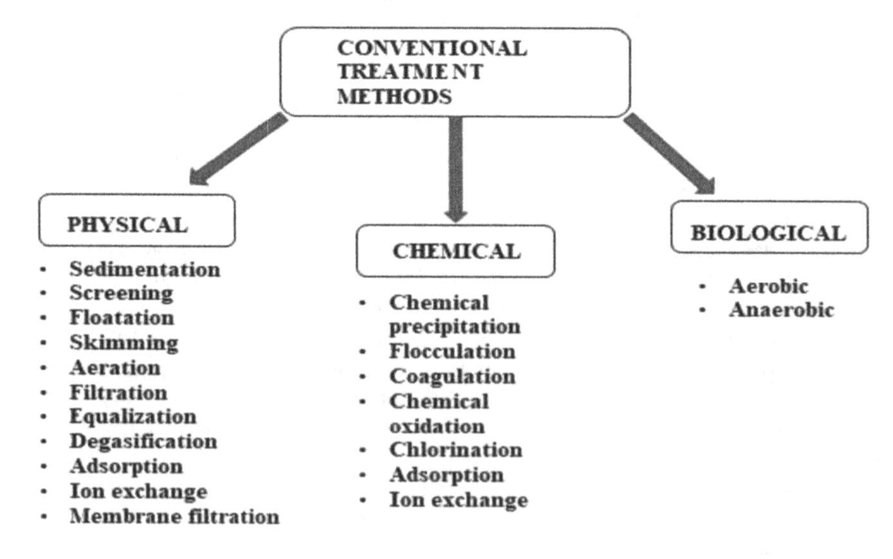

FIGURE 1.3 Various conventional treatment methods.

These positively charged particles neutralize negative charges present on pollutant surfaces. Activated carbon is also used in this process. This process is slightly effective in decolorizing effluent but not very effective against reactive and vat dyes, and it is also very costly. However, their cost-effective substitutes have also been introduced like peat, bentonite, clay, and fly ash (El-taweel et al., 2023). Adsorption is also based on the principle of the interaction of opposite charges. Its cost-effective substitute is treated ginger waste, ground nut shells, charcoal, potato plant waste, etc. But this process is effective only at low concentrations of pollutants and also generates sludge. Filtration involves ultrafiltration, nanofiltration, and reverse osmosis. This process is only effective in treating hydrolyzed dyes. Damaging and fouling of membranes have always been a concern for this process. Distillation is not suitable for the long run and is also time-consuming and has low recyclability (Holkar et al., 2016). Sedimentation eliminates solid pollutants, which are suspended in water under the force of gravity. But this process is incapable of removing dissolved pollutants. Flocculation involves the usage of clay or other particles that have specific surface charges to remove organic particles or other contaminants of low density. Their opposite charges interact with each other and make flocs. Flocculation is generally followed by sedimentation and is dependent on pH, velocity, the concentration of organic matter in waste, the charge on particles, etc. (Teh et al., 2016). Ion exchange uses zeolite, resins, clay, humus, etc. to purify, separate, and decontaminate ions and aqueous solutions. The ion exchange process is effective only for some cationic and anionic dyes. So, it cannot be used against a variety of dyes. The chemical precipitation method of treatment is based on the chemical reactions caused by some precipitating reagents, which convert soluble metals into solid particles. After that, these solid particles are removed by another process such as filtration. The efficiency of chemical precipitation mainly depends on the kind of metal and concentration of the reagent. Generally, hydroxide precipitation uses sodium or calcium hydroxide as a reagent, but this process is not much effective because of the presence of mixed metals in wastewater (Mani et al., 2019). Chemical oxidation is used to remove the desired pollutant from wastewater. The oxidative agent used in this process chemically removes, destroys, or neutralizes the pollutant's harmful effects. Chemical oxidation works by producing radicals, and these radicals oxidize or break bonds present between the chromophore and functional group, resulting in the decolorization of wastewater (Solayman et al., 2023). Chemical oxidation is divided into two types: (a) ozonation and (b) Fenton oxidation. Ozonation uses ozone gas, which produces radicals at high pH. These radicals further decolorize and inhibit the foaming of surfactant. Ozonation also increases the biodegradability of wastewater, so it is effective in water that has nonbiodegradable pollutants in large amounts. This process has the advantage that ozone gas can be used as such, as it is present in gaseous form, so it doesn't increase the volume of effluent and doesn't produce sludge. But the usage of sodium hypochlorite in this process is of concern as it produces carcinogenic amines, and due to the short life of ozone, it needs to be supplied continuously, which is very expensive. The Fenton oxidation process uses hydrogen peroxide as an oxidative agent. Radicals are produced by the interaction of hydrogen peroxide with Fe^{2+} ions at low pH. This method is also used as a pretreatment method for decolorization purposes. But this process produces sludge in large amounts (Meriç et al., 2005).

The physical and chemical processes have advantages as well as limitations. Due to the usage of chemicals and the production of many harmful toxic by-products, alternative methods of treatment are being found. Biological methods are one of the alternative methods used to treat wastewater. Microorganisms have been proven to be very effective in the degradation of various chemicals and harmful compounds. They are also found to be effective in immobilizing, mobilizing, and reducing various dyes (Lavanya et al., 2014). It has been proven that physicochemical methods give better results when combined with biological treatment.

1.5 BIOLOGICAL TREATMENT METHODS

Biological treatment methods use bacteria, fungi, algae, yeast, etc. for the treatment of wastewater in the presence or absence of air. These microbes adapt themselves according to the organic content present in the effluent due to the instinct of survival and start degrading it by using it as a source of energy and nutrients (Ghaly et al., 2014). Biological treatments are done with controlled pH, regular feeding of nutrients, and the interaction of air and oxygen (Periyasamy et al., 2018). The efficiency of a biological treatment depends on the organic contaminants and microbe load in the solution. So, the efficiency of treatment is greatly dependent on the adaptability of microorganisms, and factors like pH, temperature, presence and absence of oxygen, etc. Biological treatment methods are being proven to be cost-effective, eco-friendly, produce less sludge, and produce harmless by-products. Due to the usage of lots of chemicals and dyestuffs in textiles, discharged wastewater is full of organic contaminants. So, the usage of microbes for its treatment is one of the most effective and approachable methods. The selection of microorganisms depends on the contaminant to be treated. Since there are a variety of organic loads in wastewater, the requirement for an optimum environment for their degradation varies. Some contaminants require oxygen for their degradation, so those microbes that can work in the presence of oxygen are selected. On the other hand, some contaminants require an anaerobic environment, so the anaerobes are used for their treatment. Therefore, biological treatment methods are differentiated into two categories: (a) aerobic and (b) anaerobic. Sometimes both types of methods can be used simultaneously.

1.5.1 Aerobic and Anaerobic Methods

Oxidation processes are involved in aerobic methods of textile wastewater degradation. It helps to remove toxic compounds and suspended solids, resulting in a decrease in biological oxygen demand (BOD)/chemical oxygen demand (COD). It involves the usage of microorganisms, which act in the presence of oxygen. Aerobic methods help in decreasing smell and increasing the degradation process. Carbon dioxide is generally released during aerobic degradation (Sachidhanandham and Periyasamy, 2021). Bacterial strains have been used effectively to degrade and decolorize textile effluents containing dyes, with particular attention paid to azo dyes (Pinheiro et al., 2022; Ngo and Tischler, 2022), anthraquinone-based dyes (Jamal et al., 2022), and triphenylmethane-based dyes (Adenan et al., 2022). This is due to the hazardous nature of these dyes and their high consumption rates within the industry. Soil samples contaminated with textile waste have become the primary sources of the bacterial species utilized to decolorize dyes. Due to their prolonged exposure to the harsh environment produced by the textile effluent waste, the bacteria have adapted to them and may be able to degrade the dyes. Potential options for the breakdown and decolorization of various dyes include several distinct bacterial strains. At optimal conditions, most of the bacteria have been proven to exhibit high dye degradation efficiency. Dye degradation and decolorization rely on several different environmental and nutritional parameters (Santal et al., 2022). It's also worth noting that real-world industrial textile effluent is more complex than what can be mimicked, with a wider variety of substances influencing the interaction between the bacteria and the dye molecules. Dye degradation microorganisms should be able to decolorize and degrade the dye to its simplest molecules, making it safe for use in agriculture and the environment. Microorganisms can biodegrade dyes because their cell structures contain enzymes that break down the compounds. The biodegradation of textile dyes has uncovered a wide variety of enzymes, with laccases, lignin peroxidase, and azoreductase enzymes showing the most promise in breaking down synthetic colors of various sorts (Vishani and Shrivastav, 2022). Under anaerobic conditions, azo reductase enzymes break down azo dyes by reducing the –N-N– azo bond to form colorless poisonous aromatic amines (Ikram et al., 2023). Additional aerobic or anaerobic reduction of the amine metabolites yields correspondingly simpler, more eco-friendly metabolic molecules. Three types of azo dyes were decolored and their mineralization was boosted by employing a two-stage sequential Fenton's oxidation followed by aerobic biological treatment. First, Fenton's oxidation was used, and then, for the biological process, aerobic sequential batch reactors (SBRs) were used. The influence of pH on Fenton's oxidation was investigated. Dye decolorization and dearomatization using Fenton's process were shown to be most successful at a pH of 3. Absorbance at 200 nm was used to evaluate the degradation of dye, whereas COD was used to evaluate the degradation of aromatic amines (naphthalene chromophores). In all dyes, Fenton's oxidation technique was able to remove greater than 95% of the original hue. Naphthalene group reductions of 56%, 24.5%, and 80% were reported in RB5, RB13, and AO7 during Fenton's oxidation process; these values increased to 81.34%, 68.73%, and 92% following aerobic treatment (Tantak and Chaudhari, 2006). The *Bacillus megaterium* KY848339.1 totally breaks down the Acid red 337 azo dye into tiny aliphatic chemicals and CO_2. In 24 hours, the bacteria removed 91% of the dye from a 500 mg/L solution of dye. Due to their toxicity, azo dyes are ideally degraded by bacteria, which makes these bacteria even more desirable as dye biodegradation agents (Ewida et al., 2019). Gene sequencing was used to determine that the bacterium used to decolorize Reactive Blue 160 (RB160) was *Bacillus firmus* (Barathi et al., 2020). Therefore, it was concluded that the bacterial species were successful in decolorizing and degrading RB160 into harmless chemicals that were better for the environment than the undegraded dye. The persistent aromatic structure of anthraquinone dyes has traditionally made them resistant to biodegradation by bacteria. Recent research, however, has uncovered potential bacterial agents capable of degrading anthraquinone dyes. These bacterial isolates secrete laccases and peroxidases that efficiently break down anthraquinone dyes. *Hortaea* sp. was isolated from petroleum-contaminated soil, and its degradation of solvent green was examined by Al Farraj et al. (2019). Dye breakdown by bacteria was shown to be quickest within 24 hours when initial concentrations were as low as 10 mg/L. The biodegradation of the dye to simpler, nontoxic chemicals was also attributed to the 1,2-dioxygenase and laccase enzymes. Copper is thought to have increased laccase enzyme activity by being present in the reaction media. Previous research has demonstrated that laccases may oxidize the aromatic rings of phenolic chemicals, thereby degrading anthraquinone dyes into nontoxic metabolites. Bacteria also play a crucial role in the degradation of triphenylmethane dyes in textile wastewater. Isolates of bacteria have enzyme systems that can degrade or biosorb these colors, allowing them to be eliminated. The breakdown efficiency of four common triphenylmethane dyes – malachite green, crystal violet, methyl violet, and cotton blue – was demonstrated by a wide variety of actinobacteria isolated from soil by Adenan et al. (2020). *Bacillus* sp. AS2 was found to be responsible for the degradation of mixed dyes (Santal et al., 2022). Researchers have moved their focus from employing individual bacterial isolates to using bacterial consortia to increase the efficiency with which bacteria degrade dyes. Dye degradation efficiency is improved when a consortium of bacteria is used rather than individual bacterial cultures. This might be because metabolic networks in consortiums are more effective than those in pure single cultures. Using a consortium also has the added benefit of enabling a comprehensive therapeutic approach. Most dyes require a two-step process, beginning with anaerobic degradation and ending with either aerobic or anaerobic reduction. Having two or more bacterial species coexist in the same culture medium is an advantage of using mixed cultures. Therefore, the majority of current efforts are directed at the usage of consortia as opposed to pure single cultures in the management of textile dye. Methyl orange and Congo red were degraded into simple, nontoxic chemicals in a single step by a microaerophilic mixed bacterial consortium, as demonstrated by the work of Dissanayake et al. (2021). Biodegradation tests demonstrated that the consortium could break down and decolorize the dyes by more than 95%. The co-metabolic activities of the bacteria in the consortium were thought to have had a synergistic effect, leading to the successful biodegradation of the dyes. When compared to the 86.80% and 89.50% degradation efficiencies recorded for the pure cultures of *Zobellella taiwanensis* AT 1-3 and *Bacillus pumilus* HKG212, respectively, under similar conditions, a bacterium consortium consisting of these two bacteria was able to effectively decolorize Reactive Green 19 with a degradation efficiency above 97% (Das and Mishra, 2019). To break down azoic dyes, a two-stage anaerobic–aerobic process is often necessary; however, using mixed cultures can circumvent this requirement. Similarly, Mohanty and Kumar (2021) discovered that a bacterial consortium-BP consisting of *Bacillus flexus* TS8 (BF), *Proteus mirabilis* PMS (PM), and *Pseudomonas aeruginosa* NCH (PA) completely degraded the anthraquinone dye Indanthrene Blue RS within 9 hours, while pure cultures of the bacterial strains required an average of 19 hours. The researchers believed that the synergistic effect of the oxidoreductase enzymes in the reaction media was responsible for the rapid breakdown of the dye by the consortium bacteria. Recently, algal-bacterial consortiums have emerged as a cost-effective wastewater treatment technology capable of removing nutrients, heavy metals, minerals, and decolorizing dyes. This is part of an ongoing effort to increase the efficiency of bacterial dye wastewater remediation. Assimilative utilization of chromophores for algal biomass production, the transformation of colored molecules to noncolored compounds via CO_2 and H_2O, and adsorption of chromophores on the algal biomass are the three mechanisms responsible for dye decolorization in algal bioremediation (Li et al., 2022). Researchers have found that algal–bacterial consortiums can create an environment where bacteria can degrade dye wastewater despite fluctuations in temperature, pH, and the concentration of dissolved inorganic compounds. To remove CI Reactive Blue 40 (CI RB 40) from actual textile effluent, a consortium of algal-bacterial probiotic strains (*Pseudomonas putida*, *Chlorella*, and *Lactobacillus plantarum*). According to the results of the study, the partnership was successful in reducing COD and color in the wastewater by 89% and 99%, respectively. Phototoxicity tests verified the consortium's capacity to break down the dye into safe chemicals (Moyo et al., 2022).

1.6 MEMBRANE TECHNOLOGY

Biological treatment and membrane filtration are two halves of the membrane bioreactor's (MBR) hybrid treatment procedure. The MBR has become a popular option because the treated water generates such high quality that it may be reused. The MBR process has many advantages over traditional activated sludge processes, including a smaller physical footprint, less frequent maintenance, a constant final treated water quality regardless of sludge conditions in the bioreactor, and increased removal of nutrients, organic pollutants, and persistent organic pollutants (POPs). However, membrane fouling is the main drawback of the MBR process (Al-Asheh et al., 2021), as it decreases permeate flux or raises transmembrane pressure (TMP) with processing time, which in turn increases operating costs for membrane cleaning and shortens membrane life. The degree of fouling in an MBR depends on both the makeup of the wastewater and the biomass produced there. Multiple types of fouling, including organic fouling, inorganic fouling, particle fouling, biofouling, and combinations thereof, can damage MBRs. High-pressure membranes (nanofiltration and reverse osmosis) are particularly susceptible to inorganic fouling, while biofouling develops only after extensive use. Because of this, this analysis focuses on the mechanism of organic fouling of MBR and potential fouling control measures employed by the textile wastewater treatment industry.

1.6.1 Application of Aerobic and Anaerobic MBR in Textile Wastewater Treatment

1.6.1.1 Aerobic Treatment

The textile industry has made use of aerobic MBR. Several types of membranes have been increasingly used for wastewater treatment. To treat wastewater more efficiently, scientists have developed the MBR method. Complete physical retention of microorganisms is possible with both micro- and ultrafiltration membranes (0.05 to 0.4 μm) in the MBR (Le-Clech et al., 2006). MBR systems have been widely adopted for textile wastewater treatment because of their greater performance compared to traditional activated sludge procedures (Nataraj et al., 2009). According to a study conducted by Schoeberl et al. (2005), between 89% and 94% of COD and 65% and 91% of color were removed from textile wastewater. The MBR was effective at reducing COD by 60–95% and color by 46–98% at 525 nm. Due to the persistent nature of textile dyes in activated sludge systems, they found that the adsorption of dye molecules onto biomass was the primary method of color removal (Brik et al., 2006). The average removal effectiveness of COD and dye by aerobic MBR treatment was 94.8% and 72.9%, respectively. A synthetic azo dye cannot be biodegraded and must be extracted from sludge by first adsorbing it into microbial cells. Therefore, the adsorption of dye onto bacterial cells (activated sludge) could be largely responsible for the high efficiency with which dye is removed in an aerobic reactor (Yun et al., 2006). Huang et al. (2009) studied the efficacy of a 400 L/day submerged hollow-fiber (PVDF, 0.2 lm) MBR for treating dye-containing wastewater from a printing and dyeing facility. COD, BOD, NH_3-N, pH, and A submerged hollow fiber membrane bioreactor made up of Polyvinylidene fluoride (PVDF) with pore size of 0.2μm and capacity of 400L/day was used to treat dyeing wastewater by running it continuously for up to 100 days which in turn removed the COD by at least 90%, NH_3-N by approximately 95% and colour by about 75%. The MBR was run at a range of HRTs, from 6 to 22.5 hours. They discovered that with only 6 hours of HRT, the system was able to remove 80–90% of COD, meeting the requirements for effluent quality in the textile sector. Due to insufficient NH_3-N removal (90–95%) and color removal (60–75%), the effluent could only be classified as second-class wastewater. Water from the denim textile industry had extremely high TDS, color, and conductivity. The feasibility of using a microbial community called "IHK22" in an MBR pilot plant to decolorize reconstituted textile effluent before it is reused. At first, a low biomass concentration of 4 g_{MLVSS} L^{-1} was used in the bioreactor. For dye mass loading rates between 1.25 and 2.5 mg $g_{MLVSS}{}^{-1}$ d^{-1}, decolorization performances remained at very high rates (91–100%) throughout the study period. The decolorization capabilities of the MBR were impacted (80–87%), however, when the dye concentration was increased to 7.5 mg $g_{MLVSS}{}^{-1}$ d^{-1}. Thus, a complete decolorization of the MBR was seen once the biomass content reached 8 g_{MLVSS} L^{-1}. Most of the soluble organic matter in the treated effluent (80–90%) is due to microorganisms and the microfiltration membrane and can be eliminated at concentrations below 320 mg $g_{MLVSS}{}^{-1}$ d^{-1}, regardless of the MBR's operation duration. The results obtained show that the efficacy of dye removal through an MBR can be affected by the dye mass loading rate. Optimal MBR performance for efficient textile wastewater treatment is achieved through the use of a sufficient concentration of biomass (Khouni et al., 2020). In another treatment, maximum nitrification was observed when the biomass concentration was 8–10 gL^{-1} and the SRT was adjusted from short to long, as described by Innocenti et al. (2002). This is probably because the MBR process provides ample time for slow-growing microorganisms to exist at high SRTs. According to research by Huang et al. (2001), which compared the effects of changing SRT on the efficiency of conventional bioreactors and MBRs, the latter maintained a constant 90% removal efficiency regardless of SRT, while the former saw a decrease in COD removal efficiency with decreasing SRT (from 70% to 80% for 5 and 10 days of SRT, respectively).

1.6.1.2 Anaerobic Treatment

For wastewater treatment, anaerobic MBR technology has emerged as a promising alternative to traditional anaerobic systems in recent decades. Potential benefits of anaerobic MBR (AnMBR) include efficient removal of color from wastewater at a reduced energy cost. In one of the studies, the color and COD levels in the permeate were reduced to 70 mg/L and 150 Pt-Co. The GPC analysis showed that only low-molecular-weight organics were present in the permeate after the dynamic cake layer produced on the membrane eliminated pollutants with molecular weights larger than 15 kDa observed in the supernatant. Successful operation of an MBR was achieved at a flux of 4.10.7 LMH with a weekly cleaning cycle (Yurtsever et al., 2020). The anaerobic MBR systems (a) have minimal energy consumption, (b) produce low sludge, and (c) generate valuable methane as an end product. Anaerobic MBRs have been studied extensively for the treatment of textile wastewater, and the results demonstrate that this is an effective technique for dealing with wastewater with a high organic strength and a high particle load. The removal of COD and total suspended solids (TSS) using anaerobic MBR was reported to be very effective by Ivanovic and Leiknes (2012) for textile wastewater, but the removal of TN and TP was observed to be quite ineffective. Jegatheesan et al. (2016) found that the effluent quality from an MBR remained consistent despite variations in organic loading rate (OLR) in the feed. It was also found that, compared to the standard anaerobic method, MBR effluent quality was more consistent. Anaerobic MBRs had lower operational costs than aerobic MBRs, according to a study by Achilli et al. (2011), who compared the two methods for treating municipal wastewater. Because of poor sludge management in the aerobic MBR, this has occurred. When treating urban wastewater, Wen et al. (1999) discovered that the anaerobic MBR had a high tolerance to temperature fluctuations between 12°C and 26°C and that these processes could decrease 88% of COD at 12°C. In their analysis of the treatment of municipal wastewater, Martinez-Sosa et al. (2011) observed that anaerobic externally submerged MBR could reduce COD by 90% at both 35°C and 20°C. In a large variety of influents, microbial species at optimum temperature have efficacy in eliminating contaminants. Results showed that decolorization levels greater than 99% were possible with SAMBRs. Even though azo dye concentrations up to 3.2 gL^{-1} had no discernible effect on decolorization, they did have a profound (up to 80–85%) inhibitory effect on methane generation (Spagni et al., 2012).

1.7 HYBRID PROCESSES COMBINING MBR AND OTHER ADVANCED TREATMENT TECHNOLOGIES

Decolorization and treatment of textile wastewater utilizing MBR and other combinations of physical, chemical, and physicochemical treatment approaches have been the subject of a large number of studies in recent years. To treat textile wastewater and investigate fouling behavior, a unique submerged MBR has been designed. This research compares the effectiveness of three different types of reactors operating at a pilot scale: MBRs, ozonized MBRs, and those combined with photocatalysis. Polyvinylidene difluoride hollow-fiber membrane modules with pores measuring

0.1 m were used in the membrane filtration process. Tungsten oxide, a visible photocatalyst, was made into spongy alginate beads and used in the photocatalytic reactor. The photocatalyst dose has been optimized at 500 mg/L. Integrating ozone with MBR at the maximum dosage of 5 g/h has resulted in a membrane filterability ratio of about 10%. Color and COD elimination efficiencies of 94% and 93% were demonstrated. Optimal ozone dosage (5 g/h) significantly increased biodegradability efficiency from 0.2 to 0.4. A study on reversible and irreversible fouling has been done to understand the fouling nature (Sathya et al., 2019). An important intermediate widely used in the synthesis of anthraquinone dyes, 1-amino-4-bromoanthraquinone-2-sulfonic acid, was removed from wastewater by Qu et al. (2009) using an MBR supplemented with *Sphingomonas xenophobia* QYY and a photochemical (photocatalysis and ozonation) treatment process. This system was run alongside an identical control system, but in the absence of QYY. In a 10 L MBR, they used a hollow-fiber membrane with a 0.2 lm pore size, cycling the device on and off for 8 minutes at a time. The effectiveness of an MBR was compared to that of a traditional activated sludge process in a study by Yang et al. (2020). When compared to other treatment methods, MBR was superior in removing color, COD, and TSS. The average biomass concentrations in the study were 3 gL^{-1} for the conventional activated sludge and 2.3 gL^{-1} for the MBR. Both systems had an initial OLR of 1 kg COD/m^3 and a hydraulic retention duration of 2 days (Yang et al., 2020). Iron carbon microelectrolysis (IC-ME) despite of its complex working mechanism is still very useful because of its economical, environment friendliness and excellent performance in removal of organic contaminants, dyes, acetonitrile etc. The IE process resulted in the release of iron (Fe) ions, which had a noticeable effect on the MBR's ability to remove COD and color. The Fe ions made the sludge flocs more stable, which reduced the buildup of biofouling on the membrane. Simultaneously, the ions may have triggered inorganic fouling. The findings suggest that regulating the concentration of Fe ions is essential for getting the most out of the MBR-IE hybrid system. High levels of COD (90%) and total phosphate (97%) were removed in a study of real textile-dying wastewater conducted by Yan et al. (2012) using a pilot scale (hydraulic capacity of 1.67 m^3) hybrid coagulation (40 mg/L of poly-aluminum chloride)/MBR (PVDF membrane, pore size of 0.2 lm, and surface area of 25 m^2, where membrane flux was maintained at 7 L/m^2h).

1.8 CONCLUSIONS

The effluent water from the textile industry is difficult to treat due to the presence of hazardous chemicals with poor biodegradability. To make textiles more durable against physical, chemical, and biological elements, a wide variety of chemicals and additives are utilized, including dyes, detergents, surfactants, and biocides. The textile sector should produce eco-friendly goods, and all aspects of the textile processing chain will need to be adjusted to eliminate harmful substances to the greatest extent possible. Additionally, we need to forego the use of chemicals and dyes that have negative impacts on both biotic and abiotic components of our ecosystems. In recent decades, many cutting-edge technologies have emerged, one of which is membrane technology, which has shown great promise in the treatment of industrial effluents. Last but not least, water technologists and textile industry specialists need to do more to lower the sector's water footprint. Over the past decade, there have been numerous studies testing the efficacy of MBR technology for textile wastewater. Anaerobic MBR is a promising treatment technology for the removal of high-strength organic loading. The MBR is a promising choice for the treatment of textile wastewater. However, no research contrasting aerobic and anaerobic MBR processes has been conducted for this specific use. Therefore, it is crucial to increase the economic viability of anaerobic MBR technology by developing an efficient and cost-effective biogas recovery (especially methane) method as a future prospect.

REFERENCES

Achilli, A., Marchand, E. A., & Childress, A. E. (2011). A performance evaluation of three membrane bioreactor systems: Aerobic, anaerobic, and attached-growth. *Water Science and Technology*, *63*(12), 2999–3005.

Adenan, N. H., Lim, Y. Y., & Ting, A. S. Y. (2020). Discovering decolorization potential of triphenylmethane dyes by actinobacteria from the soil. *Water, Air, & Soil Pollution*, *231*(12), 560.

Adenan, N. H., Lim, Y. Y., & Ting, A. S. Y. (2022). Removal of triphenylmethane dyes by *Streptomyces bacillaris*: A study on decolorization, enzymatic reactions and toxicity of treated dye solutions. *Journal of Environmental Management*, *318*, 115520.

Al-Asheh, S., Bagheri, M., & Aidan, A. (2021). Membrane bioreactor for wastewater treatment: A review. *Case Studies in Chemical and Environmental Engineering*, *4*, 100109.

Al Farraj, D. A., Elshikh, M. S., Al Khulaifi, M. M., Hadibarata, T., Yuniarto, A., & Syafiuddin, A. (2019). Biotransformation and detoxification of Anthraquinone Dye Green 3 using halophilic *Hortaea* sp. *International Biodeterioration & Biodegradation*, *140*, 72–77.

Alsukaibi, A. K. (2022). Various approaches for the detoxification of toxic dyes in wastewater. *Processes*, *10*(10), 1968.

Al-Tohamy, R., Ali, S. S., Li, F., Okasha, K. M., Mahmoud, Y. A. G., Elsamahy, T., & Sun, J. (2022). A critical review on the treatment of dye-containing wastewater: Ecotoxicological and health concerns of textile dyes and possible remediation approaches for environmental safety. *Ecotoxicology and Environmental Safety*, *231*, 113160.

Asghar, A., Raman, A. A. A., & Daud, W. M. A. W. (2015). Advanced oxidation processes for in-situ production of hydrogen peroxide/hydroxyl radical for textile wastewater treatment: A review. *Journal of Cleaner Production*, *87*, 826–838.

Azanaw, A., Birlie, B., Teshome, B., & Jemberie, M. (2022). Textile effluent treatment methods and eco-friendly resolution of textile wastewater. *Case Studies in Chemical and Environmental Engineering*, 100230.

Barathi, S., Karthik, C., Nadanasabapathi, S., & Padikasan, I. A. (2020). Biodegradation of textile dye reactive blue 160 by *Bacillus firmus* (Bacillaceae: Bacillales) and non-target toxicity screening of their degraded products. *Toxicology Reports*, *7*, 16–22.

Benkhaya, S., M'rabet, S., & El Harfi, A. (2020). A review on classifications, recent synthesis and applications of textile dyes. *Inorganic Chemistry Communications*, *115*, 107891.

Bhatia, D., Sharma, N. R., Kanwar, R., & Singh, J. (2018). Physicochemical assessment of industrial textile effluents of Punjab (India). *Applied Water Science*, *8*(3), 83.

Bhuvaneswari, A. (2021). Identify the indicators for sustainability in textile effluent. *Turkish Journal of Computer and Mathematics Education (TURCOMAT)*, *12*(11), 4968–4971.

Bidu, J. M., Njau, K. N., Rwiza, M., & Van der Bruggen, B. (2023). Textile wastewater treatment in anaerobic reactor: Influence of domestic wastewater as co-substrate in color and COD removal. *South African Journal of Chemical Engineering*, *43*, 112–121.

Brik, M., Schoeberl, P., Chamam, B., Braun, R., & Fuchs, W. (2006). Advanced treatment of textile wastewater towards reuse using a membrane bioreactor. *Process Biochemistry, 41*(8), 1751–1757.

Chequer, F. D., De Oliveira, G. R., Ferraz, E. A., Cardoso, J. C., Zanoni, M. B., & de Oliveira, D. P. (2013). Textile dyes: Dyeing process and environmental impact. *Eco-friendly textile Dyeing and Finishing*, IntechOpen *6*(6), 151–176.

Chowdhary, P., Bharagava, R. N., Mishra, S., & Khan, N. (2020). Role of industries in water scarcity and its adverse effects on environment and human health. In *Environmental concerns and sustainable development: Volume 1: Air, Water and Energy Resources* (pp. 235–256).

Das, A., & Mishra, S. (2019). Complete biodegradation of azo dye in an integrated microbial fuel cell-aerobic system using novel bacterial consortium. *International Journal of Environmental Science and Technology, 16*, 1069–1078.

David Noel, S., & Rajan, M. R. (2015). Phytotoxic effect of dyeing industry effluent on seed germination and early growth of lady's finger. *Journal of Pollution Effects and Control.* 10.4172/2375-4397.1000138

De Bon, H., Parrot, L., & Moustier, P. (2010). Sustainable urban agriculture in developing countries. A review. *Agronomy for Sustainable Development, 30*, 21–32.

Dissanayake, M., Liyanage, N., Herath, C., Rathnayake, S., & Fernando, E. Y. (2021). Mineralization of persistent azo dye pollutants by a microaerophilic tropical lake sediment mixed bacterial consortium. *Environmental Advances, 3*, 100038.

Doble, M., & Kumar, A. (2005). *Biotreatment of industrial effluents.* Elsevier Butterworth-Heinemann, UK.

Dotto, J., Fagundes-Klen, M. R., Veit, M. T., Palacio, S. M., & Bergamasco, R. (2019). Performance of different coagulants in the coagulation/flocculation process of textile wastewater. *Journal of Cleaner Production, 208*, 656–665.

El-taweel, R. M., Mohamed, N., Alrefaey, K. A., Husien, S., Abdel-Aziz, A. B., Salim, A. I., & Radwan, A. G. (2023). A review of coagulation explaining its definition, mechanism, coagulant types, and optimization models; RSM, and ANN. *Current Research in Green and Sustainable Chemistry*, 100358.

Ewida, A. Y., El-Sesy, M. E., & Abou Zeid, A. (2019). Complete degradation of azo dye acid red 337 by *Bacillus megaterium* KY848339. 1 isolated from textile wastewater. *Water Science, 33*(1), 154–161.

Ghaly, A. E., Ananthashankar, R., Alhattab, M. V. V. R., & Ramakrishnan, V. V. (2014). Production, characterization and treatment of textile effluents: A critical review. *Journal of Chemical Engineering and Process Technology, 5*(1), 1–19.

GilPavas, E., Dobrosz-Gómez, I., & Gómez-García, M. (2020). Efficient treatment for textile wastewater through sequential electrocoagulation, electrochemical oxidation and adsorption processes: Optimization and toxicity assessment. *Journal of Electroanalytical Chemistry, 878*, 114578.

Holkar, C. R., Jadhav, A. J., Pinjari, D. V., Mahamuni, N. M., & Pandit, A. B. (2016). A critical review on textile wastewater treatments: Possible approaches. *Journal of Environmental Management, 182*, 351–366.

Huang, R. R., Hoinkis, J., Hu, Q., & Koch, F. (2009). Treatment of dyeing wastewater by hollow fiber membrane biological reactor. *Desalination and Water Treatment, 11*(1–3), 288–293.

Huang, X., Gui, P., & Qian, Y. (2001). Effect of sludge retention time on microbial behaviour in a submerged membrane bioreactor. *Process Biochemistry, 36*(10), 1001–1006.

Ikram, M., Zahoor, M., Naeem, M., Islam, N. U., Shah, A. B., & Shahzad, B. (2023). Bacterial oxidoreductive enzymes as molecular weapons for the degradation and metabolism of the toxic azo dyes in wastewater: A review. *Zeitschrift für Physikalische Chemie, 237*(1–2), 187–209.

Innocenti, L., Bolzonella, D., Pavan, P., & Cecchi, F. (2002). Effect of sludge age on the performance of a membrane bioreactor: Influence on nutrient and metals removal. *Desalination, 146*(1–3), 467–474.

Ivanovic, I., & Leiknes, T. O. (2012). The biofilm membrane bioreactor (BF-MBR)—A review. *Desalination and Water Treatment, 37*(1–3), 288–295.

Jamal, M., Awadasseid, A., & Su, X. (2022). Exploring potential bacterial populations for enhanced anthraquinone dyes biodegradation: A critical review. *Biotechnology Letters, 44*(9), 1011–1025.

Jegatheesan, V., Pramanik, B. K., Chen, J., Navaratna, D., Chang, C. Y., & Shu, L. (2016). Treatment of textile wastewater with membrane bioreactor: A critical review. *Bioresource Technology, 204*, 202–212.

Khan, S., & Malik, A. (2014). Environmental and health effects of textile industry wastewater. *Environmental Deterioration and Human Health: Natural and Anthropogenic Determinants*, 55–71.

Khan, S., & Malik, A. (2018). Toxicity evaluation of textile effluents and role of native soil bacterium in biodegradation of a textile dye. *Environmental Science and Pollution Research, 25*, 4446–4458.

Khouni, I., Louhichi, G., & Ghrabi, A. (2020). Assessing the performances of an aerobic membrane bioreactor for textile wastewater treatment: Influence of dye mass loading rate and biomass concentration. *Process Safety and Environmental Protection, 135*, 364–382.

Kishor, R., Purchase, D., Saratale, G. D., Saratale, R. G., Ferreira, L. F. R., Bilal, M., & Bharagava, R. N. (2021). Ecotoxicological and health concerns of persistent coloring pollutants of textile industry wastewater and treatment approaches for environmental safety. *Journal of Environmental Chemical Engineering, 9*(2), 105012.

Lavanya, C., Rajesh, D., Sunil, C., & Sarita, S. (2014). Degradation of toxic dyes: A review. *International Journal of Current Microbiology and Applied Sciences, 3*(6), 189–199.

Le-Clech, P., Chen, V., & Fane, T. A. (2006). Fouling in membrane bioreactors used in wastewater treatment. *Journal of Membrane Science, 284*(1–2), 17–53.

Lellis, B., Fávaro-Polonio, C. Z., Pamphile, J. A., & Polonio, J. C. (2019). Effects of textile dyes on health and the environment and bioremediation potential of living organisms. *Biotechnology Research and Innovation, 3*(2), 275–290.

Li, X., Jia, Y., Qin, Y., Zhou, M., & Sun, J. (2021). Iron-carbon microelectrolysis for wastewater remediation: Preparation, performance and interaction mechanisms. *Chemosphere, 278*, 130483.

Li, Y., Cao, P., Wang, S., & Xu, X. (2022). Research on the treatment mechanism of anthraquinone dye wastewater by algal-bacterial symbiotic system. *Bioresource Technology, 347*, 126691.

Mani, S., Chowdhary, P., & Bharagava, R. N. (2019). Textile wastewater dyes: Toxicity profile and treatment approaches. *Emerging and Eco-Friendly Approaches for Waste Management*, 219–244.

Manikandan, P., Palanisamy, P. N., Baskar, R., Sivakumar, P., & Sakthisharmila, P. (2015). Physico chemical analysis of textile industrial effluents from Tirupur city, TN, India. *International Journal of Advance Research in Science and Engineering, 4*(2), 93–104.

Martinez-Sosa, D., Helmreich, B., Netter, T., Paris, S., Bischof, F., & Horn, H. (2011). Pilot-scale anaerobic submerged membrane bioreactor (AnSMBR) treating municipal wastewater: The fouling phenomenon and long-term operation. *Water Science and Technology, 64*(9), 1804–1811.

Meriç, S., Selçuk, H., & Belgiorno, V. (2005). Acute toxicity removal in textile finishing wastewater by Fenton's oxidation, ozone and coagulation–flocculation processes. *Water Research, 39*(6), 1147–1153.

Mohanty, S. S., & Kumar, A. (2021). Enhanced degradation of anthraquinone dyes by microbial monoculture and developed consortium through the production of specific enzymes. *Scientific Reports*, *11*(1), 7678.

Mojsov, K. D., Andronikov, D., Janevski, A., Kuzelov, A., & Gaber, S. (2016). The application of enzymes for the removal of dyes from textile effluents. *Advanced Technologies*, *5*(1), 81–86.

Moyo, S., Makhanya, B. P., & Zwane, P. E. (2022). Use of bacterial isolates in the treatment of textile dye wastewater: A review. *Heliyon*, e09632.

Mubashar, M., Naveed, M., Mustafa, A., Ashraf, S., Shehzad Baig, K., Alamri, S., & Kalaji, H. M. (2020). Experimental investigation of *Chlorella vulgaris* and *Enterobacter* sp. MN17 for decolorization and removal of heavy metals from textile wastewater. *Water*, *12*(11), 3034.

Mullai, P., Yogeswari, M. K., Vishali, S., Namboodiri, M. T., Gebrewold, B. D., Rene, E. R., & Pakshirajan, K. (2017). Aerobic treatment of effluents from textile industry. In *Current Developments in Biotechnology and Bioengineering* (pp. 3–34). Elsevier. Netherlands

Nataraj, S. K., Hosamani, K. M., & Aminabhavi, T. M. (2009). Nanofiltration and reverse osmosis thin film composite membrane module for the removal of dye and salts from the simulated mixtures. *Desalination*, *249*(1), 12–17.

Ngo, A. C. R., & Tischler, D. (2022). Microbial degradation of azo dyes: Approaches and prospects for a hazard-free conversion by microorganisms. *International Journal of Environmental Research and Public Health*, *19*(8), 4740.

Nidheesh, P. V., Gandhimathi, R., & Ramesh, S. T. (2013). Degradation of dyes from aqueous solution by Fenton processes: A review. *Environmental Science and Pollution Research*, *20*, 2099–2132.

Patel, S., Rajor, A., Jain, B. P., & Patel, P. (2013). Performance evaluation of effluent treatment plant of textile wet processing industry: A case study of Narol textile cluster, Ahmedabad, Gujarat. *International Journal of Engineering Science and Innovative Technology*, 290–296.

Periyasamy, A. P., Ramamoorthy, S. K., Rwawiire, S., & Zhao, Y. (2018). Sustainable wastewater treatment methods for textile industry. *Sustainable Innovations in Apparel Production*, 21–87.

Pinheiro, L. R. S., Gradíssimo, D. G., Xavier, L. P., & Santos, A. V. (2022). Degradation of azo dyes: Bacterial potential for bioremediation. *Sustainability*, *14*(3), 1510.

Qu, Y. Y., Yang, Q., Zhou, J. T., Gou, M., Xing, L. L., & Ma, F. (2009). Combined MBR with photocatalysis/ozonation for bromoamine acid removal. *Applied Biochemistry and Biotechnology*, *159*, 664–672.

Roy, C., Jahan, M., & Rahman, S. (2018). Characterization and treatment of textile wastewater by aquatic plants (macrophytes) and algae. *European Journal of Sustainable Development Research*, *2*(3), 29.

Sachidhanandham, A., & Periyasamy, A. P. (2021). Environmentally friendly wastewater treatment methods for the textile industry. In *Handbook of nanomaterials and nanocomposites for energy and environmental applications* (pp. 2269–2307). Cham: Springer International Publishing.

Samer, M. (2015). Biological and chemical wastewater treatment processes. *Wastewater Treatment Engineering*, *150*, 212.

Santal, A. R., Rani, R., Kumar, A., Sharma, J. K., & Singh, N. P. (2022). Biodegradation and detoxification of textile dyes using a novel bacterium *bacillus* sp. AS2 for sustainable environmental cleanup. *Biocatalysis and Biotransformation*, 1–15.

Saratale, R. G., Gandhi, S. S., Purankar, M. V., Kurade, M. B., Govindwar, S. P., Oh, S. E., & Saratale, G. D. (2013). Decolorization and detoxification of sulfonated azo dye CI remazol red and textile effluent by isolated *Lysinibacillus* sp. RGS. *Journal of Bioscience and Bioengineering*, *115*(6), 658–667.

Sarayu, K., & Sandhya, S. (2012). Current technologies for biological treatment of textile wastewater–a review. *Applied Biochemistry and Biotechnology*, *167*, 645–661.

Sathiyaraj, G., Ravindran, K. C., & Malik, Z. H. (2017). Physico-chemical characteristics of textile effluent collected from erode, Pallipalayam and Bhavani polluted regions, Tamilnadu, India. *Journal of Ecobiotechnology*, *9*, 1–4.

Sathya, U., Nithya, M., & Balasubramanian, N. (2019). Evaluation of advanced oxidation processes (AOPs) integrated membrane bioreactor (MBR) for the real textile wastewater treatment. *Journal of Environmental Management*, *246*, 768–775.

Schoeberl, P., Brik, M., Braun, R., & Fuchs, W. (2005). Treatment and recycling of textile wastewater—Case study and development of a recycling concept. *Desalination*, *171*(2), 173–183.

Shah Maulin, P. (2020). *Microbial bioremediation & biodegradation*. Springer.

Shah Maulin, P. (2021a). *Removal of emerging contaminants through microbial processes*. Springer.

Shah Maulin, P. (2021b). *Removal of refractory pollutants from wastewater treatment plants*. CRC Press.

Sheng, S., Liu, B., Hou, X., Wu, B., Yao, F., Ding, X., & Huang, L. (2018). Aerobic biodegradation characteristic of different water-soluble Azo Dyes. *International Journal of Environmental Research and Public Health*, *15*(1), 1–11.

Solayman, H. M., Hossen, M. A., Abd Aziz, A., Yahya, N. Y., Hon, L. K., Ching, S. L., & Zoh, K. D. (2023). Performance evaluation of dye wastewater treatment technologies: A review. *Journal of Environmental Chemical Engineering*, 109610.

Spagni, A., Casu, S., & Grilli, S. (2012). Decolorization of textile wastewater in a submerged anaerobic membrane bioreactor. *Bioresource Technology*, *117*, 180–185.

Tantak, N. P., & Chaudhari, S. (2006). Degradation of azo dyes by sequential Fenton's oxidation and aerobic biological treatment. *Journal of Hazardous Materials*, *136*(3), 698–705.

Teh, C. Y., Budiman, P. M., Shak, K. P. Y., & Wu, T. Y. (2016). Recent advancement of coagulation–flocculation and its application in wastewater treatment. *Industrial & Engineering Chemistry Research*, *55*(16), 4363–4389.

Vishani, D. B., & Shrivastav, A. (2022). Enzymatic decolorization and degradation of azo dyes. *Development in Wastewater Treatment Research and Processes*, 419–432.

Wen, C., Huang, X., & Qian, Y. (1999). Domestic wastewater treatment using an anaerobic bioreactor coupled with membrane filtration. *Process Biochemistry*, *35*(3-4), 335–340.

Yan, B., Du, C., Xu, M., & Liao, W. (2012). Decolorization of azo dyes by a salt-tolerant *Staphylococcus cohnii* strain isolated from textile wastewater. *Frontiers of Environmental Science & Engineering*, *6*, 806–814.

Yang, X., López-Grimau, V., Vilaseca, M., & Crespi, M. (2020). Treatment of textile wastewater by CAS, MBR, and MBBR: A comparative study from technical, economic, and environmental perspectives. *Water*, *12*(5), 1306.

Yun, M. A., Yeon, K. M., Park, J. S., Lee, C. H., Chun, J., & Lim, D. J. (2006). Characterization of biofilm structure and its effect on membrane permeability in MBR for dye wastewater treatment. *Water Research*, *40*(1), 45–52.

Yurtsever, A., Sahinkaya, E., & Çınar, Ö (2020). Performance and foulant characteristics of an anaerobic membrane bioreactor treating real textile wastewater. *Journal of Water Process Engineering*, *33*, 101088.

Zafar, S., Bukhari, D. A., & Rehman, A. (2022). Azo dyes degradation by microorganisms-an efficient and sustainable approach. *Saudi Journal of Biological Sciences*, 103437.

2 Aerobic Biological Reactor Systems for the Industrial Wastewater Treatment Processes

Vishal Singh Thakur, Pawan Baghmare, Anuj Bora, Jitendra Singh Verma, and Sachin Rameshrao Geed

2.1 INTRODUCTION

For many years, the biochemical process of biological wastewater treatment has been utilized to treat liquid waste. However, due to the rising volume of effluent discharge and the emergence of new contaminants, wastewater treatment methods are still extensively studied and evaluated worldwide. It is often recommended to combine waste management and wastewater treatment to improve their overall economy and energy efficiency (Narayanan and Narayan, 2019). Aerobic wastewater treatment is a biological process that uses oxygen to break down organic contaminants and other pollutants such as nitrogen and phosphorus. Mechanical aeration equipment like air blowers and compressors supply more oxygen to the wastewater, allowing aerobic microbes to feed on the organic material and produce carbon dioxide or biomass that can be harvested. Typically, aerobic treatment follows anaerobic treatment to ensure complete degradation of industrial effluent, which can then be safely disposed of in compliance with strict environmental standards (Narayanan and Narayan, 2019).

Aerobic treatment methods are widely used in several industries such as the food, beverage, chemical, and municipal sectors. The treated effluent generated by these methods is of superior quality and can be used as an excellent agricultural fertilizer that is both odorless and marketable. Combining aerobic treatment systems with anaerobic treatment guarantees complete nutrient and contaminant removal, allowing for safe wastewater disposal without violating environmental regulations. The efficiency of the degradation process in an aerobic bioreactor largely depends on the size of the bubbles and the mass transfer rate from gas to liquid (Samer, 2015; Naga Vignesh, 2020). It is crucial to maintain ideal oxygen solubility, as the presence of salt can prevent oxygen from solubilizing. Basic aerobic bioreactors can be constructed with oxidation ponds or aerated lagoons, with a revolving disc holding microorganisms in a biofilm for regular mixing. Stirred tank, airlift, and inverse fluidized bed aerobic bioreactors are some of the most commonly used aerobic bioreactors on an

DOI: 10.1201/9781003381327-2

industrial scale. The stirred tank reactor is the most popular aerobic reactor as it draws air from beneath the reactor, while gas turbulence is used to provide mixing in the airlift bioreactor (ALR). Compared to stirred tank reactors, airlift reactors have a higher oxygen transfer coefficient. By feeding air from the bottom, airlift reactors ensure efficient gas transfer and optimal mixing of the contents in the reactor (Naga Vignesh, 2020).

This chapter focuses on the various types of aerobic reactors used for treating industrial wastewater, with an emphasis on the limitations, design characteristics, and performance traits of each bioreactor system. These include moving beds, packed beds, biofilm-fixed bioreactors, fluidized beds, ALRs, and sludge beds/sludge blankets. Additionally, the chapter delves into the usefulness of hybrid biological processes in treating industrial wastewater, advancements in aerobic reactor systems for industrial wastewater treatment, and the challenges and future potential in this field. The chapter also considers the possibility of combining waste treatment and waste utilization using aerobic methods.

2.2 BIOLOGICAL TREATMENT OF INDUSTRIAL WASTEWATER

Secondary treatment is a wastewater treatment process that involves biological treatment with secondary sedimentation. Wastewater is introduced into a well-designed bioreactor, where microorganisms like bacteria, algae, and fungus use organic materials under either aerobic or anaerobic conditions (Grady et al., 2011; Shah Maulin, 2020). The bioreactor offers an ideal environment for microbes to grow and consume dissolved organic materials for energy. As long as the microbes have access to oxygen and food via the settled wastewater, they continue the biological oxidation of soluble organic compounds. Bacteria in the bioreactor's fundamental trophic level are primarily responsible for carrying out the biological process, and the bioconversion of dissolved organic materials into dense bacterial biomass enables the fundamental purification of wastewater (Grady et al., 2011).

After the biological process, the treated wastewater must undergo sedimentation to remove the microbiological biomass. Unlike primary sedimentation, where the sludge comprises fecal particles, in secondary sedimentation, the sludge is made up of bacterial cells. Microorganisms, mainly heterotrophic bacteria, but also fungi, carry out the biological removal of organic debris from the settling effluent. The microbes break down the organic materials through biological oxidation and biosynthesis, resulting in the release of minerals and other byproducts of biological oxidation. Biosynthesis is the process by which colloidal and dissolved organic materials are converted into new cells, producing dense biomass that can be removed by sediment deposition (Gray, 2005).

The term "industrial wastewater" refers to water that is discharged from various industrial operations, such as manufacturing, cleaning, and other commercial activities, and contains dissolved and suspended materials. The toxins present in industrial wastewater depend on the type of plant and industry. For instance, sectors such as mining, steel/iron manufacturing, industrial laundries, power generation plants, food/beverage manufacturing, metal finishing, etc. are some examples of industries that produce wastewater. Industrial wastewater typically contains chemical compounds, toxic metals, oil, pesticides, silt, medicines, and other industrial by-products. The treatment of industrial wastewater is generally challenging as it requires an industry-specific assessment of the setups and specific treatment facilities. To tackle this issue, on-site filter presses are constructed to treat the effluent wastewater (Grady et al., 2011; Shah Maulin, 2021a,b; Ahmed et al., 2022).

Wastewater is a by-product of the production of commercial goods in various industries. Water plays a vital role in the manufacturing process of several businesses, including food and beverage, clothing, paper, and chemicals. In order to convert this wastewater into drinkable water, a biological wastewater treatment process is employed. This process utilizes microbes, protozoa, and other specialized microorganisms that work together to break down organic contaminants and separate them from the solution through a flocculation action. The result is a sludge that is easier to manage, dewater, and dispose of as solid waste. The biological wastewater treatment process can be divided into three primary categories (Chertow and Ehrenfeld, 2012; Shah Maulin, 2021a,b; Chavan et al., 2022).

- *Aerobic:* Aerobic microbial activity necessitates the presence of oxygen for microorganisms to convert organic materials into carbon dioxide and microbial biomass.
- *Anaerobic:* The anaerobic process involves the breakdown of organic matter by bacteria in the absence of oxygen. This often leads to the production of methane, carbon dioxide, or an excess amount of biomass.
- *Anoxic:* Anoxic treatment is a process where bacteria use other substances besides oxygen to grow, such as nitrate, nitrite, sulfate, selenate, and selenite, for the elimination of these compounds.

After removing the suspended solid waste in the first stage of treatment, the wastewater undergoes biological treatment in the secondary stage. This involves the use of naturally occurring bacteria to break down toxins and pollutants found in the wastewater. The difference between aerobic and anaerobic biological treatment is that the former requires oxygen to function. To facilitate aerobic bacterial consumption of organic matter, dissolved oxygen levels of around 2–3 ppm are maintained through mechanical aeration devices (Boyd, 1995; Chavan et al., 2022). The optimal method for maintaining biomass growth and achieving maximum microbial performance depends on the specific properties of the wastewater, such as dissolved oxygen, pH, temperature, nutrients, and environment. Sufficient aeration is crucial for maintaining proper wastewater treatment by ensuring the bacteria always have the necessary levels of dissolved oxygen (di Biase et al., 2019; Ahmed et al., 2022).

Biological wastewater treatment is a long-standing biochemical process that has been used for centuries. As industrial effluent discharge rates continue to rise and the variety of contaminants in effluent streams continues to expand, wastewater treatment techniques are being extensively researched and tested worldwide. Waste usage and wastewater treatment must always go hand in hand. Therefore, in such a scenario, it is necessary to suggest and develop improvements in wastewater handling and treatment systems that demonstrate both their overall economy and energy efficiency (di Biase et al., 2019).

Aerobic wastewater treatment is a biological process that uses oxygen to decompose organic pollutants and other contaminants like nitrogen and phosphorus. To continuously supply oxygen to the wastewater or sewage, a mechanical aeration device like an air blower or compressor is used. Aerobic bacteria break down the organic material in the wastewater, producing carbon dioxide and biomass that can be removed. This method is a steady, efficient, and straightforward process that produces high-quality secondary effluent. The end product is a sludge that is free of odor and can be sold as a premium agricultural fertilizer. By combining aerobic and anaerobic treatment methods, complete nutrient and contamination removal can be achieved, allowing for safe wastewater release without violating stringent environmental regulations (Paździor et al., 2019; Ahmed et al., 2022).

Industrial operations generate substantial amounts of persistent and harmful pollutants, which accumulate in the discharge water of the plant. Discharging such contaminated effluents into waterways can harm aquatic ecosystems and even pose a risk to human health. It is therefore necessary to treat these effluents effectively (Spina et al., 2012). However, the complex aromatic chemical compounds, severe physicochemical properties, and presence of a native bacterial fauna often render textile and pharmaceutical effluents resistant to standard biological treatments. Additionally, businesses must constantly upgrade their products to remain competitive, which impact the industrial process, resulting in extremely complex and diverse wastewaters whose composition may change dramatically over time (Spina et al., 2012).

Biological treatment is an essential component of every wastewater treatment facility that processes wastewater from municipalities, industries, or a combination of both sources and contains soluble organic pollutants. The economic benefits of biological treatment are clear when compared to other treatment methods, such as chemical oxidation and thermal oxidation, in terms of capital expenditure and operational costs. The aerobic activated sludge treatment method has been utilized for well over a century as the go-to biological treatment technique. However, the increasing demand to comply with stricter discharge standards or risk being prohibited from discharging treated wastewater has prompted the implementation of several advanced biological treatment methods in recent years (Mittal, 2011).

Aerobic wastewater treatment systems utilize oxygen-dependent microorganisms, including bacteria and protozoa, to cleanse water of pollutants. This process differs from anaerobic systems, which function without oxygen. Through enhancing the natural microbial breakdown process, these systems are capable of removing industrial wastewater contaminants. The bacteria's biological oxygen demand (BOD) serves as a measure for the organic pollutants that are broken down into smaller molecules. High levels of BOD may result from various contaminants, such as industrial discharges, human waste, or fertilizer runoff, and signal a high amount of biodegradable materials in the wastewater (Mittal, 2011; Englande et al., 2015; Salgot and Folch, 2018).

2.3 LIMITATIONS OF VARIOUS PROCESSES, DESIGN CHARACTERISTICS, AND PERFORMANCE TRAITS OF BIOREACTOR

2.3.1 Limitations of Various Processes

There are various limitations associated with bioreactor processes. Some of the common limitations are (Usman et al., 2021):

- *Mass transfer limitations:* Mass transfer refers to the transfer of molecules across the bioreactor's boundary layers. This process can be limited due to various factors, such as the size of the reactor, the stirring rate, and the viscosity of the medium. These limitations can lead to insufficient nutrient supply, accumulation of waste products, and reduced productivity (Reisman, 1993; Hölker and Lenz, 2005; Kadic and Heindel, 2014).
- *Oxygen limitations:* The availability of oxygen is a critical factor in many bioreactor processes. Oxygen transfer can be limited by factors such as the aeration rate, the viscosity of the medium, and the agitation rate. Oxygen limitations can lead to reduced cell growth, decreased productivity, and altered product quality.
- *pH limitations:* The pH of the bioreactor's medium can impact the growth of microorganisms and affect the product yield and quality. The pH can be affected by factors such as the buffer capacity of the medium, the metabolic activity of the cells, and the presence of inhibitors. Maintaining the desired pH level in the bioreactor can be challenging, and deviations from the optimal range can lead to reduced productivity (Reisman, 1993; Kadic and Heindel, 2014).
- *Sterilization limitations:* The bioreactor must be sterile to prevent contamination and ensure the viability of the cells. Sterilization can be limited by factors such as the size and design of the bioreactor, the type of medium used, and the type of microorganisms being cultured. Inadequate sterilization can lead to contamination and loss of cell culture.
- *Product inhibition limitations:* In some bioreactor processes, the accumulation of the product can inhibit cell growth and reduce productivity. This can be due to factors such as product toxicity or feedback inhibition. Strategies such as continuous harvesting or removal of the product can be used to overcome these limitations.

Table 2.1 gives the limitations of bioreactors for wastewater treatment (Reisman, 1993; Hölker and Lenz, 2005; Kadic and Heindel, 2014).

TABLE 2.1
Applications and Limitations of Bioreactors for Wastewater Treatment

Type of Bioreactor	Application	Limitation	Reference
Stirred tank bioreactor (STBR)	Commonly used for culturing biological agents such as cells, enzymes, and antibodies, as well as for producing various chemicals and substances.	Limitations include the need for bearings and shaft seals, size limitations, foaming, and increased power consumption due to mechanical pressure pumps.	di Biase et al. (2019)
Packed-bed bioreactor	Often used in food, beverage, and nutraceutical synthesis, as well as for waste treatment.	Limitations include unwanted temperature gradients, inadequate temperature control, complicated cleaning, difficulty in replacing catalysts, and negative side effects.	Warnock and Bratch (2005)
Fluidized-bed bioreactor (FBBR)	Used to produce gasoline and other fuels, as well as various chemicals and polymers, including rubber, vinyl chloride, polyethylene, and polypropylene.	Limitations include increased reactor vessel size, pumping requirements and pressure drop, particle entrainment, lack of current understanding, erosion of internal components, and pressure loss scenarios.	Karthikeyan and Kandasamy (2005)
Airlift bioreactor/ fermentor	Used for the production of various substances, including antibiotics, chitinolytic enzymes, exopolysaccharides, gibberellic acid, and lactic acid.	Limitations include the need for more air flow and higher pressures, as well as the need to control agitation through supply air.	Petronela and Gavrilescu (2012)
Activated sludge bioreactor	Employed in the treatment of wastewater and sewage, as well as for the production of biofuels such as biogas and bioethanol.	Limitations include high energy and capital costs.	Stott (2003)

2.3.2 Advantage of Bioreactor

Bioreactors provide a number of advantages over conventional techniques for cultivating microorganisms or cells (Betts and Baganz, 2006).

- *Controlled requirements:* Bioreactors enable accurate control of temperature, pH, oxygen levels, and other environmental factors, which can significantly improve culture growth and productivity.
- *Scalability:* Bioreactors can be scaled up or down according to the desired production volume, making it possible to easily expand or reduce production.

- *Automation:* Fully automated bioreactors allow for precise monitoring and control of the culture without the need for consistent human involvement.
- *Sterility:* Bioreactors can be sterilized, which is important for industrial and medical applications.
- *High productivity:* Bioreactors could even generate a large number of cells, microorganisms, or products per unit of volume, making them cost-effective.
- *Repeatability:* Bioreactors can provide a secure platform for reproducible experiments and manufacturing.
- *Versatility:* Even though bioreactors can be employed to grow a huge variety of microorganisms and cells, they are useful in a variety of fields.

Bioreactors have the ability to mimic the natural growth conditions of microorganisms and to study their behavior under various conditions (Betts and Baganz, 2006).

All bioreactors can be employed to enhance the efficiency, production efficiency, and repeatability of a wide range of industrial, medical, and research processes, enabling the production of high-quality and continuous products (Mitra and Murthy, 2022).

2.3.3 Design Characteristics of Bioreactors

The use of bioreactors for various purposes has a long history that dates back to prehistoric times. Through observation and experience, ancient cultures were able to overcome bioengineering design challenges to produce products like wine and beer. This laid the foundation for biotechnological processes, particularly in food production and processing (Asenjo, 1994). A good bioreactor design should prioritize increased productivity and validation of desired parameters in order to produce consistent, higher-quality products at a lower cost. The bioreactor design and operation are influenced by various factors, including the organism being produced, the specific conditions required for product formation, the product cost, and the scale of production (Asenjo, 1994; Narayanan and Narayan, 2019).

In designing a bioreactor, it is crucial to maintain monoseptic conditions, optimal mixing, and low, uniform shear rates throughout the fermentation process to control and positively influence biochemical processes and avoid contamination. The bioreactor's capital investment and operational costs should also be taken into consideration. Appropriate mass transfer (oxygen), heat transfer, and clearly defined material feeding and pH conditions should be maintained to avoid underdosing or overdosing. For aerobic processes, a proper supply of oxygen and water, particulate suspension, sufficient substrate, salts for nutrition, vitamins, etc. should be ensured (Hutmacher and Singh, 2008; Narayanan and Narayan, 2019).

A good bioreactor design must meet certain requirements, such as sterilization, simple construction and measurement, process control devices, regulating techniques, scale-up, operational flexibility, compatibility with both upstream and downstream processes, antifoaming indicators, and gas evolution product and by-product removal. The working volume of a bioreactor refers to the portion of its total capacity occupied by the medium, microorganisms, and gas bubbles, while the headspace is the remaining portion. Typically, the working volume is between 70 and 80 percent of the total bioreactor volume, determined by how quickly foam forms inside the reactor (Hutmacher and Singh, 2008; Narayanan and Narayan, 2019).

- *Agitator:* An external power drive, an impeller, and the baffles are used to increase mass transfer rates via the bulk liquid and bubble boundary layers while ensuring vigorous mixing. It provides the shear conditions required for bubbles to pop (srmuni.ac.in). The Rushton turbine impeller is used in the majority of microbial fermentation processes.
- *Air delivery:* The system includes a compressor, inlet air sterilization equipment, an air sparger, and an exit air sterilization system to avoid contamination.
- *Foam control:* The system includes a compressor, inlet air sterilization process, air sparger, and exit air sterilization system to prevent cross-contamination.
- *Temperature control:* The system includes a heat transfer system and temperature probes (jacket and coil). Electric heaters and steam produced by boilers provide heating, while cooling is accomplished by cooling towers or refrigerants like ammonia.
- *pH:* When diluted in the medium, neutralizing agents—which are used to adjust pH—should not be harmful or corrosive to cells. Small-scale bioreactors frequently use sodium carbonate.
- *Sampling ports:* These are employed for both sample collection and the injection of nutrients, water, salts, etc. into bioreactors.
- *Cleaning and sterilization:* This system is crucial for preventing contamination. The ideal method for cost-effective and extensive equipment sterilization is thermal sterilization using steam. In general, chemical sterilization is chosen for equipment that is sensitive to heat. Sterilization is accomplished using radiation, including x-rays for liquids and UV for surfaces, as well as membrane filters with consistent mycelium and depth filters with glass wool.
- *Charging and emptying:* In the bioreactor, lines are utilized for reactant input and product outflow (Singh et al., 2014).

Designing a bioreactor requires the expertise of mechanical, electrical, and bioprocess engineers. The design should ensure that the bioprocesses are compliant with regulations. Recently, new processes have been developed that use isolated enzymes instead of whole cells to perform chemical changes. This approach has advantages over traditional methods because it eliminates the need to cater to the specific requirements of living cells. However, the conditions for expressing the catalytic activity of enzymes must be identified, and the use of isolated enzymes can be expensive and time-consuming. Additionally, the removal of the enzyme from the product after the bioreaction is complete can be challenging. To address these issues, immobilized enzyme technology has been developed. In this method, the substrate solution is passed through an enzyme immobilized in a bed or tubular, and the product is continuously collected as bioreactor effluent. The design and operation of an immobilized system are similar to heterogeneous catalysis processes. Product recovery is possible with lower separation expenses than in homogeneous systems. Various immobilized enzyme systems have been used to investigate gas–liquid–solid contacting bioreactors. The enzyme immobilization process can take several forms and is dependent on the enzyme's properties, the substrate, and the bioprocess fluid. Enzymes can be supported by a mesh-type or traditional mass-transfer structure or incorporated in a film, gel, or silica-derived system, high-permeable ion-exchange resins, or other polymeric supports. One example of a bioreactor that employs this technology is the trickle bed bioreactor (Kaur and Sharma, 2021).

Include features that allow process control over reasonable ranges of process variables.

- The operation must be reliable.
- The operation must be contamination-free.
- The traditional design consists of open vessels that are cylindrical or rectangular in shape and made of wood or stone.
- In order to prevent contamination, closed systems are now commonly used for fermentation.
- To allow for repeated sterilization and cleaning, the fermenter should be constructed using non-toxic and corrosion-resistant materials.
- Fermentation vessels with a capacity of a few liters (small sizes) are typically constructed using glass and/or stainless steel materials.
- Stainless steel is commonly used for pilot-scale and many production vessels, as it has polished internal surfaces. However, to minimize costs, very large fermenters are often constructed from mild steel that is lined with glass or plastic.
- In order to maintain aseptic operation, it is essential for all pipelines used for transporting air, inoculation, and nutrients for fermentation to be sterilizable, usually through the use of steam (Jagani et al., 2010).

2.3.4 Performance Traits of Bioreactors

Aerobic bioreactors are used in a variety of industries to treat wastewater and organic waste materials. These bioreactors rely on microorganisms that use oxygen to break down organic compounds, producing carbon dioxide and water as byproducts. The performance of aerobic bioreactors is influenced by several key factors, including (Reisman, 1993; Asenjo, 1994; Kadic and Heindel, 2014):

Oxygen transfer rate: The rate at which oxygen is transferred into the bioreactor is critical to the performance of aerobic bioreactors. Oxygen is required for the microorganisms to carry out their metabolic processes, so an adequate supply of oxygen is necessary to maintain a healthy microbial population.

Mixing: Effective mixing is also essential for the performance of aerobic bioreactors. It ensures that the microorganisms are evenly distributed throughout the bioreactor and have access to nutrients and oxygen.

Nutrient supply: The availability of nutrients, such as nitrogen and phosphorus, is critical for the growth and activity of the microorganisms. A balanced supply of nutrients is necessary to maintain a healthy microbial population and to prevent the accumulation of undesirable compounds.

pH control: The pH level of the bioreactor is another critical factor that affects the performance of aerobic bioreactors. The pH level should be maintained within a specific range to ensure optimal growth and activity of the microorganisms.

Temperature control: The temperature of the bioreactor is another important factor that affects the performance of aerobic bioreactors. The temperature should be maintained within a specific range to ensure optimal microbial activity.

Overall, the performance of aerobic bioreactors depends on maintaining a healthy microbial population, which requires a balance of oxygen, nutrients, mixing, pH, and temperature. Proper control of these factors is essential to achieve optimal treatment efficiency and prevent any undesirable outcomes (Reisman, 1993; Asenjo, 1994; Kadic and Heindel, 2014).

2.4 AEROBIC BIOREACTOR SYSTEMS FOR INDUSTRIAL WASTEWATER

The presence of oxygen in wastewater plays an important role as an electron acceptor, resulting in higher energy outputs and the formation of essential mud due to the growth of aerobic bacteria. However, the use of oxygen in wastewater treatment may be limited due to the poor solubility of oxygen in water. Since the early 1900s, activated sludge has been a popular biological treatment method widely used in municipal and industrial applications. It was first developed in England and involves introducing wastewater into an aeration tank where suspended microbes help in the aeration process. The organic material is broken down and consumed, resulting in the formation of biological solids that flocculate into larger clumps or flocs. These flocs are then separated from the wastewater through sedimentation in a settling tank. The suspended solids are controlled by recycling the settled sediments into the aeration tank, while excess particles are disposed of as sludge. While activated sludge treatment systems require more area and produce large volumes of sludge compared to other options, the capital and maintenance costs are typically lower (Chan et al., 2009; Goli et al., 2019; Remmas, 2022).

Aerobic reactors support a variety of significant reactions that utilize different electron donors and oxygen as the terminal electron acceptor. These reactions include the oxidation of organic matter, which lowers BOD; the oxidation of ammonia to nitrate, which lowers ammonia concentration; and the luxury absorption of phosphate with polyphosphate synthesis, which lowers effluent phosphate concentrations (Chan et al., 2009; Goli et al., 2019; Remmas, 2022).

There are several technologies available for aerobic wastewater and sewage treatment, including:

Conventional activated sludge: Conventional activated sludge treatment is a widely used method in various industrial settings for treating wastewater. The process involves introducing the wastewater into an aeration tank, where it is mixed with activated sludge, which is a mixture of microorganisms that help in breaking down the organic matter present in the wastewater. The aeration tank is aerated to provide sufficient oxygen to the microorganisms for the oxidation process to take place. The microorganisms consume the organic matter, which leads to the formation of biological solids or flocs. The flocs are separated from the wastewater in the settling tank through sedimentation. The settled solids are returned to the aeration tank for further treatment, while the excess solids are removed as sludge. Conventional activated sludge treatment is an efficient process for the removal of organic matter, nitrogen, and phosphorus from wastewater. However, it has some limitations, including the high costs associated with sludge handling and disposal. Moreover, the process requires a significant amount of energy to maintain the aeration and mixing in the aeration tank (Valderrama et al., 2012; Thongsai et al., 2022).

Membrane bioreactor: Membrane bioreactors (MBRs) are a type of wastewater treatment technology that combines a biological process (such as activated sludge) with membrane filtration. MBRs use membranes as a physical barrier to separate the mixed liquor from the treated water, allowing for a high level of effluent quality. In MBRs, the biological process occurs in a separate tank or basin, which is aerated to promote the growth of microorganisms that break down organic matter in wastewater (Valderrama et al., 2012). The mixed liquor from this tank is then passed through a membrane module, which separates the treated water from the mixed liquor. The membrane module is typically made up of hollow fiber or flat sheet membranes and can remove suspended solids, bacteria, viruses, and other contaminants from the treated water. One advantage of MBRs is that they produce high-quality effluent that meets strict water reuse standards. MBRs also have a small footprint, making them suitable for use in areas where space is limited. Additionally, MBRs have a low sludge production rate, which reduces the costs associated with sludge disposal. However, MBRs can be expensive to operate and maintain, as the membranes require periodic cleaning and replacement. MBRs also require a constant supply of energy to power the aeration system and the membrane filtration process. Despite these challenges, MBRs are becoming an increasingly popular choice for industrial and municipal wastewater treatment (Deowan et al., 2016).

2.5 TYPES OF AEROBIC BIOREACTORS

2.5.1 Moving-Bed Bioreactor

The moving-bed bioreactor (MBBR) is a biological wastewater treatment technology that utilizes suspended plastic biofilm carriers as a substrate for the growth of bacteria. These carriers are free-floating and kept in motion by aeration in an aerated reactor tank. The bacteria grow on the surface of the carriers and remove organic matter and nutrients from the wastewater. MBBR technology is commonly used in industries as a secondary or tertiary treatment process after primary treatment has been completed (Bassin et al., 2017). The MBBR process is efficient in removing organic matter, nitrogen, and phosphorus, making it a popular choice for treating municipal and industrial wastewater. One advantage of the MBBR process is its ability to handle varying flow rates and pollutant loadings, making it a flexible and adaptable technology. It also has a small footprint compared to other treatment technologies, making it ideal for industries where space is limited. However, the MBBR process requires a constant supply of oxygen and a continuous supply of plastic biofilm carriers, which can increase operating costs. Additionally, the system can be susceptible to clogging, and the plastic carriers can break down over time, potentially leading to environmental concerns (Bassin et al., 2017). The key feature of MBBR technology is the use of specially designed carrier elements that support the growth of biomass, which moves freely inside the reactor due to mechanical mixing, liquid recirculation, or aeration (as shown in Figure 2.1). The goal of the MBBR design is to create a continuously operating biofilm reactor with low head loss, a high specific biofilm surface area, and a low risk of clogging. The carrier particles on which the biofilm grows are small and can be moved with water inside the reactor, increasing the surface area for growth. This movement is typically achieved through vigorous stirring in anaerobic reactors and aeration in aerobic reactors, which ensures that the entire reactor volume is utilized for treatment (Bassin et al., 2017; Safwat, 2019). The MBBR technology has been successfully applied to a wide range of wastewater types and is known for its simplicity, flexibility, and low environmental impact in removing pollutants such as BOD/COD, nitrogen, and ammonia from both industrial and municipal wastewater streams (Safwat, 2019).

The MBBR is a technology with a high specific surface area and low head loss that operates on the non-cloggable aerobic/anaerobic biofilm reactor concept. By combining two methods, MBBRs can fully utilize a reactor's capacity without head loss or clogging issues. Furthermore, it promotes the growth of slowly growing

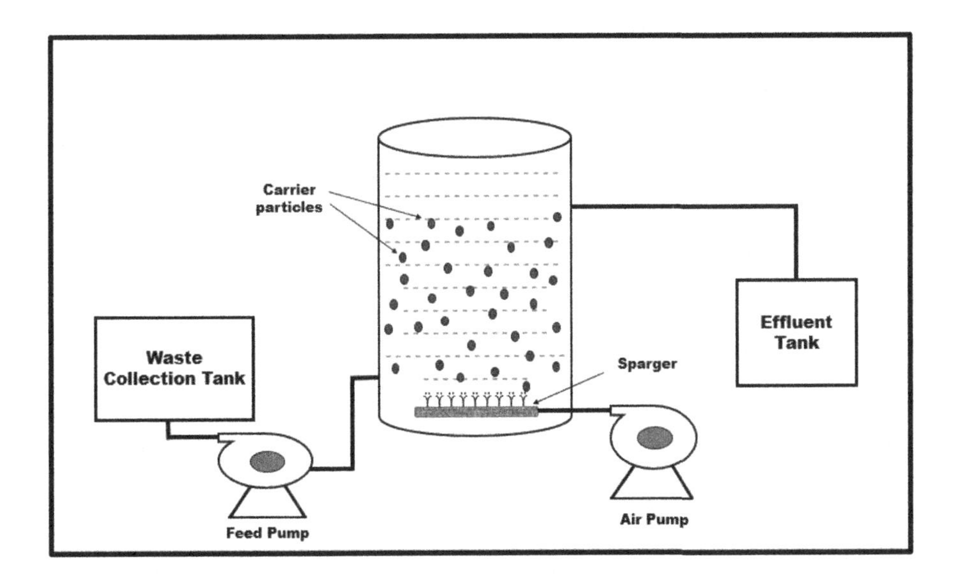

FIGURE 2.1 Schematic diagram of a moving-bed biofilm reactor.

microorganisms such as nitrifiers, which leads to nitrification in the same reactor system. Sludge recirculation is not required, resulting in longer solid retention times. These functional requirements are met by using moving biocarriers with attached biomass in an aeration tank. A thin biofilm is a layer of biomass that is formed on a carrier particle, which is mobile and can move freely around the reactor vessel. While mechanical mixing keeps the carrier materials in suspension, the aeration system in the MBBR system primarily controls the movement of the carrier particles. The biocarrier materials in the MBBR reactor primarily comprise low-density polyethylene sheets with a density slightly less than 1.0 g/cm^3, allowing them to easily remain in a floating state. MBBR can be thought of as a modular treatment system for wastewater that combines carbon oxidation, nitrification, or denitrification in a single reactor without making any substantial modifications to the reactor's configuration (Wang et al., 2020).

2.5.2 Packed-Bed Bioreactor

Packed-bed bioreactors (PBBRs) are commonly used for immobilization systems. In a PBBR, cells are immobilized inside a stationary matrix that forms the bed, enabling attached microbial growth in biofilm reactors. Activated carbon, polymer beads, and silica granules are commonly used as support particles in these heterogeneous systems (Warnock and Bratch, 2005). The biofilm surrounding each particle consists of a thin layer of microbial cells. The high response rates and ease of use make PBBRs beneficial (Narayanan and Narayan, 2019). However, one significant disadvantage of fixed-bed bioreactors is their extremely low heat and mass transfer coefficients due to their low liquid velocities. This can cause gas flooding and uneven liquid distribution, leading to stagnant gas pockets that prevent effective gas interactions and carbon dioxide elimination (Kocadagistan et al., 2005; Zhong, 2011).

Traditional chemical engineering principles for catalytic packed-bed reactors have been used to design PBBRs that use enzymes and microbiological cells as biocatalysts. These bioreactors are typically cylindrical and contain immobilized xylanase. Fluid is evenly integrated in a circular direction, but not in the axial direction, resulting in isothermal conditions for most bioreactions. Therefore, the addition or subtraction of heat is not required. Immobilization methods are used to create biocatalysts with specific dimensions. A single pass is often used to run the PBBR, with the feed recirculation channel also being an option (Sen et al., 2017). PBBRs are widely used in wastewater treatment, and their tolerance for heavy metal removal is crucial. The effects of industrial pollution on wastewater, such as Cu, Cd, Zn, and Ni, have been researched to evaluate the PBBR's tolerance for different composite heavy metal levels. The heavy metal composition of the wastewater outflow stream is compared to the source material at the optimal hydraulic retention time of 2 hours (Azizi et al., 2016).

2.5.3 Biofilm-Fixed Bioreactors

Biofilm fixed-bed bioreactors are commonly used in the treatment of industrial wastewater. In these bioreactors, microorganisms are attached to a solid support medium, which can be made from a variety of materials such as plastic, ceramic, or natural materials. The microorganisms grow as a biofilm on the surface of the media, forming a complex ecosystem of microorganisms that work together to break down the organic compounds in the wastewater (Lazarova and Manem, 1994; Córdoba et al., 2008; Chan et al., 2009; Rathour et al., 2021).

The use of biofilm fixed-bed bioreactors for industrial wastewater treatment has several advantages, including:

High treatment efficiency: The biofilm on the support media provides a large surface area for microbial attachment, allowing for high microbial densities and efficient biodegradation of the organic compounds in the wastewater.

Robustness: Biofilm fixed-bed bioreactors are known for their resilience to fluctuations in wastewater quality and quantity. The biofilm acts as a natural buffer, providing a stable and predictable microbial community that can withstand changes in influent flow or load.

Low operating costs: Biofilm fixed-bed bioreactors are relatively low-maintenance and do not require as much energy or resources as other wastewater treatment methods.

Compact design: Biofilm fixed-bed bioreactors have a small footprint, making them ideal for industrial applications where space is limited.

However, biofilm fixed-bed bioreactors also have some limitations, including the potential for clogging due to the accumulation of biomass on the support media. Regular cleaning and maintenance of the support media may be necessary to prevent clogging and ensure optimal performance.

2.5.4 Fluidized Bed Bioreactor

Fluidized bed bioreactors (FBBRs) have become a widely used method for both aerobic and anaerobic wastewater treatment. In this system, microorganisms are coated onto suspended particles that are fluidized to maintain full mixing of the phases. Compared to traditional biological treatment techniques, the support materials in FBBRs have exceptionally large specific surfaces and are able to achieve treatment levels more quickly (Figure 2.2). The increased surface contact between microbes and contaminants due to fluidization makes it a beneficial method. Some experts claim that FBBRs combine the high efficiency of an activated sludge process with the stability and ease of operation of a trickling filter. Although the use of FBBRs is well established, understanding the key variables that affect the processes is crucial for proper design, upscaling, or process optimization. Interest in FBBRs has increased since they have been shown to be more successful at treating wastewater than other contacting technologies such as fixed-bed columns and activated sludge (Shieh and Keenan, 2005; Bello et al., 2017). Fluidized-bed reactors can provide a level of mixing between the two extremes of packed-bed and stirred tank reactors. These biological reactors may run either as a single-column system or a double-column system, depending on the treatment method being employed. The FBBR is

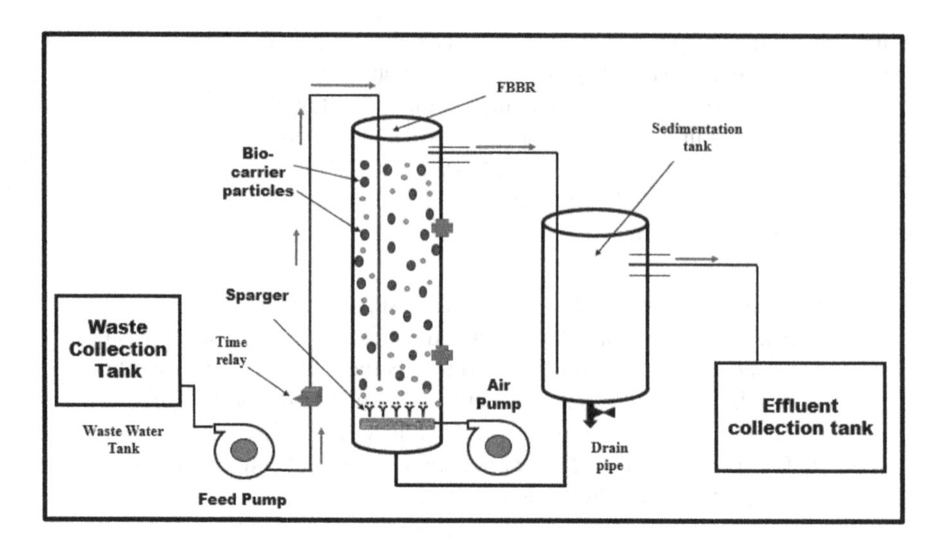

FIGURE 2.2 Schematic diagram of fluidized bed bioreactor.

an associated growth process in which fluid flows upward, lifting stationary cells with it. As the particles rise, gravity causes them to fall, and fluidization is the result of particles moving both vertically and horizontally at the same time (Shieh and Keenan, 2005; Bello et al., 2017). The following are some of the major benefits of a FBBR over a fixed-bed reactor: higher heat transfer and mass transfer rates are anticipated; the system is homogeneous, making it simpler to monitor and manage the operational parameters; and scaling up is possible without affecting concentration gradients. However, operational challenges do exist, as it can be difficult to forecast fluidization patterns or back-mixing. The fluidized-bed reactor containing immobilized cells has mostly been used for wastewater treatment, and microcarrier cultures have also been grown in fluidized reactors (Zhong, 2011).

A "fluidized-bed bioreactor" is a type of liquid–solid FBBR that has been developed. The microbes attach to the fluidized medium, resulting in the formation of a biofilm on the surface. When aeration is involved in the process, the air stream or recirculating wastewater induces fluidization in the column, as described by Zhu et al. (2000). The FBBR process, like all fluidization processes, benefits from superior mixing, enhanced mass transfer, and an expanded surface area. FBBRs utilize smaller particles than other attached-growth systems, such as MBBR and integrated fixed-film activated sludge (IFAS) systems, which provide exceptional microbial attachment properties. As a result, much thicker biofilms are formed, and the surface area of the film exposed to the water is much larger than in traditional attached-growth processes. Due to the greater interaction between the biofilm and the wastewater substrates, larger chemicals that are typically more difficult to treat can be broken down using this method. Furthermore, compared to a normal bioreactor, the FBBR has demonstrated its ability to handle higher loadings and run at shorter hydraulic retention durations, as demonstrated by Nelson et al. (2017).

2.5.5 Airlift Bioreactors

An ALR is a type of gas–liquid bioreactor that operates based on the draught tube principle. It is similar to stirred tank reactors, except for the impeller. This reactor employs compressed air for aeration and agitation, which is introduced through a glass grid aerator that allows humidified air to pass through for mixing and oxygenation (Figure 2.3). The draught tube or external loop in an ALR offers several advantages, including improved fluid circular activity, faster mass and heat transfer rates, reduced bubble coalescence, even distribution of shear stresses, and improved cell growth conditions (Wang and Zhong, 2007).

The ALR typically has an internal loop with a sparger band that regulates the passage of air bubbles and liquid into the bottom of the central draught tube. Air bubbles rise inside the central draught tube, while some condense and climb to the highest point of a column, and others circulate downhill from the area outside the draught tube by following the degassed liquid (Wang and Zhong, 2007). The ALR offers improved fluidization properties compared to traditional FBBRs. In loop reactor systems, the particles carrying the immobile cells are more easily fluidized and can be kept suspended even when faced with high fluid phases or gas-phase velocities, thanks to a circulating liquid phase. The FBBRs can be shielded against the

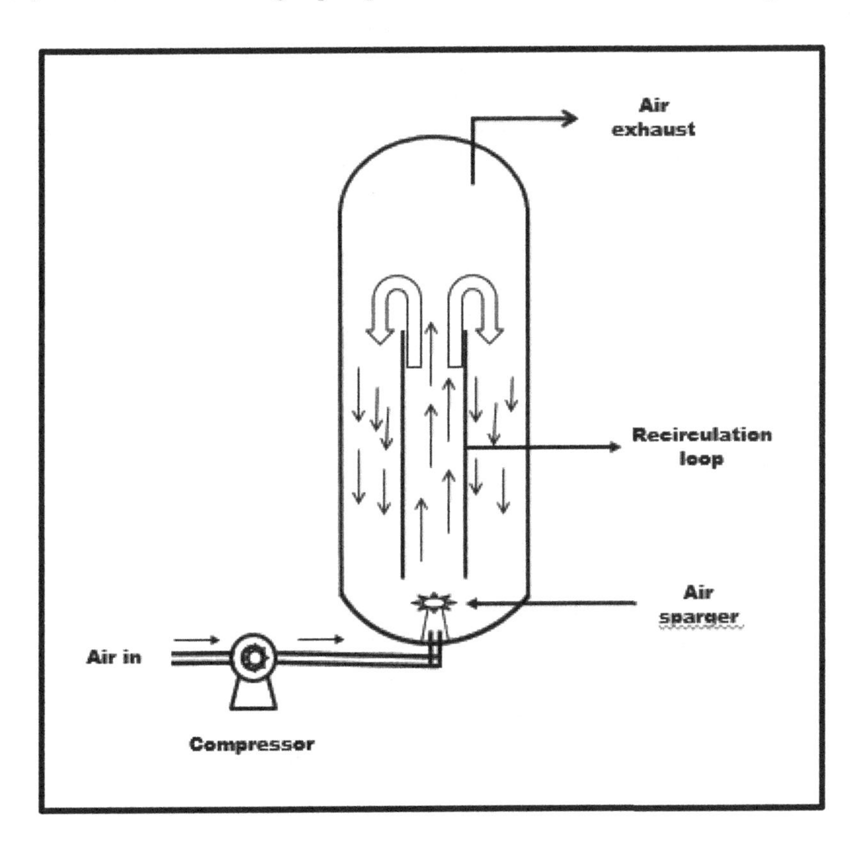

FIGURE 2.3 Schematic diagram of an airlift bioreactor.

unfavorable washout brought on by bed expansion. Gas flow-derived mixing in the ALR results in low shear rates and effective liquid–solid as well as gas–liquid mass transfers (Zhu, 2006).

The ALR has been used for wastewater treatment due to its low shear rate, effective mixing, low operating expenses, and straightforward operating system. It is an excellent alternative to other bioreactors for wastewater treatment.

2.5.6 Membrane Bioreactors

MBRs are designed specifically for aerobic biological treatment of wastewater systems that employ membrane filtration and activated sludge process elements. By eliminating the need for conventional sedimentation techniques, MBRs effectively recycle and separate suspended materials, as shown in Figure 2.4. The result is a high-quality effluent that concentrates BOD. Although MBRs have a high initial cost and ongoing maintenance expenses, their efficiency is unmatched.

The aerobic wastewater treatment process employing aerobic reactors has been significantly improved, and advanced microbial strains have evolved as a viable solution. While mixed liquor solids are suspended in an aeration tank in both the MBR and the classic activated sludge bioreactor, the biosolids are separated differently in the two reactors. In the MBR, a polymeric membrane based on a microfiltration or ultrafiltration unit is used to separate the biosolids, whereas a clarifier uses the gravity settling process in a typical activated sludge bioreactor (Lakatos, 2018).

For industrial waste streams that have high organic loadings and specific biorefractory, inhibitory, and difficult-to-treat chemicals, alternate treatment methods like the MBR are preferred. Due to the process's resilience, the application of MBR for the treatment of industrial wastewater has gained interest. The reactor's technological

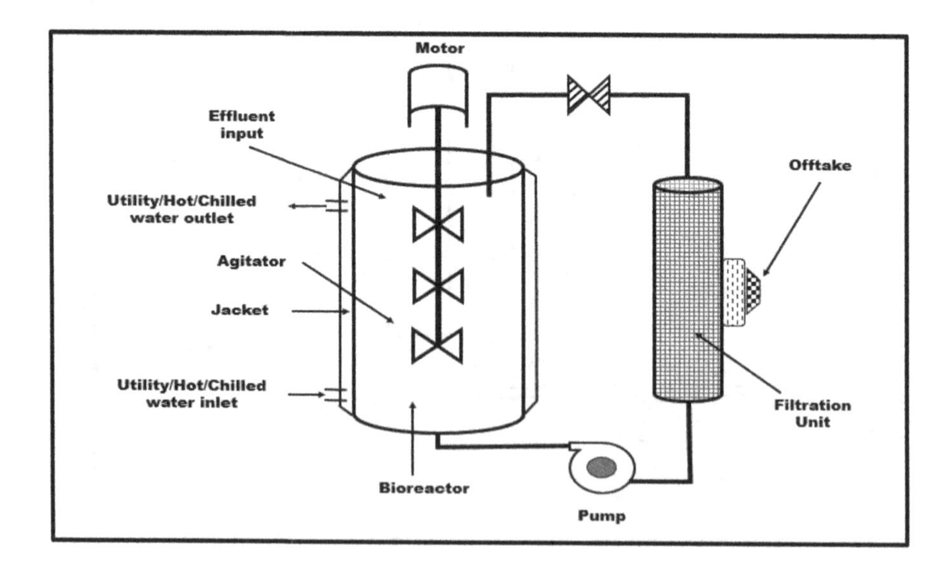

FIGURE 2.4 Schematic diagram of a membrane bioreactor.

characteristics significantly influence the choice of biomass and solid separation. However, it is crucial to emphasize that industrial wastewater's refractory substances may significantly impact the microbial screening process within an MBR. The type of industrial method employed and the quantity of nonbiodegradable chemicals resulting from it affect the efficiency of organic pollutant removal (Fazal et al., 2015).

Depending on how they are configured, MBR systems may be divided into two main categories. The submerged MBR has a membrane built within the bioreactor itself and applies negative pressure to the membrane's permeate side to supply the driving force across it. This device was created to make the system simpler and use less energy, but it produces a lower permeation flow due to its lower transmembrane pressure. The second type, the side stream MBR, circulates mixed liquor into an exterior membrane module and is known for its high-effluent fluxes, easier membrane maintenance, and less challenging scale-up, despite the high pumping costs associated with the recirculation of mixed liquors (Baek and Pagilla, 2006).

The widespread use of MBRs in the treatment of industrial wastewater is hindered by low microbial activity and substantial membrane fouling issues brought on by harsh wastewater conditions such as high temperature, toxicity, and high salinity. pH is a crucial factor that specifically impacts membrane fouling behavior in MBR by influencing both the membrane surface and the characteristics of the mixed liquor suspension. Because many industrial sectors, such as pulp, paper, and textile manufacturing, produce high-temperature streams, temperature is another aspect that needs to be taken into consideration. Despite the increase in membrane flow caused by temperature rise due to the decrease in liquid viscosity, long-term studies have shown that cake layers created at high temperatures have considerably better filtration resistance and are also more compact or less porous (Baek and Pagilla, 2006; Song et al., 2020). Overall, MBRs are an innovative solution to wastewater treatment challenges that provide high-quality effluent while addressing complex industrial waste streams.

2.5.7 Sludge Bed/Sludge Blanket

A sludge bed/sludge blanket aerobic bioreactor is a type of wastewater treatment system that uses microorganisms to remove pollutants from wastewater. This type of bioreactor consists of a tank filled with a bed of small plastic or ceramic media, onto which a layer of activated sludge is settled. The sludge blanket is maintained by the continuous circulation of wastewater through the tank. The microorganisms in the sludge bed/sludge blanket bioreactor are typically aerobic bacteria, which require oxygen to survive. As the wastewater flows through the bed, the microorganisms attach to the media and use the organic pollutants as a food source. The pollutants are converted into carbon dioxide, water, and biomass.

One of the unique features of the sludge bed/sludge blanket bioreactor is the stratification of the biomass. The settled biomass forms a blanket, which separates the upper, oxygenated layer from the lower, anoxic layer. This stratification allows for the simultaneous treatment of organic matter and nitrogen, as denitrification can occur in the anoxic layer (Baek and Pagilla, 2006).

Sludge bed/sludge blanket bioreactors offer several advantages over other types of aerobic bioreactors. They are highly effective in removing organic pollutants and nitrogen, and they have a smaller footprint than conventional activated sludge systems. They also have a lower energy requirement for mixing and a reduced sludge production rate. However, sludge bed/sludge blanket bioreactors also have some limitations. They are sensitive to shock loads and require careful monitoring to prevent the formation of dead zones and clogging. They also require regular maintenance to prevent the accumulation of excess biomass and maintain the sludge blanket.

2.6 HYBRID BIOLOGICAL PROCESSES FOR INDUSTRIAL WASTEWATER TREATMENT

Before the development of moving-bed biofilm reactors, traditional wastewater treatment techniques such as rotating biological reactors, trickling filters, fixed-film reactors, MBRs, and aerated submerged reactors were commonly used (Madan et al., 2022). However, these systems had various drawbacks such as the need for a large amount of space, mechanical breakdown or repair, higher spending and operating expenses, and odor issues. Furthermore, submerged reactors were prone to hydraulic instability, fixed film reactors experienced uneven biofilm distribution and biofilm growth medium blockage, and existing biofilm treatment systems encountered operational challenges such as inappropriate biofilm formation and limited mass transfer (Samal et al., 2022). Over the past two decades, the value of biofilm methods for wastewater treatment has increased significantly, and the moving bed technique, which combines elements from biofilter and activated sludge processes, has been developed to remove considerable contaminants, organic matter, and nutrients from industrial and municipal wastewater (Samal et al., 2022).

Although traditional activated sludge techniques are frequently insufficient to provide high reductions of most organic contaminants, combining two techniques, such as biofilm and suspended biomass, might enhance the efficiency of the biological degradation of organic contaminants due to the increased biodiversity in the systems. Hybrid methods based on membranes have emerged as a possible solution for treating complicated industrial effluent, which contains many different chemicals such as heavy metals, salts, and nutrients that make treatment difficult (Madan et al., 2022).

Activated carbon can help with the adsorption of organic compounds that are only somewhat biodegradable. The inclusion of biofilm surface inside a hybrid process looks promising for enhancing the removal of organic micropollutants in small wastewater treatment facilities, especially if they need to be enlarged to boost nutrient removal. Additionally, the cost of such a method should be lower than that of an extra treatment such as activated carbon adsorption. However, further research is required to enhance these water treatment processes, especially in terms of technological and economic competitiveness (Samal et al., 2022; Madan et al., 2022).

2.7 ADVANCEMENT OF AEROBIC REACTOR SYSTEMS FOR INDUSTRIAL WASTEWATER TREATMENT

Aerobic bioreactor systems are widely used for the treatment of industrial wastewater. These systems rely on microorganisms to break down organic compounds in wastewater, converting them into less harmful by-products such as carbon dioxide and water. Recent advancements in aerobic bioreactor technology have significantly improved the efficiency and effectiveness of wastewater treatment (Narayanan and Narayan, 2019). One key advancement in aerobic bioreactor systems is the use of MBRs. MBRs combine biological treatment with physical separation through the use of a membrane filtration system, which allows for the removal of small particles, bacteria, and viruses. This technology results in a higher quality of treated water and reduces the need for additional treatment steps. The advancement is the use of sequencing batch reactors (SBRs), which allow for more precise control of the biological treatment process. SBRs operate in cycles, with each cycle consisting of several phases such as aeration, settling, and decanting. This system can be programmed to adjust to changes in wastewater quality and flow rate, leading to improved treatment efficiency and flexibility (Narayanan and Narayan, 2019). Additionally, the use of advanced sensors and automation systems in aerobic bioreactor systems has improved monitoring and control of the treatment process. Sensors can measure key parameters such as dissolved oxygen, pH, and temperature, allowing for real-time adjustments to the aeration rate and nutrient supply. This results in optimal conditions for the microorganisms and improved treatment performance. There have been several recent advancements in aerobic bioreactor systems, including:

High-throughput sequencing technology: This technology enables the rapid and cost-effective identification and quantification of microbial communities within aerobic bioreactors. It provides insights into the microbial diversity and dynamics of the system, which can help optimize the bioreactor's performance and stability.

Biofilm-based aerobic bioreactors: Biofilm-based systems use microorganisms attached to a support material to treat wastewater. They offer several advantages over suspended-growth aerobic bioreactors, including higher treatment efficiency, lower sludge production, and better resistance to shock loads.

Membrane-aerated biofilm reactors: These systems combine the benefits of biofilm-based systems with membrane technology. They use a membrane to supply oxygen to the microorganisms attached to the support material, which can improve treatment efficiency and reduce energy consumption.

Electrochemical aerated biofilm reactors: These systems combine biofilm-based treatment with electrochemical processes, which can enhance the removal of pollutants and improve the bioreactor's stability.

Integration with renewable energy sources: Aerobic bioreactor systems can be integrated with renewable energy sources such as solar or wind power to reduce their carbon footprint and energy costs (Narayanan and Narayan, 2019).

2.8 CURRENT STATUS OF AEROBIC BIOREACTOR SYSTEMS

Aerobic bioreactor systems are widely used for the treatment of various types of wastewater, including municipal, agricultural, and industrial wastewater. These systems continue to evolve and improve, with ongoing research and development efforts aimed at optimizing their performance and efficiency. One area of focus is the use of advanced materials and membranes in aerobic bioreactor systems. New types of membranes with improved properties such as higher permeability, fouling resistance, and selectivity are being developed, which can enhance the separation and filtration of wastewater. Another area of advancement is the use of artificial intelligence (AI) and machine learning in aerobic bioreactor systems. These technologies can help optimize the operation of the bioreactor by predicting changes in wastewater quality and flow rate, adjusting the treatment process in real time, and predicting system failures before they occur. In addition, there is growing interest in the use of microbial fuel cells (MFCs) in aerobic bioreactor systems. MFCs use electrochemical reactions to convert organic matter in wastewater into electrical energy, which can be used to power the bioreactor system. This technology has the potential to make aerobic bioreactor systems more energy-efficient and sustainable.

2.9 CHALLENGES AND FUTURE SCOPE

Aerobic bioreactor systems have emerged as a promising technology for the treatment of various types of wastewater, including industrial and municipal wastewater. Aerobic bioreactor systems are widely used, but they still face several challenges that need to be addressed for improved performance and efficiency. Some of these challenges include the following:

High energy consumption: Aerobic bioreactor systems require a constant supply of oxygen, which can be costly and energy-intensive. Finding alternative, more energy-efficient methods of providing oxygen to the microorganisms is a challenge that needs to be addressed.

Incomplete removal of contaminants: While aerobic bioreactor systems are effective at removing a wide range of contaminants from wastewater, some compounds, such as pharmaceuticals and personal care products, can be difficult to remove using this technology. Research is needed to develop new methods for removing these compounds from wastewater.

Nutrient removal: In addition to organic compounds, wastewater also contains nutrients such as nitrogen and phosphorus. These nutrients can contribute to eutrophication and other environmental problems if not removed from the wastewater. Aerobic bioreactor systems can be designed to remove these nutrients, but doing so can be challenging and requires careful optimization of the system.

Improving reactor design: Innovative reactor designs that enhance mass transfer, provide optimal mixing, and ensure uniform biomass distribution will increase treatment efficiency.

Optimizing nutrient supply: The use of innovative nutrient delivery systems such as in situ nutrient recovery, nutrient recycling, and nutrient extraction from wastewater will reduce nutrient costs and improve treatment efficiency.

Bioreactor automation: The integration of process control technologies, such as sensors and advanced algorithms, will enable real-time monitoring and control and reduce operating costs.

Bioreactor integration: Integration of aerobic bioreactors with other technologies such as anaerobic digestion, membrane filtration, and advanced oxidation processes will improve treatment efficiency and enable resource recovery.

Scaling up: Scaling up aerobic bioreactor systems for large-scale applications can be challenging, particularly when it comes to ensuring efficient oxygen transfer and maintaining a healthy microbial population. The design and operation of these systems need to be optimized to ensure reliable and efficient performance at larger scales.

In terms of future scope, there is a growing interest in using aerobic bioreactor systems for resource recovery and waste-to-energy applications. For example, aerobic bioreactors can be used to produce biogas from organic waste materials, which can be used to generate electricity or heat. There is also a growing interest in using aerobic bioreactors to recover nutrients such as nitrogen and phosphorus from wastewater for use in agriculture and other applications. As research continues in these areas, it is likely that we will see further development and optimization of aerobic bioreactor systems for wastewater treatment and resource recovery.

REFERENCES

Ahmed, S.F., Mofijur, M., Parisa, T.A., Islam, N., Kusumo, F., Inayat, A., Badruddin, I.A., Khan, T.Y. and Ong, H.C., 2022. Progress and challenges of contaminate removal from wastewater using microalgae biomass. Chemosphere, 286, p.131656.

Asenjo, J.A., 1994. Bioreactor System Design. CRC Press.

Azizi, S., Kamika, I. and Tekere, M., 2016. Evaluation of heavy metal removal from wastewater in a modified packed bed biofilm reactor. PloS One, 11(5), p.e0155462.

Baek, S.H. and Pagilla, K.R., 2006. Aerobic and anaerobic membrane bioreactors for municipal wastewater treatment. Water Environment Research, 78(2), pp.133–140.

Bassin, J.P., Rachid, C.T., Vilela, C., Cao, S.M., Peixoto, R.S. and Dezotti, M., 2017. Revealing the bacterial profile of an anoxic-aerobic moving-bed biofilm reactor system treating a chemical industry wastewater. International Biodeterioration & Biodegradation, 120, pp.152–160.

Bello, M.M., Raman, A.A.A. and Purushothaman, M., 2017. Applications of fluidized bed reactors in wastewater treatment–a review of the major design and operational parameters. Journal of Cleaner Production, 141, pp.1492–1514.

Betts, J.I. and Baganz, F., 2006. Miniature bioreactors: Current practices and future opportunities. Microbial Cell Factories, 5, pp.1–14.

Boyd, C.E., 1995. Bottom Soils, Sediment, and Pond Aquaculture. Springer Science & Business Media.

Chan, Y.J., Chong, M.F., Law, C.L. and Hassell, D.G., 2009. A review on anaerobic–aerobic treatment of industrial and municipal wastewater. Chemical Engineering Journal, 155(1-2), pp.1–18.

Chavan, S., Yadav, B., Atmakuri, A., Tyagi, R.D., Wong, J.W. and Drogui, P., 2022. Bioconversion of organic wastes into value-added products: A review. Bioresource Technology, 344, p.126398.

Chertow, M. and Ehrenfeld, J., 2012. Organizing self-organizing systems: Toward a theory of industrial symbiosis. Journal of Industrial Ecology, 16(1), pp.13–27.

Córdoba, A., Vargas, P. and Dussan, J., 2008. Chromate reduction by Arthrobacter CR47 in biofilm packed bed reactors. Journal of Hazardous Materials, 151(1), pp.274–279.

Deowan, S.A., Galiano, F., Hoinkis, J., Johnson, D., Altinkaya, S.A., Gabriele, B., Hilal, N., Drioli, E. and Figoli, A., 2016. Novel low-fouling membrane bioreactor (MBR) for industrial wastewater treatment. Journal of Membrane Science, 510, pp.524–532.

di Biase, A., Kowalski, M.S., Devlin, T.R. and Oleszkiewicz, J.A., 2019. Moving bed biofilm reactor technology in municipal wastewater treatment: A review. Journal of Environmental Management, 247, pp.849–866.

Englande, A.J. Jr, Krenkel, P. and Shamas, J., 2015. Wastewater treatment &water reclamation. Reference module in earth systems and environmental sciences.

Fazal, S., Zhang, B., Zhong, Z., Gao, L. and Chen, X., 2015. Industrial wastewater treatment by using MBR (membrane bioreactor) review study. Journal of Environmental Protection, 6(06), p.584.

Grady, C.L. Jr, Daigger, G.T., Love, N.G. and Filipe, C.D., 2011. Biological Wastewater Treatment. CRC press.

Gray, N.F., 2005. Water Technology: An Introduction for Environmental Scientists and Engineers (2nd edition), Elsevier Science & Technology Books, ISBN 0750666331.

Goli, A., Shamiri, A., Khosroyar, S., Talaiekhozani, A., Sanaye, R., Azizi, K., 2019. A review on different aerobic and anaerobic treatment methods in dairy industry wastewater. J. Environ. Treat. Tech. 7, 113–141.

Hölker, U. and Lenz, J., 2005. Solid-state fermentation—Are there any biotechnological advantages? Current Opinion in Microbiology, 8(3), pp.301–306.

Hutmacher, D.W. and Singh, H., 2008. Computational fluid dynamics for improved bioreactor design and 3D culture. Trends in Biotechnology, 26(4), pp.166–172.

Jagani, N., Jagani, H., Hebbar, K., Gang, S.S., Vasanth Raj, P., Chandrashekhar, R.H. and Rao, J., 2010. An overview of fermenter and the design considerations to enhance its productivity. Pharmacologyonline, 1, pp.261–301.

Kadic, E. and Heindel, T.J., 2014. An Introduction to Bioreactor Hydrodynamics and Gas-Liquid Mass Transfer. John Wiley & Sons.

Karthikeyan, K. and Kandasamy, J., 2005. Up-flow anaerobic sludge blanket (UASB) in wastewater treatment. Encycl. Life Support Syst., 1, pp.315–333.

Kaur, I. and Sharma, A.D., 2021. Bioreactor: Design, functions and fermentation innovations. Research and Reviews in Biotechnology and Biosciences, 8, pp.116–125.

Kocadagistan, B., Kocadagistan, E., Topcu, N. and Demircioğlu, N., 2005. Wastewater treatment with combined upflow anaerobic fixed-bed and suspended aerobic reactor equipped with a membrane unit. Process Biochemistry, 40(1), pp.177–182.

Lakatos, G., 2018. Biological wastewater treatment, in: Wastewater and Water Contamination: Sources, Assessment and Remediation. pp. 105–128.

Lazarova, V. and Manem, J., 1994. Advances in biofilm aerobic reactors ensuring effective biofilm activity control. Water Science and Technology, 29(10), pp.319–328.

Madan, S., Madan, R. and Hussain, A., 2022. Advancement in biological wastewater treatment using hybrid moving bed biofilm reactor (MBBR): A review. Applied Water Science, 12(6), p.141.

Mitra, S. and Murthy, G.S., 2022. Bioreactor control systems in the biopharmaceutical industry: A critical perspective. Systems Microbiology and Biomanufacturing, pp.1–22. https://doi.org/10.1007/s43393-021-00048-6

Mittal, A., 2011. Biological wastewater treatment. Water Today, 1, pp.32–44.

Naga Vignesh, S., 2020. Working principle of typical bioreactors, Bioreactors: Sustainable Design and Industrial Applications in Mitigation of GHG Emissions. INC.

Narayanan, C.M. and Narayan, V., 2019. Biological wastewater treatment and bioreactor design: A review. Sustainable Environment Research, 1, pp.1–17.

Nelson, M.J., Nakhla, G. and Zhu, J., 2017. Fluidized-bed bioreactor applications for biological wastewater treatment: A review of research and developments. Engineering, 3, pp.330–342.

Paździor, K., Bilińska, L. and Ledakowicz, S., 2019. A review of the existing and emerging technologies in the combination of AOPs and biological processes in industrial textile wastewater treatment. Chemical Engineering Journal, 376, p.120597.

Petronela, C. and Gavrilescu, M., 2012. Airlift reactors: Applications in wastewater treatment. Environmental Engineering and Management Journal, 11, pp.1505–1515.

Rathour, R., Jain, K., Madamwar, D. and Desai, C., 2021. Performance and biofilm-associated bacterial community dynamics of an upflow fixed-film microaerophilic-aerobic bioreactor system treating raw textile effluent. Journal of Cleaner Production, 295, p.126380.

Reisman, H.B., 1993. Problems in scale-up of biotechnology production processes. Critical Reviews in Biotechnology, 13(3), pp.195–253.

Remmas, N., 2022. Biotreatment potential and microbial communities in aerobic bioreactor systems treating agro-industrial wastewaters. Processes, 10(10), p.1913.

Safwat, S.M., 2019. Moving bed biofilm reactors for wastewater treatment: A review of basic concepts. International Journal of Research, 6(10), pp.85–90.

Salgot, M. and Folch, M., 2018. Wastewater treatment and water reuse. Current Opinion in Environmental Science & Health, 2, pp.64–74.

Samal, K., Geed, S.R. and Mohanty, K., 2022. Hybrid biological processes for the treatment of oily wastewater. In Advances in Oil-Water Separation (pp. 423–435). Elsevier.

Samer, M., 2015. Biological and chemical wastewater treatment processes. Wastewater Treatment Engineering, 150, p.212.

Shah Maulin, P., 2020. Microbial Bioremediation & Biodegradation. Springer.

Shah Maulin, P., 2021a. Removal of Refractory Pollutants from Wastewater Treatment Plants. CRC Press.

Shah Maulin, P., 2021b. Removal of Emerging Contaminants Through Microbial Processes. Springer.

Sen, P., Nath, A. and Bhattacharjee, C., 2017. Packed-bed bioreactor and its application in dairy, food, and beverage industry. In Current Developments in Biotechnology and Bioengineering (pp. 235–277). Elsevier.

Shieh, W.K. and Keenan, J.D., 2005. Fluidized bed biofilm reactor for wastewater treatment. In Bioproducts (pp. 131–169). Springer Berlin Heidelberg.

Singh, J., Kaushik, N. and Biswas, S., 2014. Bioreactors–technology & design analysis. The Scitech Journal, 1(6), pp.28–36.

Song, W., Xie, B., Huang, S., Zhao, F., and Shi, X., 2020. 6 - Aerobic membrane bioreactors for industrial wastewater treatment. Current Developments in Biotechnology and Bioengineering. Elsevier B.V, pp.129–145.

Spina, F., Anastasi, A.E., Prigione, V.P., Tigini, V. and Varese, G., 2012. Biological treatment of industrial wastewaters: A fungal approach. Chemical Engineering Transactions, 27, pp.175–180.

Stott, R., 2003. 31 - Fate and behaviour of parasites in wastewater treatment systems. In Mara, D., Horan, N. (Eds.), Handbook of Water and Wastewater Microbiology (pp. 491–521). Academic Press

Thongsai, A., Phuttaro, C., Saritpongteeraka, K., Charnnok, B., Bae, J., Noophan, P.L. and Chaiprapat, S., 2022. Efficacy of anaerobic membrane bioreactor under intermittent liquid circulation and its potential energy saving against a conventional activated sludge for industrial wastewater treatment. Energy, 244, p.122556.

Usman, M., Kavitha, S., Kannah, Y., Yogalakshmi, K.N., Sivashanmugam, P., Bhatnagar, A. and Kumar, G., 2021. A critical review on limitations and enhancement strategies associated with biohydrogen production. International Journal of Hydrogen Energy, 46(31), pp.16565–16590.

Valderrama, C., Ribera, G., Bahí, N., Rovira, M., Giménez, T., Nomen, R., Lluch, S., Yuste, M. and Martinez-Lladó, X., 2012. Winery wastewater treatment for water reuse purpose: Conventional activated sludge versus membrane bioreactor (MBR): A comparative case study. Desalination, 306, pp.1–7.7.

Wang, T., Wu, T., Wang, H., Dong, W., Zhao, Y., Chu, Z., Yan, G. and Chang, Y., 2020. Comparative study of denitrifying-MBBRs with different polyethylene carriers for advanced nitrogen removal of real reverse osmosis concentrate. International Journal of Environmental Research and Public Health, 17(8), p.2667.

Warnock, J.N., Bratch, K., Al-Rubeai, M., 2005. Packed Bed Bioreactors. Bioreactors for Tissue Engineering: Principles, Design and Operation. Springer Netherlands, Dordrecht, pp.87–113.

Warnock, J.N. and Bratch, K., 2005. Chapter 4 packed bed bioreactors. Culture, pp.87–113.

Zhong, J.J., 2011. Bioreactor engineering. In Comprehensive Biotechnology (2nd Edition). pp. 165–177, Elsevier.

Zhu, Y., 2006. Immobilized Cell Fermentation for Production of Chemicals and Fuels. Bioprocess. Value-Added Prod. from Renew. Resour. New Technol. Appl. pp. 373–396.

Zhu, J., Kang, W., Wolfe, J.H. and Fraser, N.W., 2000. Significantly increased expression of β-glucuronidase in the central nervous system of mucopolysaccharidosis type VII mice from the latency-associated transcript promoter in a nonpathogenic herpes simplex virus type 1 vector. Molecular Therapy, 2, pp.82–94.

3 Energetic Valorization of Industrial Wastes Through Microbial Digestion

Djaber Tazdaït and Rym Salah-Tazdaït

3.1 INTRODUCTION

Over the past 60 years, continuous demographic and economic growth has resulted in numerous forms of acute pollution, including wastes derived from agriculture, industry, and household use. A significant portion of these wastes consists of hazardous materials with adverse effects on the environment in all areas and on humans (Salah-Tazdaït and Tazdaït 2019). For instance, it is expected that by 2025, 2.2 billion tons of solid waste will be produced globally (Kawai and Tasaki 2016). A large number of industrial compounds are involved in chemical pollution, including antibiotics, pesticides, pharmaceuticals, plastics, fertilizers, heavy metals, personal care products, and other substances. Several non-food materials and waste products generated as effluents from food processing industries are also known as important sources of pollution. These waste products can have different pollution levels and different biodegradation characteristics. The dairy industry, for example, produces a large number of readily biodegradable products. In contrast, effluents containing high levels of cellulosic matter are less biodegradable, if not nondegradable. Xenobiotics generated from pharmaceutical and fine chemical industries are much more difficult to dispose of through biological treatment because of their complex chemical structures and toxic and inhibitory effects on the degradative and growth potential of the microorganisms involved. However, thanks to the physiological and biochemical diversity of the indigenous flora of the environment, it would be possible to degrade virtually every xenobiotic released with more or less efficiency. Hazardous xenobiotic levels in different environmental media deeply influence microorganisms' biodegradation of these contaminants. There is a particular concentration of contaminants needed for biodegradation to take place. This concentration is known as the minimum substrate concentration (S_{min}). Here, the contaminant would serve as a substrate for the growth of microorganisms (Becker and Seagren 2010; Shah 2020). Based on the degradation extent, two mechanisms – mineralization and biotransformation – are involved in the biodegradation of organic contaminants. The mineralization process (complete biodegradation) refers to the breakdown of organic pollutants into simple mineral substances (CO_2, H_2O, SO_4^{2-}, NH_3, etc.) through a series of catabolic reactions

DOI: 10.1201/9781003381327-3

without producing any toxic metabolite. Consequently, in this case, the metabolized pollutant is used as a source of nutrients (nitrogen, carbon, phosphorus, or sulphur) and/or as an energy source for the growth of the degrading biomass (Tazdaït et al. 2013). During the biotransformation process (partial biodegradation), the pollutants are partially catabolized by undergoing minor chemical modifications, which produce compounds that are usually less toxic than the parent pollutants. In contrast, in some cases, the released metabolites may be as toxic as or more toxic than the parent compounds. Growth substrates for this process include simple (glycerol, glucose, citrate, acetate, etc.) as well as complex (molasses, syrup date, whey, etc.) (Salah-Tazdaït et al. 2018; Tazdaït and Salah-Tazdaït 2021).

The production of energy in the form of heat through incineration is still the predominant method of waste valorization. However, this type of treatment suffers a significant drawback: generating several toxic compounds, mainly in gaseous forms, which can cause different serious diseases (cancers, respiratory diseases, heart diseases, dermatological diseases, etc.). A promising alternative to this method is the biological approach, which permits efficient energy extraction from various organic wastes through microbial metabolic reactions. In fact, the biological method involving microorganisms offers a more effective and eco-friendly alternative by coupling the biodegradation of toxic or non-toxic wastes with chemical (methane, hydrogen, and biofuels) and physical (electricity) energy production (Salah-Tazdaït and Tazdaït 2019; Shah 2021b). In fact, thanks to the biochemical, genetic, and physiological diversity of the microorganisms, different microbial metabolic pathways could yield very interesting energetic metabolites. Thus, the hydrogenotrophic methanogenic (*Methanogenium*, *Methanobacterium*, *Methanococcus*, *Methanobrevibacter*, etc.) and acetoclastic methanogenic (*Methanothrix* and *Methanosarcina*) Archaea are capable of producing methane through anaerobic digestion of various organic wastes in association with a variety of heterotrophic microorganisms, which accomplish the hydrolysis (hydrolytic bacteria), acidogenesis, and acetogenesis (fermentative bacteria) processes of the wastes. Besides, the microbial production of hydrogen has attracted considerable attention in recent years. Many microbial species that vary taxonomically, biochemically, and physiologically can produce hydrogen as a metabolic by-product. These microorganisms are classified into strictly anaerobic heterotrophic bacteria (*Ruminococcus albus*, *Clostridium pasteurianum*, *Clostridium felsineum*, methylotrophic bacteria, etc.), heterotrophic cytochrome-containing facultative anaerobic bacteria (*Escherichia coli*, *Enterobacter aerogenes*, etc.), and photosynthetic cyanobacteria, including those fixing nitrogen. Purple bacteria, in particular, combine hydrogen production with the consumption of organic matter for their growth. Several types of waste are convenient for biological hydrogen production, including sewage sludge, municipal solid waste, solid recovered fuel, etc. (Lui et al. 2020; Shah 2021a). The advancement of knowledge related to microbial bioenergetics has led to the design of different hybrid devices, coupling biological and physicochemical systems, capable of producing electricity using the microorganisms' respiratory systems. These devices, called microbial fuel cells (MFCs), offer an option for directly producing electricity from oxidized electron donors, including organic wastes, by entire microbial cells or free enzymes used as biocatalysts (Salah-Tazdaït and Tazdaït 2019). Various industrial, agro-industrial, and urban organic wastes have been used as electron donors for microorganisms to produce electricity in different MFC configurations. Examples of wastes include animal manure and cheese (Michalopoulos et al. 2017), potato waste (Du and Li 2016), and remazol brilliant blue R dye (Khan et al. 2015), with maximum power density values ranging from 2.23 to 50 mW/m^2. This chapter provides a comprehensive overview of exploiting microbial metabolic diversity in the treatment of industrial wastes in conjunction with the production of electricity in MFCs and the production of energetic compounds, including biohydrogen, biomethane, and biofuels, in different bioprocess configurations.

3.2 TYPES OF WASTE FROM THE INDUSTRIAL SECTOR

As the name implies, industrial wastes are wastes generated by companies engaged in different sectors of economic activity, such as construction, manufacturing, agriculture, and services. In 2011, it was estimated that the worldwide production of industrial waste was 9.2 billion tons (Vignesh et al. 2021). Due to their diverse origins, industrial wastes are very diverse in their characteristics and nature. Depending on their nature, they can be hazardous or non-hazardous. Industrial hazardous wastes contain toxic elements that can seriously affect the environment and public health (Ishaq and Dincer 2021). They can be of different chemical natures, including organic (solvents, dyes, hydrocarbons, etc.), mineral (acids, alkalis, oxidants, etc.), or gaseous (fluorinated gases, nitrogenated gases, carbon monoxide, etc.). In contrast, non-hazardous industrial wastes have no hazardous, explosive, corrosive, or toxic effects. These wastes include, among others, food wastes, plastic, fly ash, paper, and cardboard. A classification of industrial wastes is represented in Figure 3.1.

3.3 BIOMETHANE PRODUCTION COUPLED WITH INDUSTRIAL WASTE BIODEGRADATION

Anaerobic digestion is now considered the best process for converting organic waste products into renewable energy or biofuel, driven by a favourable regulatory and economic context, linked in particular to the taking into account, on the scale of industrialized countries, the depletion of fossil resources and the necessary protection of the environment. The treatment channels for organic waste products integrating anaerobic digestion are experiencing significant growth. In addition to the production of renewable energy, and depending on the nature of the product treated, methanation also makes it possible to achieve two additional objectives: to reduce the polluting load of waste or organic effluents and to produce a stabilized digestate that can be used as a fertilizer or organic amendment in agriculture. Anaerobic digestion thus constitutes an essential technological building block for the development of the bioeconomy, in close connection with the territories and their agrosystems (Uddin et al. 2021).

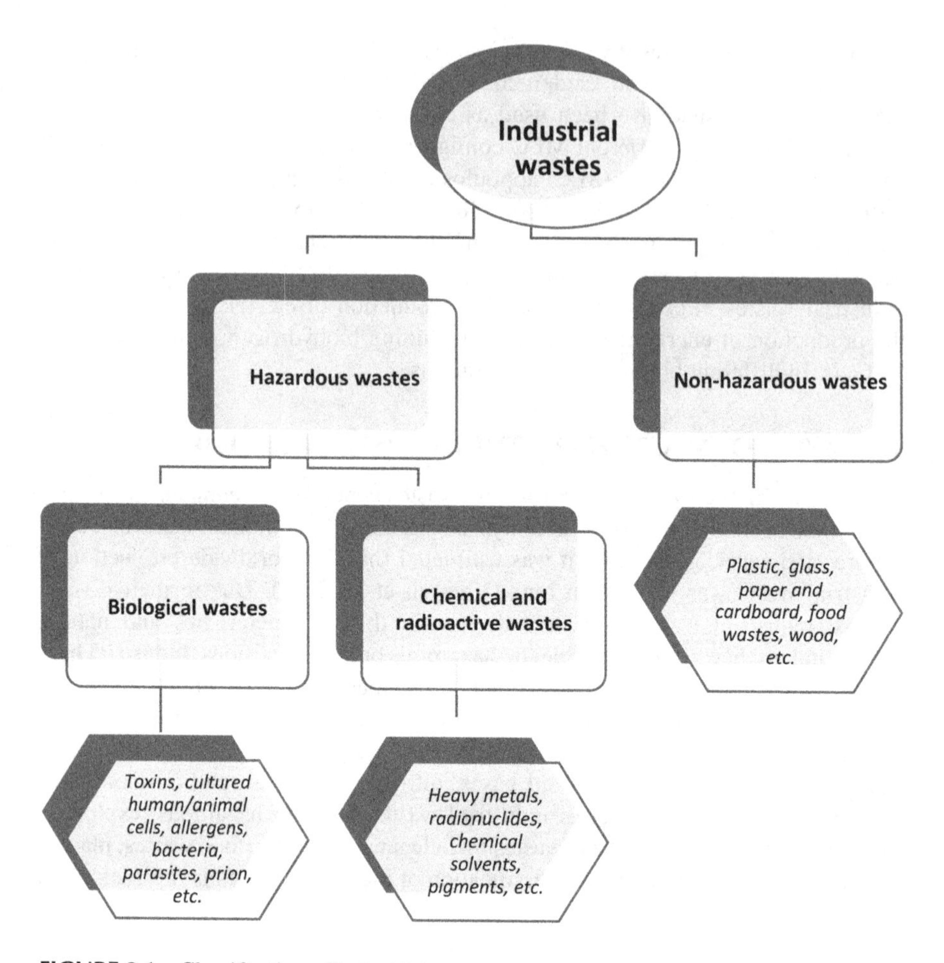

FIGURE 3.1 Classification of industrial wastes.

3.3.1 Wastes Used in the Production of Methane

The organic substrates that can be used in methanization processes can be of various origins:

- waste related to domestic activities, such as residual household waste;
- bio-waste from selective collection from individuals or produced by large producers linked to the distribution of foodstuffs (fairground markets, wholesale markets, retailers, etc.) and catering (commercial and collective catering);
- green waste collected mainly in selective sorting centres;
- waste produced by agricultural activities such as crop residues and organic waste from livestock manure (slurry, manure);
- co-products or by-products resulting from the transformation of animal and plant materials by the agri-food industries (Table 3.1);

TABLE 3.1
Some Examples of Industrial Wastes Used in Methane Production

Microorganisms	Substrates	Yields	References
Anaerobic digestion	Sugarcane bagasse (agroindustrial waste)	1,290×10^6 m^3 CH_4/year	Posso and Mantilla (2019)
Anaerobic digestion (consortium dominated by *Methanosarcina* species) at 37°C	Industrial paper wastes: pulp residues	323 (± 4.96) mL CH_4 g^{-1}	Walter et al. (2016)
Anaerobic digestion (consortium dominated by *Methanosarcina* species) at 55°C	Industrial paper wastes: wood residues	368 (± 7.89) mL CH_4 g^{-1}	Walter et al. (2016)
Anaerobic digestion	Fruit-juice industrial waste	918 mL CH_4 g^{-1}	Akbay et al. (2022)

- solid residues produced during collective or non-collective sanitation, such as sludge from wastewater treatment plants, grease, or waste matter (Zahan et al. 2018; do Carmo Precci Lopes et al. 2022; Pongsopon et al. 2023).

3.3.2 Biological Processes of Anaerobic Digestion of Organic Matter

During the process of anaerobic digestion, the conversion of complex organic compounds into methane and carbon dioxide is carried out by the complementary action of microorganisms belonging to a complex microbial community, both from a taxonomic and functional point of view. The more complex the chemicals to be degraded, the greater the number of interacting microbial species. There is consensus on the simplified model of Zeikus (1980) developed to describe the methanization process; it includes four steps (hydrolysis, acidogenesis, acetogenesis, and methanogenesis) carried out by different microbial groups. Each step leads to the formation of intermediate compounds, which in turn serve as substrates for the next step (Andreides et al. 2022).

3.3.2.1 Hydrolysis and Acidogenesis

During hydrolysis, macromolecules such as polysaccharides, lipids, proteins, and nucleic acids are cleaved, generally by specific extracellular enzymes, until monomers are obtained (simple monosaccharides, fatty acids, amino acids, and base nitrogen), which are transported inside the cell where they are degraded. The bacteria participating in this stage have a strict or facultative anaerobic-type metabolism and form a heterogeneous phylogenetic set comprising many bacterial groups.

During acidogenesis, these monomers are metabolized by fermentative microorganisms to produce mainly volatile fatty acids (VFAs) (acetate, propionate, butyrate, isobutyrate, valerate, and isovalerate), but also alcohols, sulphide of hydrogen (H_2S), carbon dioxide (CO_2), and hydrogen (H_2). This step thus leads to simplified fermentation products. The bacteria participating in this step can be strict or facultative anaerobes with a very short duplication time (30 minutes to a few hours). Strict anaerobic bacteria of the genus *Clostridium* often constitute a significant fraction of the anaerobic population that participates in the acidogenesis step, although other bacterial genera, such as *Acetobacterium,* may be involved. The speed of this step is very high compared to the following steps, where the microorganisms have a longer duplication time. Under organic overload (massive intake of rapidly biodegradable compounds), the faster metabolism of this trophic group can lead to an accumulation of intermediates, particularly hydrogen and acetate. These metabolites inhibit acetogenic and methanogenic microorganisms and may be responsible for dysfunctions or even the cessation of anaerobic digestion (Patakova et al. 2019; Salah-Tazdaït and Tazdaït 2019).

3.3.2.2 Acetogenesis

The acetogenesis stage transforms the various compounds resulting from the previous phase into direct methane precursors: acetate, carbon dioxide, and hydrogen. There are two groups of acetogenic bacteria: obligate, hydrogen-producing, and non-syntrophic acetogenic (Salah-Tazdaït and Tazdaït 2019).

3.3.2.2.1 Obligated Hydrogen-Producing Acetogen

OHPAs (obligate hydrogen-producing acetogens) are strict anaerobes capable of producing acetate and hydrogen from reduced metabolites resulting from acidogenesis, such as propionate and butyrate. These biological reactions only become possible at very low hydrogen partial pressures. The accumulation of hydrogen reflects a dysfunction and inevitably leads to the cessation of acetogenesis. This implies the need for constant removal of the hydrogen produced. This elimination can be achieved through the syntrophic association of these bacteria with hydrogenotrophic microorganisms, ideally methanogenic archaea, which will consume the hydrogen produced (Su et al. 2018).

3.3.2.2.2 Non-Syntrophic Acetogenic Bacteria

The metabolism of these bacteria is mainly oriented towards the production of acetate. They develop in environments rich in CO_2, which is often the case in anaerobic niches. They are primarily autotrophic. Non-syntrophic acetogenic bacteria are conventionally divided into two groups. The first group comprises bacteria that produce acetate, butyrate, and other compounds from simple sugars. The bacteria in this group are classified into several genera, such as *Acetobacterium, Acetogenium, Clostridium, Sporomusa,* etc. “Homo-acetogenic” bacteria constitute the second group. They use hydrogen and carbon dioxide to produce acetate according to the reaction (3.1). The bacteria in this second group essentially belong to the genus *Clostridium.* They do not seem to compete for hydrogen with hydrogenotrophic methanogenic archaea and are present in much lower quantities in anaerobic biotopes. They could, however, be identified as hydrogenotrophic partners of syntrophs (Karekar et al. 2022).

$$2\,HCO_3^- + 4\,H_2 + H^+ \rightarrow CH_3COO^- + 4\,H_2O \tag{3.1}$$

3.3.2.3 Methanogenesis

During this last step, the products of acetogenesis are converted into methane by strict anaerobic microorganisms belonging to the archaea domain. They basically use acetate, formate, carbon dioxide, and hydrogen as substrates to produce methane. Their doubling time is shorter overall than that of acidogenic bacteria. Methane production in a digester is classically described as being carried out by two main categories of methanogenic archaea.

3.3.2.3.1 Hydrogenotrophic Methanogens

These microorganisms derive their energy from methane production by reducing carbon dioxide or formic acid with hydrogen, according to reactions (3.2) and (3.3).

$$4H_2 + HCO_3^- + H^+ \rightarrow CH_4 + 3H_2O \quad (3.2)$$

$$HCOO^- + H^+ + 3H_2 \rightarrow CH_4 + 2H_2O \quad (3.3)$$

Hydrogenotrophic methanogenic archaea live in syntrophic associations with fermentative bacteria, continuously supplying them with carbon dioxide and hydrogen. They thus play an essential role in maintaining a low hydrogen pressure, promoting the growth of acetogenic bacteria.

The most represented genera are *Methanobacterium*, *Methanobrevibacter*, *Methanospirillum*, *Methanogenium*, and *Methanocorspusculum* (Muñoz-Tamayo et al. 2019).

3.3.2.3.2 Acetoclastic Methanogens

The two genera most frequently encountered in a digester are *Methanosarcina* and *Methanosaeta* (or *Methanothrix*). Archaea of the genus *Methanosaeta* use acetate as the sole carbon source to produce methane, according to reaction (3.4).

$$CH_3COO^- + H^+ \rightarrow CH_4 + CO_2 \quad (3.4)$$

Archaea of the genus *Methanosarcina* have a broader spectrum of substrates: they can use, in addition to acetate, carbon dioxide, hydrogen, methanol, and methylamines, to form methane. These acetoclastic methanogenesis reactions are slow and not very exothermic. Nevertheless, they generate more than 70% of methane (Nikitina et al. 2023).

3.4 BIOHYDROGEN PRODUCTION COUPLED WITH INDUSTRIAL WASTE BIODEGRADATION

The reactive nature of hydrogen means that, in the industrial world, dihydrogen (H_2) is widely used as a reagent in many fine chemical, petrochemical, and food processing processes. In the current energy transition context, H_2 is an energy carrier of interest. The production of "green" or carbon-free hydrogen is an auspicious sector for the future. However, hydrogen is an omnipresent biochemical reaction intermediate in the living world, playing a significant role as a carrier of electrons between microbial species, particularly under fermentation conditions.

So-called "dark" fermentation, as opposed to photo-bioprocesses dependent on a light source, is a good process geared towards hydrogen production. Long considered an undesirable process of degradation of organic matter because it generates olfactory nuisances and by-products with little economic interest, namely acetate and butyrate, it has recently become particularly attractive for its hydrogen production. In addition, the advantage of producing hydrogen by fermentation lies in using a wide range of organic substrates, whether pure or not, pure carbohydrates, organic waste, or other agricultural residues. From an industrial point of view, this production sector of "green" hydrogen by fermentation has not yet experienced real growth, given the still-emerging hydrogen market. Therefore, the development of this biotechnology type remains at the pilot-scale stage. These biohydrogen production processes still have certain limits regarding their immediate industrialization. In order to optimize the yields of the sector, couplings with other biological or chemical processes must be considered, such as, for example, photofermentation or microbial electrolysis processes. This "integrated" approach is essential to the success of the "biohydrogen" sector. Certain incompatibilities between processes nevertheless remain to be removed via, for example, the presence of inhibiting compounds. In the short term, the fermentation processes for the production of H_2 seem to be easily integrated into the waste treatment/recovery channels by methanization (Sittijunda et al. 2022).

3.4.1 Wastes Used in Hydrogen Production

Microbial species of the genus *Clostridium* sp., which produce hydrogen by fermentation, have a great affinity for simple sugars. So far, the most studied synthetic substrates have been based on glucose and sucrose. More complex sugars such as cellulose, starch, yeast effluents, rice distillery effluents, industrial bioethanol production waste, vinasses, molasses, and glycerol-rich effluents have also been used, etc. Generally, the more the substrate is rich in easily accessible sugars or glycerol, the more it is suitable for hydrogen production by fermentation (Gomez-Flores et al. 2017).

Recently, there is a growing interest in using agricultural residues because they are an abundant, cheap, and readily biodegradable organic material, except for woody compounds. Agricultural residues include not only straw, fruit and vegetable peels, stalks and ears of energy or non-energy crops, and bagasse but also livestock residues such as slurry and manure. Due to its highly fermentable nature, bio-waste (community waste, catering, etc.) is also interesting for biological recovery.

The production of hydrogen by fermentation process from waste (Table 3.2), effluents, and other residues has attained an ever-increasing interest in recent years, with a large number of studies done in their valorization in dark fermentation (Wang and Yin 2018).

3.4.2 Processes Involved in the Production of Hydrogen

Hydrogen can be produced by the so-called "dark" fermentation or the "photosensitive" pathways.

TABLE 3.2
Some Examples of Industrial Wastes Used in Hydrogen Production

Microorganisms	Substrates	Yields	References
Clostridium beijerinckii ATCC 8260	Soft drink wastewater (agroindustrial wastewater)	18.5±1.68 mL	Martinez-Burgos et al. (2022)
Clostridium butyricum DEBB-B348	Soft drink wastewater (agroindustrial wastewater)	27.4±1.84 mL	Martinez-Burgos et al. (2022)
Anaerobic digester sludge	Cheese whey	8.66 L	Islam et al. (2021)
Anaerobic sludge	Plastic industry waste	0.28 L	Islam et al. (2021)
Clostridium beijerinckii DSM 791	Rice mill wastewater	214.9 mL	Rambabu et al. (2021)

3.4.2.1 Dark Fermentation

Biomass fermentation in the dark is by far the bioprocess allowing the highest volume density of H_2 production. It cannot be considered an effective means of dedicatedly converting solar energy into H_2 on a large scale. However, it remains a potentially effective way of producing H_2 in configurations where biomass is locally available (agricultural residues, organic waste, etc.).

3.4.2.1.1 Hydrogen-Producing Pathways

3.4.2.1.1.1 Anaerobic Metabolisms in the Environment In nature, when conditions become strictly anaerobic, organic matter is biologically degraded to produce methane and carbon dioxide by a microbial process called anaerobic digestion or methanation. The first steps in this process are:

1. a phase of hydrolysis of the most complex organic compounds (carbohydrates, proteins, and lipids) into simpler monomers;
2. an acidogenesis or fermentation phase during which the monomers obtained previously are metabolized and transformed into metabolic by-products (organic acids, alcohols). If the conditions are favourable, it is during the acidogenesis stage that hydrogen production can be observed.

However, during the anaerobic fermentation of these compounds, various microbial metabolisms enter into a competition; the oxidation of organic compounds leads either to the enzymatic reduction of protons in the form of hydrogen when these are in excess in the bacterial cell or to the production of metabolic by-products without excess protons (Moussa et al. 2022).

3.4.2.1.1.2 Microbial Metabolisms Related to Hydrogen Production A large number of metabolic pathways are likely to be involved in the production of hydrogen by fermentation. In the glycolysis pathway (or Embden–Meyerhof–Parnas pathway), glucose is converted into pyruvate while producing adenosine triphosphate (ATP), which serves as an energy store for the cell. ATP will later be used in the synthesis pathways of the various cellular elements (proteins, DNA, etc.). Reducing elements such as nicotinamide adenine dinucleotide (NADH) or reduced ferredoxin (FdH_2) are also produced. As a central product of metabolism, pyruvate is transformed into acetyl coenzyme A (acetyl-CoA), carbon dioxide, and hydrogen by the action of pyruvate-ferredoxin oxidoreductase and hydrogenase (Wang et al. 2022).

Pyruvate can also be converted into acetyl-CoA and formate, the latter of which can be directly transformed into CO_2 and H_2 under the action of a formate-hydrogen lyase. This "formate" pathway is found in certain facultative anaerobes, such as *E. coli* (Roger et al. 2018).

Finally, acetyl-CoA is converted into acetate, butyrate, or ethanol depending on the microorganisms, their oxidation state, the substrate's load, and the environmental conditions influencing the metabolic fluxes. Part of the NADH is used to form butyrate and ethanol, and the excess NADH is recycled back to NAD while producing hydrogen.

Thus, among the wide range of by-products resulting from the different microbial metabolisms, the two pathways producing hydrogen from carbohydrates are essentially associated with the "acetate" pathway (3.5) and the "butyrate" pathway (3.6), well-known in many species of the genus *Clostridium* or the family *Enterobacteriaceae.*

The production of acetate results in a theoretical stoichiometric yield of 4 moles of H_2 per mole of hexose, equivalent to 498 mL of H_2 per gram of hexose (0°C, 1 atm), while the molar amount of hydrogen is lower via the butyrate route with only 2 moles of H_2 per mole of hexose, equivalent to 249 mL of H_2 per gram of hexose (0°C, 1 atm).

$$C_6H_{12}O_6 + 2H_2O \rightarrow 2CH_3COOH + 2CO_2 + 4H_2 \quad (3.5)$$

$$C_6H_{12}O_6 \rightarrow CH_3CH_2CH_2COOH + 2CO_2 + 2H_2 \quad (3.6)$$

Another pathway has been observed but less studied, with simultaneous acetate and ethanol production yielding 2 moles of hydrogen per mole of glucose (3.7).

$$C_6H_{12}O_6 + 3H_2O \rightarrow CH_3CH_2OH + CH_3COOH + 2CO_2 + 2H_2 \quad (3.7)$$

Thermodynamic limitations are essential in the case of hydrogen production. Indeed, even if, from a stoichiometric point of view, it is possible to produce up to 4 moles of hydrogen per mole of glucose consumed via acetic fermentation, there is a strong thermodynamic limit to these reactions. This limit is known as the Thauer limit. It states that hydrogenases, key enzymes catalysing the oxidation of NADH and the reduction of two H^+ ions to H_2, cannot function above hydrogen partial pressures of 10^{-3} atm. This implies that low gas flow rates are observed for high yields of glucose-to-hydrogen conversion. In order to optimize the production of hydrogen, it is sometimes necessary to shift the equilibrium by using separation techniques to eliminate the hydrogen (Ergal et al. 2020; Lo et al. 2020).

3.4.2.2 Photosensitive Pathways

Photosynthetic bacteria constitute several families, some of which can produce H_2. Unlike algae or higher plants, they do not produce O_2 in light because they do not oxidize water as a source of electrons but oxidize organic substrates. In these micro-organisms, hydrogenases are found, but most of the H_2 production is catalysed by nitrogenases. This enzyme is produced in large quantities in certain strains, such as *Rhodobacter capsulatus*, leading to relatively high H_2 production. Many organic substrates, such as those present in waste from agri-food processes, can be used. The photosynthetic bacteria can extract H_2 from organic acids such as acetate, constituting the ultimate fermentation residues in the dark, thanks to their photosynthetic system. The use of light energy allows them to oxidize organic compounds while giving a thermodynamic "boost" to bring the electrons to a potential compatible with the production of H_2. However, a research effort is to be developed to improve the efficiency of H_2 production from substrates that are more difficult to metabolize by these organisms and which are often present in the available resources (Hoffmann et al. 2015).

Some cyanobacteria and microalgae possess a hydrogenase capable of accepting electrons from the "oxygenic" photosynthetic chain. There are also cyanobacteria capable of fixing atmospheric nitrogen and therefore possessing a nitrogenase. These different species thus exhibit a production of H_2 in the light, which is fleeting because it is rapidly inhibited due to the inactivation of hydrogenase by the oxygen produced by photosynthesis through the oxidation of water at the Photosystem II (PSII) level. To obtain a sustainable production of H_2, it is necessary to either modify the enzyme (hydrogenase or nitrogenase) to make it insensitive to O_2 or physically or temporally separate the reactions producing H_2 from those which produce O_2 (Li et al. 2022).

3.4.2.2.1 Production of "Photofermentary" Hydrogen by Bacteria with Anoxygenic Photosynthesis

In photosynthetic bacteria, light absorption leads to a cyclic transfer of electrons, creating a gradient of protons and therefore allowing the synthesis of ATP. The photosynthetic bacteria that have been most studied for H_2 production are the purple bacteria; they have interesting potential for hydrogen production, and the fact that they are not associated with oxygen production makes them reasonably easy to implement. They oxidize various organic substrates (mainly organic acids and sugars) and, for some of them, inorganic substrates such as $S_2O_3^{2-}$, H_2S, and Fe^{2+} to ultimately produce biomass, H_2, and CO_2. For the production of H_2, part of the electrons resulting from the oxidation of these substrates is injected at the level of carriers of the photosynthetic chain, where the electrochemical gradient created by the cyclic transfer of electrons allows, directly or via the synthesis of ATP, the reduction of highly electronegative compounds like ferredoxin by a mechanism that is not yet fully understood. Under favourable conditions, the reduced ferredoxin and the ATP thus produced make it possible to feed a nitrogenase. One can theoretically approach the maximum yields of conversion of organic substrates into H_2 for cells without growth or at least for which the energy investment in the phenomena linked to growth and maintenance is negligible compared to the flow of substrate (cells in stationary phase or immobilized in matrices, for example). As with any photoprocess, the performance remains dependent on the light transfer conditions, which must be optimized to approach optimal yields (Higuchi-Takeuchi and Numata 2019).

Photosynthetic bacteria can use fermentation products found in agricultural or food industry waste (such as acetate and butyrate). Some may also use carbohydrate sources or like *Rhodopseudomonas palustris* aromatic compounds (e.g. lignin monomers) (Brown et al. 2022).

3.4.2.2.2 Hydrogen Production by Organisms with Oxygenic Photosynthesis

Organisms with oxygenic photosynthesis (microalgae, cyanobacteria) draw electrons from water oxidation, providing access to an almost inexhaustible source for producing H_2. However, in this case, the problem of the sensitivity of the catalysts to O_2 for the production of H_2 becomes crucial.

3.4.2.2.2.1 Cyanobacteria

3.4.2.2.2.1.1 *Production via Hydrogenase* In cyanobacteria, there are two types of hydrogenases: hydrogenases functioning only in the direction of H_2 uptake and quinone reduction (Hup type for "hydrogen uptake") and reversible hydrogenases possessing a diaphorase module coupling the oxidation-reduction of H_2 to that of NAD(P)H (Hox type for "hydrogen oxidoreduction"). The latter are those used for the production of H_2. However, because the potential of the $NADH/NAD^+$ couple is higher than that of the H_2/H^+ couple, the production of H_2 requires the substrate to be extremely reduced. In fact, thermodynamically, the uptake of H_2 is favoured, and the production of H_2 by cyanobacteria possessing this enzyme is very low (at least in wild-type strains).

3.4.2.2.2.1.2 *Production via Nitrogenase* Some cyanobacteria also possess nitrogenase. In filamentous cyanobacteria, nitrogenase is localized in heterocysts, specialized cells lacking PSII in which the O_2 tension is low enough to allow the functioning of these enzymes. Nevertheless, "Hup"-type hydrogenases are also frequently found in heterocysts. These hydrogenases recycle the H_2 produced by the nitrogenases and minimize the H_2 released. It has been shown that the suppression of this hydrogenase makes it possible to sustainably stimulate the production of H_2 by this system (Khanna and Lindblad 2015).

3.4.2.2.2.2 Microalgae

3.4.2.2.2.2.1 *Main Mechanisms of Hydrogen Production by Microalgae* Hydrogen production by microalgae was observed for the first time in 1942 in the species *Scenedesmus obliquus*, and then in various green algae such as *Chlorella pyrenoidosa* and *Chlamydomonas reinhardtii*.

Two metabolic pathways are implemented, depending on the origin of the electrons involved in the reduction of protons to hydrogen:

- *Dependent PSII pathway:* the electrons come directly from the oxidation of water, these being transported via photosystem II (PSII) and photosystem I (PSI) to ferredoxin, which is the donor of hydrogenase.
- *Independent PSII pathway:* it involves the degradation of an endogenous substrate (mainly starch and chloroplast proteins). The injection of electrons into the photosynthetic chain occurs between the two photosystems at the level of the plastoquinone pool, in particular via NADH dehydrogenase type 2.

Under natural conditions, hydrogen photoproduction is transient and does not last more than a few seconds to a few minutes. This phenomenon is due to the inhibition of the process by oxygen produced by photosynthesis. Oxygen strongly and irreversibly inhibits algal hydrogenase. One possibility is to create a sulphur deficiency, which gradually inhibits oxygen production by photosynthesis by significantly reducing the activity of PSII (Fan et al. 2016; Ruiz-Marin et al. 2020; Redding et al. 2022).

3.5 BIOFUEL PRODUCTION COUPLED WITH INDUSTRIAL WASTE BIODEGRADATION

Biofuels are fuels produced from non-fossil plant or animal matter, called biomass. Biofuels are renewable energies and, unlike fossil fuels, do not contribute to aggravating certain global environmental impacts, such as the greenhouse effect, provided that their production is carried out under efficient energy conditions and is low in fossil fuels and durability.

The biomass used for biofuel production covers the organic or fermentable by-products of industrial, agri-food, paper, or wood processing activities (Table 3.3).

TABLE 3.3
Some Examples of Industrial Wastes Used in Biofuel Production

Biofuel	Microorganisms	Substrates	Yields	References
Bioethanol	*Pichia kudriavzevii*	Industrial Cocoa Waste	13.8 g/L	Mendoza-Meneses et al. 2021
Bioethanol	Co-culture of *Trichoderma harzianum*, *Aspergillus sojae*, and *Saccharomyces cerevisiae*	Apple pomace	0.945 g/g	Evcan and Tari 2015
Biobutanol	*Clostridium acetobutylicum*	Cassava bagasse hydrolysate	76.40 g/L	Huzir et al. 2018
Biodiesel	*Rhodosporidium toruloides*	Food waste saccharified liquid	5 g/L	Zhang et al. 2021

3.5.1 Ethanol

The ethanol produced for biofuel use comes mainly from sugar beets and wheat. Bioethanol production technology involves:

3.5.1.1 Hydrolysis

Hydrolysis is applied industrially to cereal starch with two processes:

- *The wet process:* the grains are soaked in an aqueous solution containing sulphuric acid to facilitate the separation of the components. The grain is then ground, and its constituents (bran, gluten, starch, etc.) are separated conventionally for the bran and partly by washing. Only the starch is treated during enzymatic hydrolysis and then fermented;
- *The dry process:* the grain is also ground, and all of its constituents undergo enzymatic hydrolysis and fermentation.

3.5.1.2 Fermentation

All fermentable hexoses, mainly glucose and sucrose, can be converted to ethanol and carbon dioxide after fermentation. The reaction is catalysed by an enzyme produced by the yeast *Saccharomyces cerevisiae*. The fermentation reaction equations are as follows (3.8 and 3.9):

$$\underset{\text{Glucose}}{C_6H_{12}O_6} \rightarrow 2\underset{\text{Ethanol}}{C_2H_5OH} + 2CO_2 \tag{3.8}$$

$$\underset{\text{Saccharose}}{C_{12}H_{22}O_{11}} + H_2O \rightarrow 4\underset{\text{Ethanol}}{C_2H_5OH} + 4CO_2 \tag{3.9}$$

The traditional fermentation of hexoses by yeast remains the primary industrial way of producing ethanol. Many research works have approached different fermentations with bacteria, other yeasts, even fungi, and the fermentation of pentoses, but they have not yet found industrial concretizations allowing them to produce ethanol under competitive and efficient conditions.

3.5.1.3 Separation of Ethanol

The fractionation of fermentation products leads to the separation of ethanol from the rest of the constituents. The technology applied consists of two stages: distillation and dehydration (Alfonsín et al. 2019).

3.5.2 Butanol

Butanol is obtained by acetone–butanol–ethanol (ABE) fermentation by the bacterium *Clostridium acetobutylicum* according to the following reactions (3.10, 3.11, and 3.12):

$$\underset{\text{Glucose}}{C_6H_{12}O_6} \rightarrow \underset{\text{Butanol}}{C_4H_9OH} + 2CO_2 + H_2O \tag{3.10}$$

$$\underset{\text{Glucose}}{C_6H_{12}O_6} + H_2O \rightarrow \underset{\text{Acetone}}{CH_3COCH_3} + 3CO_2 + 4H_2 \tag{3.11}$$

$$\underset{\text{Glucose}}{C_6H_{12}O_6} \rightarrow 2\underset{\text{Ethanol}}{C_2H_5OH} + 2CO_2 \tag{3.12}$$

Butanol has several isomers and can be used in a mixture with gasoline. It would reduce the impact of ethanol on vapour pressure, but this product is not without health risks (Moon et al. 2016).

3.5.3 Biodiesel

Depending on their carbon chain length, many esters can be obtained with different characteristics. Triglycerides from vegetable oils are commonly esterified with methanol. The reaction used in this case is transesterification (3.13):

$$\underset{\text{Triglyceride}}{RCOOCH_2CH(OOCR)CH_2OOCR} + 3\underset{\text{Methanol}}{CH_3OH} \rightarrow 3\underset{\text{Fatty ester}}{RCOOCH_3} + \underset{\text{Glycerin}}{HOCH_2CH(OH)CH_2OH} \tag{3.13}$$

(Paladino and Neviani2022).

3.6 PRODUCTION OF ELECTRICITY FROM INDUSTRIAL WASTES

The biological waste-to-energy concept has gained significant attention recently because of its economical and eco-friendly advantages. The production of electricity from the microbial elimination of organic wastes is based on the direct biological conversion of the chemical energy contained in organic compounds into electrical energy. In these systems, called MFCs, microorganisms form a biofilm by colonizing a conductive surface, acting as a catalyst for oxidation, and allowing the production of electrons. These microorganisms have different ecosystem origins, including marine sediments, activated sludge, and rhizosphere. The MFCs have a dual function of producing energy in the form of electricity and biotreatment of various biodegradative organic compound wastes, including agroindustrial waste (whey, molasses, etc.), domestic wastewater, and industrial waste loaded with different xenobiotics (pesticides, dioxins, polycyclic aromatic hydrocarbons, dyes, etc.). In MFCs, two compartments are present: an anode compartment holding an anode on which microorganisms are immobilized, forming a biofilm structure, as well as a cathode compartment containing a cathode, which, in the case of full microbial fuel cells (FMFCs), is colonized by microorganisms, and free of biomass in the case of semi-microbial fuel cells (SMFCs) (Salah-Tazdaït and Tazdaït 2019). In FMFCs, the adhered microorganisms act as catalysts on the two electrodes. An external circuit transports the electrons to the cathode, where final electron acceptor compounds, usually dioxygen, trap them along with the protons to produce reduced compounds such as water. The protons result from the oxidation of the initial electron donors through catabolic reactions and attain the cathode by passing through a proton-exchange membrane separating the two compartments. The protons and electrons at the cathode of SFMCs reduce dioxygen to water, while FMFCs' cathode uses adsorbed biomass to operate as a catalyst for reducing dioxygen into water (Figure 3.2). The electron transfer from

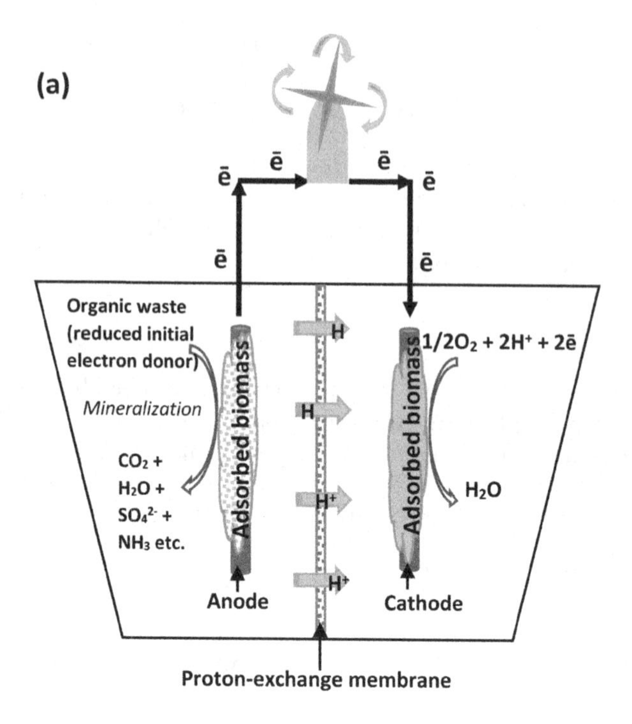

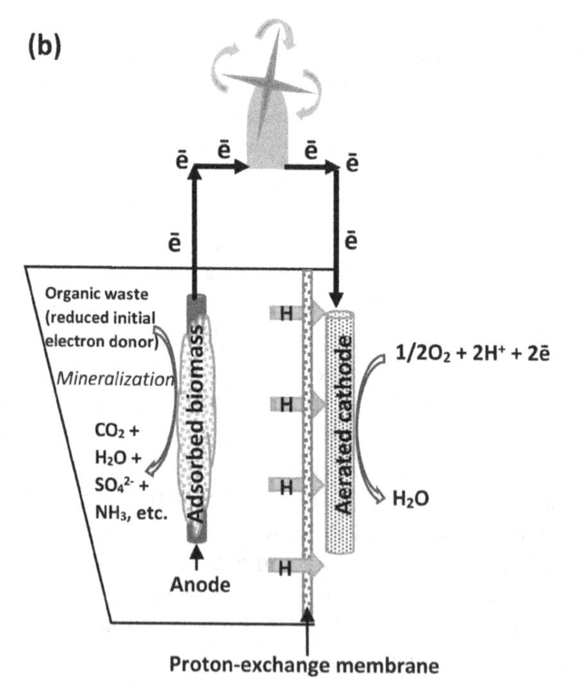

FIGURE 3.2 Schematic diagram of (a) a full microbial fuel cell (FMFC) and (b) a semi-microbial fuel cell (SMFC).

the microorganisms to the anode can be described in two ways: (1) by direct electron transfer through microbially produced filaments called bacterial nanowires (pili, cytochrome polymers) (Veerubhotla et al. 2017; Wang et al. 2023), (2) or mediated by extracellular secreted soluble compounds such as phenazines and pyocyanin (Rabaey et al. 2005). As anodes in MFCs, various electrodes, such as carbon veil, carbon mesh, carbon fibre paper, carbon cloth, graphite plate, cylindrical graphite rod, and reticulated vitrified carbon, are currently used. The choice of anode is primarily determined by the microorganism to be attached. On the other hand, it is worth noting that some process factors, such as ohmic losses, activation losses, polarization losses, concentration losses, and microbial metabolic losses, have a negative impact on MFCs performance (Mahadevan et al. 2014).

Many efforts are being made to improve the yield of the electrical energy harvested from MCFs. This section of the chapter includes descriptions of some of the most current studies. Thus, Zhao et al. (2021) undertook a study on pollutants (methyl orange, Cr(VI), and nitrate) removal combined with electricity generation. This study was conducted in a hybrid system consisting of a cylindrical plexiglass solid-phase denitrification bioreactor containing an anode and an aerated cathode made of carbon felt and activated carbon with a 2.7 cm^3 volume each. The bioreactor was embedded with polybutylene succinate serving as a solid carbon source and inoculated with two types of sludge, one inoculating the anode and the other inoculating the denitrifying part of the system. The system was fed with synthetic wastewater composed of 30 mg/L methyl orange, 100 mg/L NO_3^-, 20 mg/L Cr(VI), and other mineral elements. The results showed that the hybrid system successfully performed both electrogenesis and denitrification. Interestingly, after 60 days of operation, the average removal efficiencies of methyl orange, NO_3^-, and Cr(VI) (93.0%, 95.5%, and 98.6%, respectively) were significantly enhanced by the MFC process as compared to the control experiments (53.1%, 72.7%, and 72.1%). Besides, it was found that the genus *Pseudomonas* colonizing the anode was responsible for bioelectricity generation. The system achieved maximum power densities of 61.2 and 16.1 mW/m^2 at 48 h and 24 h HRT (hydraulic retention time), respectively.

In their study, Umar et al. (2021) employed double-chamber benthic MFCs to investigate their efficiency in degrading benzene using sugarcane waste as a carbon and energy source. The system consisted of two plexiglass chambers (size: 20 cm × 14 cm (length × width)) equipped with a cylindrical graphite anode and cathode with the following characteristics: 7 cm length, 1 cm diameter, and 23.5 cm^2 surface area. The anode and the cathode were immersed in sugarcane waste and phosphate buffer solutions, respectively. The cathode compartment was aerated using an air pump, while the anode compartment was maintained under anoxic conditions. The results showed that the system could generate a current density of 95 mA/m^2 and a maximum power density of 24.2 mW/m^2 after 40 days of operation. Besides, within 155 days of operation, benzene was found to be bioremediated through the mineralization process, with a maximum removal efficiency of about 82.6%. Lastly, the analysis of the biofilm attached to the surface of the electrodes (cathode and anode) revealed the presence of *Bacillus* sp. and *Pseudomonas* sp. The study results demonstrated that the system had good application possibilities for electricity generation and benzene bioremediation.

More recently, a study reported the performance of an MFC in degrading organic compounds contained in pulp and paper mill wastewater and converting them into different value-added products, including glycerol, vanillin, alcohols, and organic acids (Srivastava et al. 2023). The reactor consisted of a dual-chamber MFC equipped with an anode and a cathode made of aluminium with surfaces of 150 and 80 cm^2, respectively. In order to favour its colonization by the microorganisms (bacteria and fungi) present in the wastewater used, the anode was attached to graphite rods and immersed into the wastewater to be treated, while the cathode was immersed in a sodium chloride solution at 100 g/L and operated under aerobic conditions. The system exhibited a maximum voltage of 702 mV within 60 hours of the experiment and allowed for efficient degradation of the various organic compounds present in the industrial wastewater and the liberation of plenty of interesting value-added by-products. These by-products included, among many others, glycerol, propionic acid, acetic acid, oleic acid, ethanol, and propanol.

In another study, Yaghmaeian et al. (2023) investigated the treatment of a vegetable oil industry wastewater using a double-chamber MFC endowed with an anode composed of carbon paper and TiO_2-HX@MWCNT-COOH-Al_2O_3 composite and a carbon felt cathode made of activated carbon powder obtained from Bambuseae. The experimental setup enabled maximum elimination rates of 94%, 89%, 87%, 74%, 79%, and 65% for chemical oxygen demand, biochemical oxygen demand, ammonium, nitrate, total suspended solids, and volatile suspended solids, respectively. Besides, the device reached maximum average power density, energy efficiency, and Coulombic efficiency of 30 W/m^3, 35%, and 85%, respectively.

Alonso-Lemus et al. (2022) studied the treatment of pharmaceutical wastewater with the following major physicochemical characteristics: 27.6 mg/L COD, 30.54 mg/L total dissolved solids, 93.91 mg/L total phosphorous, 6.13 mg/L nitrates, pH 9.2 using a double-chamber MFC equipped with a bioanode immersed in 240 mL of the pharmaceutical wastewater saturated with nitrogen gas and a platinum-based cathode immersed in potassium hydroxide solution saturated with oxygen. The bioanode consisted of three layers: a layer of carbon cloth serving as a support, a catalytic layer made of biocarbon prepared from leather waste, and a biofilm layer of *Bacillus subtilis* or *E. coli* ATCC 25922 tested separately. With this experimental device, *Bacillus subtilis*-based biofilm exhibited the best performance in electricity generation, with a power density and voltage of 77 mW/m^2 and 602 mV, respectively. On the other hand, removal efficiencies of 45% and 15% regarding COD and BOD, respectively, were reached; however, no nitrate and phosphorous removal were observed.

3.7 CONCLUSION

The general propensity to achieve economic prosperity and rapid industrialization resulted in problems, such as the discharge of wastewater contaminated by different organic and inorganic pollutants. Besides, the ever-growing need for energy by all human societies worldwide incites us to find alternative energy sources to non-renewable energy. Thus, efforts are being made to treat and dispose of polluting wastes in an environmentally safe way in combination with their energetic valorization using the biological approach, which allows biofuel production (hydrogen, methane, bioethanol, biodiesel, etc.) and bioelectricity in an eco-friendly manner. Despite the limitations of using some biological processes, such as MFCs, intensive industrial and academic research efforts focus on optimizing the operational parameters and the process configurations. Besides, there is increasing work in screening new and more effective bacterial strains and their enzymes capable of generating high energetic yields with efficient waste degradation.

REFERENCES

Akbay, H. E. G., N. Dizge, and H. Kumbur. 2022. 'Evaluation of Electro-Oxidation and Fenton Pretreatments on Industrial Fruit Waste and Municipal Sewage Sludge to Enhance Biogas Production by Anaerobic Co-Digestion'. Journal of Environmental Management 319: 115711. https://doi.org/10.1016/j.jenvman.2022.115711.

Alfonsín, V., R. Maceiras, and C. Gutiérrez. 2019. 'Bioethanol Production From Industrial Algae Waste'. Waste Management 87: 791–797. https://doi.org/10.1016/j.wasman.2019.03.019.

Alonso-Lemus, Ivonne L., Carlos Cobos-Reyes, Mayra Figueroa-Torres, Beatriz Escobar-Morales, K. Kunhiraman Aruna, Prabhu Akash, Fabian Fernández-Luqueño, and Javier Rodríguez-Varela. 2022. 'Green Power Generation by Microbial Fuel Cells Using Pharmaceutical Wastewater as Substrate and Electroactive Biofilms (Bacteria/Biocarbon)'. Journal of Chemistry 2022 (August): 1–11. https://doi.org/10.1155/2022/1963973.

Andreides, Dominik, Dana Pokorna, and Jana Zabranska. 2022. 'Assessing the syngas biomethanation in anaerobic sludge digestion under different syngas loading rates and homogenisation'. Fuel 320 : 123929. https://doi.org/10.1016/j.fuel.2022.123929

Becker, Jennifer G., and Eric A. Seagren. 2010. 'Bioremediation of Hazardous Organics'. In Environmental Microbiology, edited by Ralph Mitchell and Ji-Dong Gu, 177–212. John Wiley & Sons, Inc. https://doi.org/10.1002/9780470495117.ch8.

Brown, Brandi, Mark Wilkins, and Rajib Saha. 2022. '*Rhodopseudomonas palustris*: A Biotechnology Chassis'. Biotechnology Advances 60: 108001. https://doi.org/10.1016/j.biotechadv.2022.108001.

Du, Haixia, and Fusheng Li. 2016. 'Size Effects of Potato Waste on Its Treatment by Microbial Fuel Cell'. Environmental Technology 37 (10): 1305–1313. https://doi.org/10.1080/09593330.2015.1114027.

Ergal, İpek, Oliver Gräf, Benedikt Hasibar, Michael Steiner, Sonja Vukotić, Günther Bochmann, Werner Fuchs, and Simon K. -M. R. Rittmann. 2020. 'Biohydrogen Production Beyond the Thauer Limit by Precision Design of Artificial Microbial Consortia'. Communications Biology 3, 443. https://doi.org/10.1038/s42003-020-01159-x.

Evcan, Ezgi, and Canan Tari. 2015. 'Production of Bioethanol From Apple Pomace by Using Cocultures: Conversion of Agro-Industrial Waste to Value Added Product'. Energy 88: 775–782. http://doi.org/10.1016/j.energy.2015.05.090.

Fan, Xiaolei, Huanyu Wang, Rongbo Guo, Dawei Yang, Yanting Zhang, Xianzheng Yuan, Yanling Qiu, Zhiman Yang, and Xiaoxian Zhao. 2016. 'Comparative Study of the Oxygen Tolerance of *Chlorella pyrenoidosa* and *Chlamydomonas reinhardtii* CC124 in Photobiological Hydrogen Production'. Algal Research 16: 240–244. https://doi.org/10.1016/j.algal.2016.03.025.

Gomez-Flores, M., G. Nakhla, and H. Hafez. 2017. 'Hydrogen Production and Microbial Kinetics of *Clostridium termitidis* in Mono-Culture and Co-Culture With *Clostridium beijerinckii* on Cellulose'. AMB Express 7 (1): 84. https://doi.org/10.1186/s13568-016-0256-2.

Higuchi-Takeuchi, Mieko, and Keiji Numata. 2019. 'Marine Purple Photosynthetic Bacteria as Sustainable Microbial Production Hosts'. Frontiers in Bioengineering and Biotechnology 7: 258. https://doi.org/10.3389/fbioe.2019.00258.

Hoffmann, M. C., E. Wagner, S. Langklotz, Y. Pfänder, S. Hött, J. E. Bandow, and B. Masepohl. 2015. 'Proteome Profiling of the *Rhodobacter capsulatus* Molybdenum Response Reveals a Role of ISCN in Nitrogen Fixation by Fe-Nitrogenase'. Journal of Bacteriology 198 (4): 633–43. https://doi.org/10.1128/JB.00750-15.

Huzir, Nurhamieza Md, Md Maniruzzaman A. Aziz, S. B. Ismail, Bawadi Abdullah, Nik Azmi Nik Mahmood, N. A. Umor, and Syed Anuar Faua'ad Syed Muhammad. 2018. 'Agro-Industrial Waste to Biobutanol Production: Eco-Friendly Biofuels for Next Generation'. Renewable and Sustainable Energy Reviews 94: 476–485. https://doi.org/10.1016/j.rser.2018.06.036.

Ishaq, H., and I. Dincer. 2021. 'A New Approach in Treating Industrial Hazardous Wastes for Energy Generation and Thermochemical Hydrogen Production'. Journal of Cleaner Production 290 (March): 125303. https://doi.org/10.1016/j.jclepro.2020.125303.

Islam, A. K. M. K., P. S. M. Dunlop, N. J. Hewitt, R. Lenihan, and C. Brandoni. 2021. 'Bio-Hydrogen Production From Wastewater: A Comparative Study of Low Energy Intensive Production Processes'. Clean Technology 3: 156–182. https://doi.org/10.3390/cleantechnol3010010.

Karekar, S., R. Stefanini, and B. Ahring. 2022. 'Homo-Acetogens: Their Metabolism and Competitive Relationship With Hydrogenotrophic Methanogens'. Microorganisms 10 (2): 397. https://doi.org/10.3390/microorganisms10020397.

Kawai, Kosuke, and Tomohiro Tasaki. 2016. 'Revisiting Estimates of Municipal Solid Waste Generation Per Capita and Their Reliability'. Journal of Material Cycles and Waste Management 18 (1): 1–13. https://doi.org/10.1007/s10163-015-0355-1.

Khan, Mohammad Danish, Huda Abdulateif, Iqbal M. Ismail, Suhail Sabir, and Mohammad Zain Khan. 2015. 'Bioelectricity Generation and Bioremediation of an Azo-Dye in a Microbial Fuel Cell Coupled Activated Sludge Process'. PLoS One 10 (10): e0138448. https://doi.org/10.1371/journal.pone.0138448.

Khanna, N., and P. Lindblad. 2015. 'Cyanobacterial Hydrogenases and Hydrogen Metabolism Revisited: Recent Progress and Future Prospects'. International Journal of Molecular Sciences 16 (5): 10537–10561. https://doi.org/10.3390/ijms160510537.

Li, Shengnan, Fanghua Li, Xun Zhu, Qiang Liao, Jo-Shu Chang, and Shih-Hsin Ho. 2022. 'Biohydrogen Production From Microalgae for Environmental Sustainability', Chemosphere 291: 132717. https://doi.org/10.1016/j.chemosphere.2021.132717.

Lo, J., J. R. Humphreys, J. Jack, C. Urban, L. Magnusson, W. Xiong, Y. Gu, Z. J. Ren, and P.-C. Maness. 2020. 'The Metabolism of *Clostridium ljungdahlii* in Phosphotransacetylase Negative Strains and Development of an Ethanologenic Strain'. Frontiers in Bioengineering and Biotechnology 8: 560726. https://doi.org/10.3389/fbioe.2020.560726.

Lopes, Alice do Carmo Precci, Christian Ebner, Frédéric Gerke, Marco Wehner, Sabine Robra, Sebastian Hupfauf, and Anke Bockreis. 2022. 'Residual municipal solid waste as co-substrate at wastewater treatment plants: An assessment of methane yield, dewatering potential and microbial diversity'. Science of The Total Environment 804: 149936. https://doi.org/10.1016/j.scitotenv.2021.149936.

Lui, Jade, Wei-Hsin Chen, Daniel C. W. Tsang, and Siming You. 2020. 'A Critical Review on the Principles, Applications, and Challenges of Waste-to-Hydrogen Technologies'. Renewable and Sustainable Energy Reviews 134 (December): 110365. https://doi.org/10.1016/j.rser.2020.110365.

Mahadevan, Aishwarya, Duminda A. Gunawardena, and Sandun Fernando. 2014. 'Biochemical and Electrochemical Perspectives of the Anode of a Microbial Fuel Cell'. In Technology and Application of Microbial Fuel Cells, edited by Chin-Tsan Wang. InTech. https://doi.org/10.5772/58755.

Martinez-Burgos, Walter J., Jair Rosário do Nascimento Junior, Adriane Bianchi Pedroni Medeiros, Leonardo Wedderhof Herrmann, Eduardo Bittencourt Sydney, and Carlos Ricardo Socco. 2022. 'Biohydrogen Production From Agro-Industrial Wastes Using *Clostridium beijerinckii* and Isolated Bacteria as Inoculum'. BioEnergy Research 15: 987–997. https://doi.org/10.1007/s12155-021-10358-1.

Mendoza-Meneses, C. J., A. A. Feregrino-Pérez, and C. Gutiérrez-Antonio. 2021. 'Potential Use of Industrial Cocoa Waste in Biofuel Production'. Journal of Chemistry 2021: 3388067. https://doi.org/10.1155/2021/3388067.

Michalopoulos, I., D. Chatzikonstantinou, D. Mathioudakis, I. Vaiopoulos, A. Tremouli, M. Georgiopoulou, K. Papadopoulou, and G. Lyberatos. 2017. 'Valorization of the Liquid Fraction of a Mixture of Livestock Waste and Cheese Whey for Biogas Production Through High-Rate Anaerobic Co-Digestion and for Electricity Production in a Microbial Fuel Cell (MFC)'. Waste and Biomass Valorization 8 (5): 1759–1769. https://doi.org/10.1007/s12649-017-9974-1.

Moon, Hyeon Gi, Yu-Sin Jang, Changhee Cho, Joungmin Lee, Robert Binkley, and Sang Yup Lee. 2016. 'One Hundred Years of Clostridial Butanol fermentation'. FEMS Microbiology Letters 363(3): fnw001. https://doi.org/10.1093/femsle/fnw001.

Moussa, Rita Noelle, Najah Moussa, and Davide Dionisi. 2022. 'Hydrogen Production From Biomass and Organic Waste Using Dark Fermentation: An Analysis of Literature Data on the Effect of Operating Parameters on Process Performance'. Processes 10(1): 156. https://doi.org/10.3390/pr10010156.

Muñoz-Tamayo, R., M. Popova, M. Tillier, D. P. Morgavi, J. P. Morel, G. Fonty, and N. Morel-Desrosiers. 2019. 'Hydrogenotrophic Methanogens of the Mammalian Gut: Functionally Similar, Thermodynamically Different-A Modelling Approach'. PLoS One 14 (12): e0226243. https://doi.org/10.1371/journal.pone.0226243.

Nikitina, Anna A., Anna Y. Kallistova, Denis S. Grouzdev, Tat'yana V. Kolganova, Andrey A. Kovalev, Dmitriy A. Kovalev, Vladimir Panchenko, Ivar Zekker, Alla N. Nozhevnikova, and Yuriy V. Litti. 2023. 'Syntrophic Butyrate-Oxidizing Consortium Mitigates Acetate Inhibition Through a Shift From Acetoclastic to Hydrogenotrophic Methanogenesis and Alleviates VFA Stress in Thermophilic Anaerobic Digestion'. Applied Sciences 13(1): 173. https://doi.org/10.3390/app13010173.

Paladino, Ombretta, and Matteo Neviani. 2022. 'Sustainable Biodiesel Production by Transesterification of Waste Cooking Oil and Recycling of Wastewater Rich in Glycerol as a Feed to Microalgae'. Sustainability 14(1): 273. https://doi.org/10.3390/su14010273.

Patakova, P., B. Branska, K. Sedlar, M. Vasylkivska, K. Jureckova, J. Kolek, P. Koscova, and I. Provaznik. 2019. 'Acidogenesis, Solventogenesis, Metabolic Stress Response and Life Cycle Changes in *Clostridium beijerinckii* NRRL B-598 at the Transcriptomic Level'. Scientific Reports 9 (1): 1371. https://doi.org/10.1038/s41598-018-37679-0.

Pongsopon, Mattana, Thamonwan Woraruthai, Piyanuch Anuwan, Thanyaphat Amawatjana, Charndanai Tirapanampai, Photchanathorn Prombun, Kanthida Kusonmano, Nopphon Weeranoppanant, Pimchai Chaiyen, and Thanyaporn Wongnate. 2023. 'Anaerobic co-digestion of yard waste, food waste, and pig slurry in a batch experiment: An investigation on methane potential, performance, and microbial community'. Bioresource Technology Reports 21: 101364. https://doi.org/10.1016/j.biteb.2023.101364.

Posso, F., and N. Mantilla 2019. 'Biomethane Production Potential From Selected Agro-Industrial Waste in Colombia and Perspectives of Its Use in Vehicular transport'. Journal of Physics: Conference Series 1386. https://doi.org/10.1088/1742-6596/1386/1/012100.

Rabaey, Korneel, Nico Boon, Monica Höfte, and Willy Verstraete. 2005. 'Microbial Phenazine Production Enhances Electron Transfer in Biofuel Cells'. Environmental Science & Technology 39 (9): 3401–3408. https://doi.org/10.1021/es048563o.

Rambabu, K., G. Bharath, A. Thanigaivelan, D. B. Das, Pau Loke Show, and Fawzi Banat. 2021. 'Augmented Biohydrogen Production From Rice Mill Wastewater Through Nano-Metal Oxides Assisted Dark Fermentation'. Bioresource Technology 319: 124243. https://doi.org/10.1016/j.biortech.2020.124243.

Redding, Kevin E., Jens Appel, Marko Boehm, Wolfgang Schuhmann, Marc M. Nowaczyk, Iftach Yacoby, and Kirstin Gutekunst. 2022. 'Advances and Challenges in Photosynthetic Hydrogen Production'. Trends in Biotechnology 40(11): 1313–1325. https://doi.org/10.1016/j.tibtech.2022.04.007.

Roger, Magali, Fraser Brown, William Gabrielli, and Frank Sargent. 2018. 'Efficient Hydrogen-Dependent Carbon Dioxide Reduction by *Escherichia coli*'. Current Biology 28(1): 140–145. https://doi.org/10.1016/j.cub.2017.11.050.

Ruiz-Marin, A., Y. Canedo-López, and P. Chávez-Fuentes. 2020. 'Biohydrogen Production by *Chlorella vulgaris* and *Scenedesmus obliquus* Immobilized Cultivated in Artificial Wastewater Under Different Light quality'. AMB Express 10 (1): 191. https://doi.org/10.1186/s13568-020-01129-w.

Salah-Tazdaït, Rym, and Djaber Tazdaït. 2019. 'Biological Systems of Waste Management and Treatment'. In Advances in Waste-to-Energy Technologies, edited by Rajeev Pratap Singh, Vishal Prasad, and Barkha Vaish, 1st ed., 115–130. CRC Press. https://doi.org/10.1201/9780429423376-7.

Salah-Tazdaït, Rym, Djaber Tazdaït, Rouchdi Berrahma, Nadia Abdi, Hocine Grib, and Nabil Mameri. 2018. 'Isolation and Characterization of Bacterial Strains Capable of Growing on Malathion and Fenitrothion and the Use of Date Syrup as an Additional Substrate'. International Journal of Environmental Studies 75 (3): 466–483. https://doi.org/10.1080/00207233.2017.1380981.

Shah, M. P. 2020. Microbial Bioremediation & Biodegradation, Springer.

Shah, M. P. 2021a. Removal of Emerging Contaminants Through Microbial Processes, Springer.

Shah, M. P. 2021b. Removal of Refractory Pollutants From Wastewater Treatment Plants, CRC Press.

Sittijunda, S., S. Baka, R. Jariyaboon, A. Reungsang, T. Imai, and P. Kongjan. 2022. 'Integration of Dark Fermentation With Microbial Electrolysis Cells for Biohydrogen and Methane Production From Distillery Wastewater and Glycerol Waste Co-Digestion'. Fermentation 8: 537. https://doi.org/10.3390/fermentation8100537.

Srivastava, Ashima, Pratibha Singh, Shaili Srivastava, and Shyni Singh. 2023. 'Reclamation and Characterization of Value-Added Products From Pulp and Paper Mill Effluent Using Microbial Fuel Cell'. Materials Today: Proceedings, ISSN 2214-7853, https://doi.org/10.1016/j.matpr.2023.02.415.

Su, X., W. Zhao, and D Xia. 2018. 'The Diversity of Hydrogen-Producing Bacteria and Methanogens Within an in Situ Coal Seam'. Biotechnology for Biofuels 11: 245. https://doi.org/10.1186/s13068-018-1237-2.

Tazdaït, Djaber, Nadia Abdi, Hocine Grib, Hakim Lounici, André Pauss, and Nabil Mameri. 2013. 'Comparison of Different Models of Substrate Inhibition in Aerobic Batch Biodegradation of Malathion'. Turkish Journal of Engineering and Environmental Sciences 37: 221–230. https://doi.org/10.3906/muh-1211-7.

Tazdaït, Djaber, and Rym Salah-Tazdaït. 2021. 'Polycyclic Aromatic Hydrocarbons: Toxicity and Bioremediation Approaches'. In Biotechnology for Sustainable Environment, edited by Sanket J. Joshi, Arvind Deshmukh, and Hemen Sarma, 289–316. Springer Singapore. https://doi.org/10.1007/978-981-16-1955-7_12.

Uddin, M. N., S. Y. A. Siddiki, M. Mofijur, F. Djavanroodi, M. A. Hazrat, P. L. Show, S. F. Ahmed, and Y. -M. Chu. 2021. 'Prospects of Bioenergy Production From Organic Waste Using Anaerobic Digestion Technology: A Mini Review'. Frontiers in Energy Research 9: 627093, https://doi.org/10.3389/fenrg.2021.627093.

Umar, Mohammad Faisal, Mohd Rafatullah, Syed Zaghum Abbas, Mohamad Nasir Mohamad Ibrahim, and Norli Ismail. 2021. 'Enhanced Benzene Bioremediation and Power Generation by Double Chamber Benthic Microbial Fuel Cells Fed With Sugarcane Waste as a Substrate'. Journal of Cleaner Production 310 (August): 127583. https://doi.org/10.1016/j.jclepro.2021.127583.

Veerubhotla, Ramya, Debabrata Das, and Debabrata Pradhan. 2017. 'A Flexible and Disposable Battery Powered by Bacteria Using Eyeliner Coated Paper Electrodes'. Biosensors and Bioelectronics 94 (August): 464–470. https://doi.org/10.1016/j.bios.2017.03.020.

Vignesh, K. S., Suriyaprakash Rajadesingu, and Kantha Deivi Arunachalam. 2021. 'Challenges, Issues, and Problems With Zero-Waste Tools'. In Concepts of Advanced Zero Waste Tools, edited by Chaudhery Mustansar Hussain, 69–90. Elsevier. https://doi.org/10.1016/B978-0-12-822183-9.00004-0.

Walter, A., S. Silberberger, M. F. D. Juárez, H. Insam, and I. H. Franke-Whittle. 2016. 'Biomethane Potential of Industrial Paper Wastes and Investigation of the Methanogenic Communities Involved'. Biotechnology for Biofuels 9: 21. https://doi.org/10.1186/s13068-016-0435-z.

Wang, Fengbin, Lisa Craig, Xing Liu, Christopher Rensing, and Edward H. Egelman. 2023. 'Microbial Nanowires: Type IV Pili or Cytochrome Filaments?. Trends in Microbiology 31 (4): 384–392. https://doi.org/10.1016/j.tim.2022.11.004.

Wang, J., and Y. Yin. 2018. 'Fermentative Hydrogen Production Using Various Biomass-Based Materials as Feedstock'. Renewable and Sustainable Energy Reviews 92: 284–306. https://doi.org/10.1016/j.rser.2018.04.033.

Wang, Yingying, Xi Chen, Katharina Spengler, Karoline Terberger, Marko Boehm, Jens Appel, Thomas Barske, Stefan Timm, Natalia Battchikova, Martin Hagemann, and Kirstin Gutekunst 2022. 'Pyruvate:Ferredoxin Oxidoreductase and Low Abundant Ferredoxins Support Aerobic Photomixotrophic Growth in Cyanobacteria'. eLife 11: e71339. https://doi.org/10.7554/eLife.71339.

Yaghmaeian, Kamyar, Ahmad Rajabizadeh, Farshid Jaberi Ansari, Sebastià Puig, Roohallah Sajjadipoya, Abbas Norouzian Baghani, Narges Khanjani, and Hossein Jafari Mansoorian. 2023. 'Treatment of Vegetable Oil Industry Wastewater and Bioelectricity Generation Using Microbial Fuel Cell via Modification and Surface Area Expansion of Electrodes'. Journal of Chemical Technology & Biotechnology 98 (4): 978–989. https://doi.org/10.1002/jctb.7301.

Zahan, Z., M. Z. Othman, and T. H. Muster. 2018. 'Anaerobic Digestion/Co-Digestion Kinetic Potentials of Different Agro-Industrial Wastes: A Comparative Batch Study for C/N Optimisation'. Waste Management 71: 663–674. https://doi.org/10.1016/j.wasman.2017.08.014.

Zeikus, J. G. 1980. 'Microbial populations in digesters'. In: Anaerobic digestion:[proceedings of the first International Symposium on Anaerobic Digestion, held at University College, Cardiff, Wales, September 1979]/edited by DA Stafford, BI Wheatley and DE Hughes. London, Applied Science Publishers,[1980]

Zhang, L., K. -C. Loh, A. Kuroki, Y. Dai, and Y. W. Tong 2021. 'Microbial Biodiesel Production From Industrial Organic Wastes by Oleaginous Microorganisms: Current Status and Prospects'. Journal of Hazardous Materials 402: 123543. https://doi.org/10.1016/j.jhazmat.2020.123543.

Zhao, Chuanfu, Bing Liu, Shuangyu Meng, Yihua Wang, Liangguo Yan, Xinwen Zhang, and Dong Wei. 2021. 'Microbial Fuel Cell Enhanced Pollutants Removal in a Solid-Phase Biological Denitrification Reactor: System Performance, Bioelectricity Generation and Microbial Community Analysis'. Bioresource Technology 341 (December): 125909. https://doi.org/10.1016/j.biortech.2021.125909.

4 Recent Advances in Reactor Design for the Treatment of Industrial Wastewater

Rajasri Yadavalli, Chunduru S. H. Sudheshna, Kolluru Naga Venkata Sujatha, Peri Shreenija, C. Nagendranatha Reddy, Bishwambhar Mishra, and Sanjeeb Kumar Mandal

4.1 INTRODUCTION

Everyone needs clean water to survive. Polluted water has negative effects on human health, wildlife, including fish, and the economy. Pollution of lakes and rivers is one of the most important environmental issues. Chemicals entering water bodies cause major destruction. Although water pollution occurs naturally, it is mostly caused by human activities (1). Chemical water pollution is a big problem – a problem and a priority for society and the government, but more significantly for the sector as a whole. There are various categories of water pollution. The first concerns the contaminants that come with it, which refer to pollutants from single sources, such as industrial emissions into water, as well as from multiple sources (2).

In this chapter, we will be learning about the sources of wastewater and recent advancements in treating wastewater to minimize pollution and make water reusable.

4.2 SOURCES OF WASTEWATER

We must comprehend the sources of wastewater and its pollutants if we are to protect water resources. The three main sectors of wastewater are the living, industrial, and agricultural sectors. Water used in domestic activities is included in domestic wastewater. Industries like food, chemicals, paper/pulp, nuclear/thermal power, laundry, pharmaceuticals, mining, and steel all produce industrial wastewater. Both organic and inorganic materials are abundant in these wastewaters. The enrichment of minerals and algae from an excessive release of these nutrients into the water, on the other hand, causes oxygen-depleted water bodies and eutrophication. With proper treatment, this recycled wastewater can be used for the removal of various contaminants. Here, contaminants from various industries have been identified, and ways to treat and reuse wastewater have been explored. First, we need to understand the composition of the wastewater. This will allow us to develop appropriate technologies to treat water before it is returned to bodies of

DOI: 10.1201/9781003381327-4

water or reused for irrigation and landscaping (3). Wastewater can be broadly divided into two types, namely industrial and domestic wastewater.

4.2.1 Domestic Wastewater

Household wastewater may include water from hospitals, industries, and other commercial establishments. Water collected during storms is also acceptable as domestic wastewater. The most common sources include fluid discharge from normal hygienic uses such as bathing, cooking, and washing meat, vegetables, and clothes. Domestic wastewater can be properly treated in spite of serious pollution.

4.2.2 Blackwater

Human waste is a type of domestic wastewater. This type of sewage is contaminated with human waste. Such wastewater is therefore also generated from flushing toilets and bidets. Common contaminants include faeces, urine, body wipes, toilet paper, soaps, shampoos, and cleaning products. This water is considered highly pathogenic due to the high content of dissolved chemicals.

4.2.3 Grey Water

Also called sullage, it is not contaminated with faecal matter. It originates from washing machines, wash basins, showers, bathtubs, and spa tubs. This is treated to make it suitable for human use.

4.2.4 Industrial Wastewater

As the name suggests, this wastewater usually arises from production or commercial activities. This water is laden with contaminants such as oil, silt, toxic chemicals, ink, pharmaceuticals, pesticides, sand, silt, and chemicals that are not as easily treated as domestic wastewater.

4.3 MAJOR EFFLUENTS OF WASTEWATER FROM VARIOUS INDUSTRIES: FOOD INDUSTRY, TEXTILE INDUSTRY, AND PHARMACEUTICAL INDUSTRY

4.3.1 Food Industry

Dairy manufacturing waste contains milk solids in various concentrations and dilutions. Almost all of these solids end up as waste operations. Temperature, colour, pH (6.5–8.0), BOD, COD, dissolved solids, suspended particles, chlorides, sulphates, oils, and fats are some of the features of dairy effluent. This is highly dependent on how much milk is processed and what kind of product is made. Casein, inorganic salts, detergents, and cleaning agents used in dairy farms are among the many milk components found in effluent from these operations. The usage of caustic soda for cleaning is to blame for the high sodium concentration.

A slaughterhouse's effluent may comprise blood, faeces, dung, hair, fat, feathers, and bones. Starch, proteins, COD, TSS, and total Kjeldahl nitrogen (TKN) are all present in significant proportions in the wastewater from the processing of potatoes. It is claimed that vegetative water, soft tissues from the olive fruit, and water used throughout various stages of oil production make up olive mill wastewater (OMW). The composition of the vegetative water, the method used to extract the oil, and the length of storage time all affect the qualitative and quantitative features of OMW (4).

4.3.2 Textile Industry

The textile manufacturing process uses large amounts of water. The main contaminants in textile waste water are persistent organics, dyes, toxins, inhibitory compounds, surfactants, soaps, detergents, chlorine compounds, and salts. Dyes are the most difficult components of textile waste water to treat. Depending on the campaign, the type of dye in the drain may change daily or hourly. The use of conventional synthetic dyes has seriously threatened the global environment. Their presence in textile effluents has seriously polluted the environment in recent decades. Their toxic and non-biodegradable nature poses a serious threat to soil fertility, crop production, and human health. At present, bio-dyes or natural dyes are attracting attention. Plants, animals, and microorganisms can produce inexpensive, non-toxic, and environmentally friendly dyes that can be applied as textile dyes. Textile dye effluents are among the most dangerous, mostly belonging to the class of contaminants in the current environmental problem and easily identifiable by the human eye. Discharge of textile waste water into water bodies should be avoided by various treatment techniques. A great demand for inexpensive and sustainable treatment methods has led to sorbents (5).

4.3.3 Pharmaceutical Industry

Pharmaceuticals are typically manufactured in batch processes, resulting in a wide variety of compounds. Wastewater is produced by a variety of operations that use a lot of water for cleaning, extraction, washing equipment, or cleaning solid cake. Consequently, the presence of pharmaceuticals in drinking water comes from two different sources: the manufacturing processes used in the pharmaceutical industry and the pharmaceutical compounds itself. As a result, it is found in both municipal and agricultural wastewater. Wastewater is generated in various manufacturing processes. Pharmaceuticals contain various chemical compounds. Active pharmaceutical ingredients (API), active pharmaceutical ingredients, and related fields all produce waste water. Medications that use a lot of water are assessed, recovery strategies for the majority of the water's valuable compounds are suggested, and treatment options for highly diluted but toxic effluents are discussed. No single technology can completely remove pharmaceuticals from wastewater. Conventional processing methods using membrane reactors and advanced post-treatment methods leading to hybrid wastewater treatment technologies seem to be the best (6). The current study claims that pharmaceutical effluents discharged into sewage contain a variety of heavy metals, phenols, drugs, and other organic compounds. These pharmaceutical effluent ingredients have a toxic nature and may be bad for the environment. It is necessary to swap out or replace the various raw materials used with other non-toxic ones. Also, various kinds of technical methods should be used to reduce manufacturing designs and sources of harmful substances. New zero-liquid waste concepts must be incorporated into waste disposal methods (7).

4.3.4 Paper and Pulp Industry

Diverse organic and inorganic substances, including those originating from wood, process chemicals, and compounds created during chemical reactions with raw materials, can be found in the wastewater from various pulp-making processes. The two primary classes of substances that contribute to COD in wastewater are carbohydrates and chemicals derived from lignin; however, they are not very harmful. Out of all substances derived from wood, resin acids, such as abietic acid and dehydroabietic acid, are generally recognized as the main cause of acute toxicity in pulp and paper industry effluents. During chemical pulping, wood typically loses 90% of its lignin, 50% of its hemicellulose, and 20% of its cellulose. The majority of this dissolved material ends up in the black liquor and is burned for disposal, though some of it is released into the process wastewater streams. The bleaching process is much more precise than pulping because the lignin is removed, leaving only the cellulose and hemicellulose. The wastewater from bleaching contains these chemicals, which are thought to have persistent, poisonous, and bioaccumulative properties because they have historically been used as bleaching chemicals (Table 4.1).

TABLE 4.1
Composition of Wastewater from Various Industries

S. No.	Industry		Composition of Effluents
1.	Food Industry		
		Dairy Industry	1. Suspended particles
			2. Chlorides
			3. Sulphates
			4. Oils
			5. Fats
			6. Casein
			7. Inorganic salts
			8. Detergents
			9. Cleaning agents
		Slaughterhouse	1. Blood
			2. Faeces
			3. Dung
			4. Hair
			5. Fat
			6. Feathers
			7. Bones
		Potato Processing Industry	1. Starch
			2. Proteins
		Olive Oil Industry	1. Vegetative water
			2. Soft tissues from the olive fruit

(Continued)

TABLE 4.1 *(Continued)*
Composition of Wastewater from Various Industries

S. No.	Industry	Composition of Effluents
2.	Textile Industry	1. Persistent organics 2. Dyes 3. Toxins 4. Inhibitory compounds 5. Surfactants 6. Soaps 7. Detergents 8. Chlorine compounds 9. Salts
3.	Pharmaceutical Industry	1. Heavy metals 2. Phenols 3. Drugs 4. Organic compounds
4.	Paper and Pulp Industry	1. Chemicals derived from lignin 2. Resin acids, which include abietic acid and dehydroabietic acid 3. Hemicellulose 4. Cellulose 5. Bleaching chemicals

4.4 TYPES OF WASTEWATER TREATMENT

Waterborne solids and liquids, or wastewater, are an example of municipal waste and are dumped into sewage systems. Wastewater contains dissolved and suspended organic materials that can decompose biologically or are putrescible. Domestic and industrial wastewaters are regarded as falling into two main types that are not fully distinct. During the wastewater treatment process, the highly complex, putrescible organic materials in wastewater are partially eliminated and partially converted by breakdown into mineral or comparatively stable organic solids. Most of the BOD and suspended particles present in wastewater are removed during primary and secondary treatments. However, it has become clearer that this level of treatment is insufficient to create reusable water for industrial and/or domestic recycling or to protect the receiving waters (8).

Because of this, additional treatment procedures have been added to wastewater treatment facilities to ensure the removal of potentially harmful or nutrient-containing substances, as well as additional organic and solids removal. In the past several years, the subject of water treatment has seen a number of innovations. Alternatives to traditional and conventional water treatment techniques have emerged. With everyone trying to find solutions to maintain important resources accessible and usable, advanced wastewater treatments have gained attention on a global scale. The loss of useful water is unavoidable, but with the help of cutting back on wastewater production and water recycling operations, there is optimism that it can be slowed or even stopped. Wastewater recycling and reuse function well with membrane technologies. High selectivity separation of substances across a range of particle sizes and molecular weights is possible with membranes. Over the centuries, membrane technology has evolved into a respectable method of separation. The major strength of membrane technology is its ability to perform processes easily and efficiently while using minimal energy and no additional chemicals (1).

Several wastewater treatment methods have been used to remove the dangerous pollutants from the water. When choosing a wastewater treatment technique, the following elements must be taken into account.

- The amount of technique required to raise the effluent water quality to a permissible level
- Adaptability of the control method
- Cost of the process
- Environmentally friendly

4.4.1 Biological Methods

The management of organic waste and sewage sludges and the optimization of wastewater treatment are two areas where bio-based solutions are particularly useful. The achievements to date include both pilot- or laboratory-scale testing and large-scale implementations (9). In addition to the standard biological treatment with activated sludge, wastewaters may directly biomethanate or serve as sources for a range of bioproducts (10). Like organic residues, municipal sewage sludge and crops grown specifically for biomass can be used as substrates for biorefineries. However, anaerobic digestion – a still-evolving technology that combines the production of biogas with the treatment of garbage – is the method most frequently used to process organic waste. Waste-utilizing biogas plants should be viewed as essential elements capable of closing the product/material loops in a modern bioeconomy because they depend on the circulation of organic matter (Table 4.2) (9).

TABLE 4.2
Biological Treatments, Their Advantages and Disadvantages

S. No.	Biological Method	Advantages	Disadvantages
1.	Anaerobic Treatment	• Creates sources of renewable energy • Lesser pollution to environment	• High capital cost • Odour nuisance
2.	Aerobic Treatment	• Simple • Limits odour generation • Reduces pathogens	• Expensive • Maintenance issues

4.4.1.1 Anaerobic Digestion of Organic Waste

During the anaerobic digestion process, which is based on microbial methane fermentation and the production of biogas, the biodegradable portion of organic content goes through partial mineralization (11). Because of its resiliency, environmental friendliness, and financial benefits, it is thought to have a significant advantage over conventional treatments for organic waste and wastewater. These benefits come from both the energy produced by the fermentation process and the quick recovery of vital nutrients like phosphorus or nitrogen, which are released from their organically bound forms. The diagram depicts the overall flow of organic matter that results from the fermentation of various feedstocks (Figure 4.1). It also shows how a digestate by-product is produced and suggests potential processing techniques for this material (9).

Digestate is a nutrient-rich waste product that is grouped according to the kind of input feedstock. It is also known as the pulp, post-ferment, or post-fermentation mass (9). This waste product is viewed as an opportunity by the biogas producer. Due to the increased administration and use that are required, the process will cost more. On the other hand, this information on biogas, may have significant value for farmers as a possible fertilizer product (12). Separations between solid and liquid municipal waste and between animal and vegetable waste are the most frequent for post-fermentation

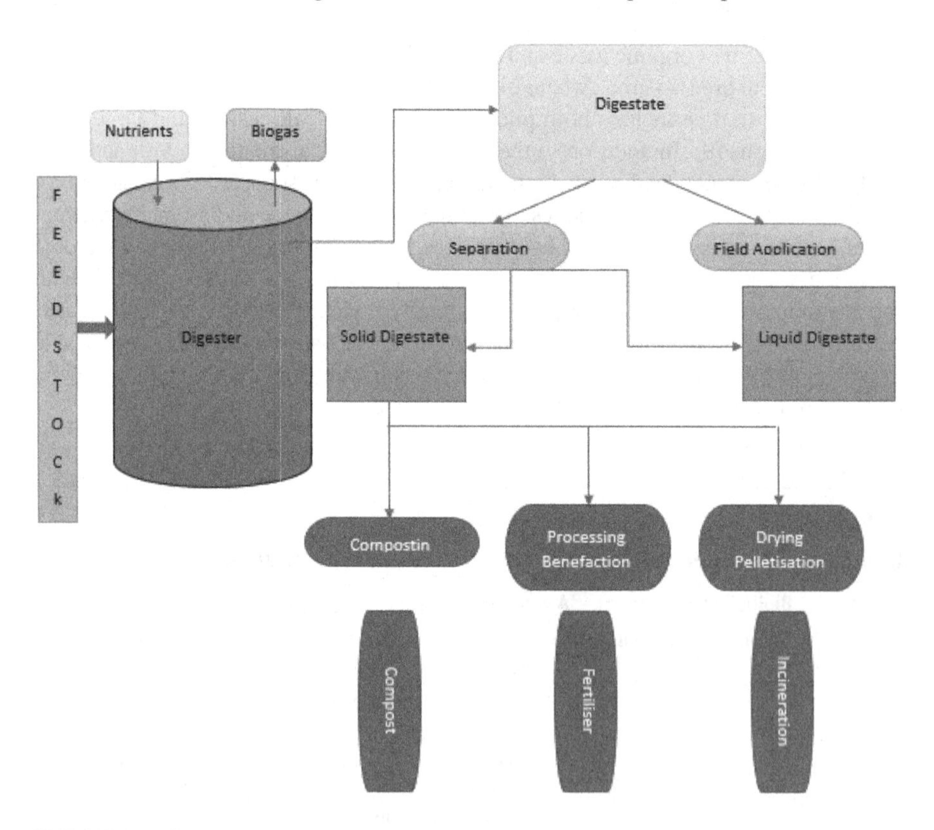

FIGURE 4.1 Flow diagram depicting the fate of fermentation of feedstocks.

pulp, respectively. It is interesting to note that, in the case of agricultural feedstock fermentation, legal changes have been made to support direct digestate soil applications, market commercialization, and the simplification of management processes.

4.4.1.2 Anaerobic Digestion of Sewage Sludges

Separable organic-mineral phases are large quantities of sewage sludge of various types that are produced during the biological treatment of municipal wastewaters. To remove ecological concerns and microbiological threats, they must go through further multistage processing (13). These complex procedures lead to reductions in volume, odour, cleanliness, chemical content stability, and susceptibility to putrefaction and breakdown in the future. For these purposes, a variety of techniques can be used, including sludge thickening, dewatering, drying, pasteurization, irradiation, lime treatment, thermal stabilization by incineration, pyrolysis, or gasification, conditioning, recovery of biogenic components, aerobic composting, and ultimately anaerobic digestion (14).

4.4.2 Physical Methods

Physical processes are used to remove contaminants in water and wastewater treatment; no biological or chemical modifications are made during this procedure (15). The most popular physical techniques for reducing water contamination include (Table 4.3):

- Sedimentation
- Degasification
- Filtration

4.4.2.1 Sedimentation

One of the important fundamental techniques for treating wastewater is sedimentation. It is a method that uses gravity to separate particles from a fluid. During the water treatment process, the water velocity is reduced, causing the particles in suspension to remain stable under still circumstances until they are gravitationally forced to settle (16). Particle size plays a significant role in the sedimentation process. To determine settling velocity, one could use the remaining time required for

TABLE 4.3
Physical Treatments, Their Advantages and Disadvantages

S. No.	Physical Method	Advantages	Disadvantages
1.	Sedimentation	• No energy • Excellent reproducibility	• Selective process • No precision
2.	Degasification	• Mitigates chemicals required for further processes	• Limited time for pollutant removal
3.	Filtration	• Autoclaving can be done for few instances	• Clogging of filters may occur

particle settling. Sedimentation is used to lower the concentration of solids prior to coagulation, thereby requiring fewer coagulants during the coagulation process. Sedimentation in dirty water treatment can come in three different forms:

- When particles settle, they cling together, or no flocculation takes place.
- The settling velocity is impacted by the steady change in particle size, and the particles flocculate.
- Particles with a high concentration settle down.

Some chemicals become adsorbed on the suspended material during the sedimentation of organic compounds. Pollutants with a higher density veer away from fluid-streamlined flow and sink to the bottom. The process of aggregation and the efficiency of solid removal are both impacted by the physical and hydrodynamic processes that the solid particles experience as a result of shear forces in the water stream. The effectiveness of the sedimentation process is also influenced by flocculation, thickening, and gravity sedimentation (17).

4.4.2.2 Degasification

The act of removing dissolved gases from a solution is known as degasification. Degasification is supported by Henry's law, which states that the amount of dissolved gas in a liquid is proportional to the partial pressure of the gas. Degasification is a process for removing carbon dioxide gas from wastewater that is both efficient and inexpensive (15). By removing the gas, it raises the pH of the water. The following factors determine how long the degasification process will last:

- Temperature of the wastewater solution
- Tank capacity
- Ultrasonic power and frequency

Typically, liquid degasification takes place at a certain temperature. Further processing cannot be done at any higher temperature when the liquid is first degasified at a specific temperature. Spraying and pumping a liquid after the operation is complete may have a big impact on the degasification process. Degassing can be accelerated by heating the wastewater solution or by intermittently applying ultrasonic energy. Small gas bubbles start to form, mix, and rise to the top once they have enough buoyancy (15).

4.4.2.3 Filtration

According to their size, pollutants are eliminated by filtration. Reusing water for various purposes is made possible by pollution removal from wastewater. Depending on the kinds of contaminants in the water, different filters are used in the procedure. The two primary types of membrane filtration are particle filtration and wastewater filtering (18). One of the key processes in the procedure for treating wastewater is particle filtering. Getting rid of materials larger than one micron in size is its intended use. The form, size, texture, density, and number of particles all influence the filters that are employed in the filtration process. The two main filter types utilized in the process of filtering contaminated water are bag and cartridge filters (19).

Solid wastes are encased in bags and filtered using bag filters. Activated carbon, nylon, etc. are examples of filtration mediums. Drainage water is allowed into the bags, keeping the solid particles in place while allowing liquid to pass through. For small-scale uses, bag filters are perfect. Wastewater treatment reduces the amount of solid waste produced, as contrasted with cartridge filters. Less so than other filter systems, bag filters hold all of the trash. In cartridge filters, solid particles are trapped outside the filter media. Polluted water enters the filter vessel from the top, travels through the filter medium, which traps particles, and then exits via the lower part. There are three types of cartridge filters: disposable, back-washable, and cleanable. The drawbacks of cartridge filters include difficulties with air reversal and a limited selection of filter materials. Additionally, particulate matter, suspended particles, germs, viruses, and other chemical pollutants may also be removed via filtration (20).

4.4.3 Chemical Methods

In addition to physical or biological methods, chemical treatments are used to lessen the discharge of pollutants and wastewater into water bodies. To transform impurities into finished goods or remove pollutants for safe disposal, various chemical processes are used. The following are some chemical methods (Table 4.4):

- Flocculation and coagulation
- Ozonation
- Chemical precipitation

TABLE 4.4
Chemical Treatments, Their Advantages and Disadvantages

S. No.	Chemical Method	Advantages	Disadvantages
1.	Flocculation and Coagulation	• Fine particles can be removed • Metals, turbidity, and colours can also be removed	• Multiple steps involved • Can lead to toxicity, if not used properly • High cost of operations
2.	Ozonation	• No chemical required • Elimination of a large variety of components • No need to alter pH and temperature • Strong activity	• Demands specialized mixing techniques • Expensive • Toxicity issues
3.	Chemical Precipitation	• Easy • Low cost • pH can be controlled	• Large quantity of sludge produced • Disposal of sludge is a challenge

4.4.3.1 Flocculation and Coagulation

The separation of solids from liquids in the treatment of industrial wastewater depends on flocculation and coagulation. By neutralizing the charges of the colloidal suspensions during the coagulation process, some compounds known as coagulants destabilize the suspensions, which causes the aggregation of smaller particles (17). To further accelerate particle aggregation and increase settling efficiency, flocculants are occasionally added. The structural characteristics of different coagulants and flocculants, such as functional groups, charge, ionic strength, and molecular weight, result in a variety of properties for each (21). Since the majority of the dispersed particles are negatively charged, long-chain anionic or non-ionic polymers are used as flocculants, whereas cationic inorganic salts are used as coagulants (22). Metal salts are quickly hydrolysed after the coagulant is added to produce cations, which are then quickly absorbed by negatively charged colloidal particles in wastewater at the isoelectric point. As a result, both the production of microflocs and the lowering of surface charge occur simultaneously. When flocculants are added, the microflocs clump together to produce denser, bigger flocs, making it simple to remove them by following physical processes like filtration or sedimentation. This process' primary mechanisms are bridging and charge neutralization (23).

4.4.3.2 Ozonation

Due to ozone's strong capacity for oxidation and disinfection, the ozonation process has attracted more interest in industrial water treatment technologies. Ozone is used, among other things, to oxidize wastewater.

- Elimination of chemicals that produce odour, taste, and colour
- Inorganic substances' oxidation states being raised
- Rarely biodegradable compound cleavage
- Organic pollutants oxidizing
- Disinfection

Ozone interacts with the components of wastewater either immediately through molecular ozone or afterwards via hydroxyl radical's emergence (24).

4.4.3.3 Chemical Precipitation

One of the best methods for removing heavy metals from industrial effluent is chemical precipitation. Through chemical formation between dissolving metal components and precipitating agents, ionic metals become insoluble particles. The elimination of cationic metals frequently involves the use of chemical compounds.

It occasionally also eliminates active molecules and anions. In the process of treating water, three main forms of precipitation are used:

- Hydroxide precipitation
- Carbonate precipitation
- Sulphide precipitation

4.5 TYPES OF BIOREACTORS FOR TREATING WASTEWATER

4.5.1 Activated Sludge Process

This method of wastewater treatment breaks down organic materials in sewage using aerobic microorganisms. The liquid created via this process is then devoid of any other solids and organic material. The process for treating activated sludge includes an aeration tank where air is injected into the liquid, a settling tank that aids in the sedimentation of water by separating the sludge from clear water treated in this way, and a treatment step that removes any remaining dissolved oxygen (DO) with nitrogenous matter or phosphate (25).

Since a long time ago, especially in local towns with dense populations, unpleasant aromas released from wastewater treatment plants (WWTPs) have been a concern. The discharge of offensive scents is linked to the formation of photochemical smog and the emission of secondary pollutants, in addition to its annoyance effects and health risks for nearby residential populations. Traditional biotechnologies for media-based odour management (biofilters, biotrickling filters, bioscrubbers, etc.) include drawbacks such as clogged media, shorted gas, humidity control, and the accumulation of harmful metabolites in the system. In contrast, the activated sludge (AS) process used in liquid-based systems can avoid these drawbacks while treating odour and wastewater. Additionally, AS-based procedures can utilize the infrastructure already in place at WWTPs, resulting in cheap operating costs and great odour reduction effectiveness. Numerous bench-scale (26, 27), pilot-scale (28), or full-scale (29, 30) studies have been carried out, and promising results have confirmed the robustness and efficiency of the AS system. There are two ways to deodorize with the AS process in WWTPs: (1) preventing odour generation or evaporation into the air, and (2) collecting and treating odours generated within the facility (31).

The suspended and attached growth systems are combined in the integrated solid membrane-activated sludge process (IFAS). IFAS systems provide significant levels of nitrification and denitrification and are excellent at removing dissolved organic carbon. A forward-looking alternative for wastewater treatment is IFAS. The moving bed biofilm reactor (MBBR), which grows biofilm on suspended media, is further developed by IFAS. By starting to create floating bio-carriers on which biofilm might build, IFAS integrates the biofilm and CAS in this manner. The media is kept in the bioreactor or is fastened to a stationary platform. The nitrifiers linked to the carrier media, which also support system compactness and resilience, enable IFAS to achieve increased biological nitrogen removal (BNR). Both autotrophic and heterotrophic bacteria can coexist in a biofilm or floc, which improves phosphate removal and BNR. In contrast to other connected growth procedures like RBC, which uses only a fraction of the bioreactor volume, IFAS maximizes the use of the bioreactor volume. As a result, IFAS combines the beneficial properties of both connected and suspended growing systems. Two common microbial communities that make up distinguishing traits in IFAS are flocs and biofilm. Their physicochemical and structural characteristics shed light on how they contribute to the oxidation of contaminants. For instance, the compact (denser than water) floc in the CAS settles easily, producing effluent that is almost sludge-free. In contrast, the biofilm's detached sludge's physicochemical characteristics make it difficult to settle. Thus, traits of both floc and biofilm are crucial to understanding their settleability carrier media/fill ratio, operating conditions, and biology are the three most important IFAS parameters. Two common microbial communities that distinguish IFAS features are flocs and biofilms. Their physicochemical and structural characteristics shed light on how they contribute to the oxidation of contaminants. For instance, the compact (denser than water) floc in the CAS settles easily, producing effluent that is almost sludge-free. In contrast, the biofilm's detached sludge's physicochemical characteristics make it difficult to settle. Thus, traits of both floc and biofilm are crucial to understanding their settleability. Carrier media/fill ratio, operating conditions, and biology are the three most important IFAS parameters. Parameters such as DO content are also important to help microorganisms digest organic compounds. Ventilation is also required to keep the carrier medium in suspension. Despite the low DO, high SRT, low hydraulic retention time (HRT), low temperature, and low C/N ratio, the IFAS study results show that this system has great potential for wastewater treatment (32).

4.5.2 The Membrane Bioreactor (MBR) Process

The goal of a two-stage anaerobic digestion process consists of separating the acid fermentation stage and the methane fermentation stage in order to provide each type of microbial population with the ideal environmental conditions in two different reactors. Compared to a single-stage method, a two-step process supports the growth and performance of a wider variety of archaea. The rapid formation of volatile fatty acids (VFAs) is a major drawback of the one-step anaerobic digestion of solid waste. Conditions for acetogenic, methanogenic, and hydrolysing acidogenic groups can be improved by a two-phase approach. The pH is typically kept between 5.5 and 6.0 throughout the acidogenic phase, and the HRT is less than 5 days. To increase treatment effectiveness and process stability, the pH in the methanogenic phase of the two-phase process is maintained above 7.0. A significant flaw that causes membrane fouling can lead to erratic and slow reactor operation. These issues have hindered the widespread commercial deployment of MBR. Researchers have developed a new type of MBR that uses inexpensive non-woven fabrics or meshes as filtration support media. A dynamic membrane (DM) was created to serve as an MBR filtration system by unfolding a cake layer on a support membrane. High-quality wastewater production and strong operational stability are two advantages of DM. The use of DM in anaerobic systems has evolved due to the treatment of wastewater under harsh conditions such as significant expansion of Suspended Solids (SS) concentrations including high salinity, organic matter content, lipids, and industrial effluents. DM replaces expensive and maintenance-intensive traditional membranes such as ultrafiltration (UF) and microfiltration (MF) membranes. DAnMBR successfully removed organics from high-strength FPW using non-woven filter cloth as a support material for DM deployment. The amount of TSS removed from the feed effluent was over 99%, which is significantly higher than standard A effluent requirements (33). Previously, several studies focused on MBR treatment of domestic/municipal wastewater. The potential use of MBRs in various forms of industrial wastewater treatment has received much attention recently.

Pressure-based membrane filtration processes have replaced secondary clarifiers, resulting in several improvements (34).

4.6 RECENT ADVANCES IN THE TREATMENT OF WASTEWATER

The world's freshwater supply has been under stress in recent decades due to rising population demands. It is anticipated that the water shortage would worsen in the years to come. According to statistical statistics, the yearly water use in the three main industries of agriculture, manufacturing, and urban development is around 4 trillion cubic meters, with a sharp increase. However, there are only around 0.01 trillion cubic meters of freshwater readily available for human use. Due to a lack of fresh water supplies for human consumption, unorthodox water sources, including wastewater, precipitation, and salty water, have been used. Less than 20% of the world's water consumption is currently attributed to industrial use, but that percentage is predicted to double by 2050. Each year, sizable amounts of untreated industrial wastewater are released into the environment, threatening both ecosystems and human health. The need for effective water management methods is driven by problems caused by increasing water consumption and wastewater from the industrial sector (35).

4.6.1 Food Industry

In the food processing industry, wastewater remedy is important for the maintenance of the environment, resource recovery, and water reuse and recycling. White, grey, and black water are the three primary categories of industrial wastewater based on their characteristics and possibilities for reuse. White water has a property that is essentially identical to that of freshwater, with a few minor exceptions like temperature. The food processing businesses that have the best potential for reuse and recycling without any necessary treatment are bakeries and the edible oil industry. White water is created in tiny volumes in these industries. After modest treatment to eliminate particulate matter (TDS or TSS), the most difficult contaminant in this wastewater category, it can be reused in grey water (36).

Increased amounts of COD, BOD, or nutrients in a water stream make it more difficult to treat gray water. Powered by an identifiable set of outputs, the best technology for gray water treatment in food processing is physical treatment. The techniques used in physical processes are simple, affordable, and highly effective in removing solids. Membrane-based technologies (such as RO, electrodialysis (ED), UF, and NF) are highly effective (in terms of water, energy, and land requirements) for treating grey water in the food industry. It has been proven that in contrast to other common physical methods such as sedimentation, flotation, and crystallization (37).

These processes separate suspended and dissolved particles very efficiently and with low energy consumption, which makes them particularly effective for recovering value-added resources from wastewater, such as phenolic compounds, antioxidants, glycosides, etc. However, there are a few practical restrictions on the use of membrane approaches, with fouling being the most important. Membrane fouling is made more difficult by the wastewater characterizations and pre-existing pollutants (meat and bone fragments, dirt and grime, straw and plant biomass, leaves, etc.) for food processing. Although recent developments in membrane fabrication, activation, application, and regeneration have greatly reduced the destructive effects of fouling, new developments are constantly taking place in this area (38).

The wastewater from the food processing sector contains a broad variety of chemical and biological pollutants, which has increased the requirement for the creation and development of selective membranes with high removal efficiency for particulate contaminants. Given that their separation using traditional treatment techniques is not possible, heavy metals constitute a crucial group of pollutants. The method advised for removing heavy metals from wastewater is membrane treatment. The physical properties of the membrane materials and their careful selection are what determine whether a rejection is effective. With excellent penetration rates, using selective nanocomposite membranes is a unique strategy that can virtually completely remove heavy metals. Data show that nanocomposite membranes are being used to remove a wide range of contaminants, including bacteria, pathogens, and ions. Its uses are not limited to removing heavy metals (39).

The third tough group of pollutants in wastewater treatment for the food processing industry is nutrients (N and P), which mostly come from protein compounds and agricultural fertilizers. While phosphorus is often recorded as PO_4, nitrogen can be found in wastewater in the forms of $N\text{-}NO_3$, $N\text{-}NH_3$, and TN. Various nitrogen forms play different roles in the design and implementation of biological treatments for the sequence of aerobic and anaerobic processes. Therefore, qualitative wastewater analysis separates them. A major cause of dangerous algal blooms is an oversupply of nutrients released primarily from untreated wastewater. HAB is the rapid and uncontrolled growth of wild algal species in aquatic environments (especially lakes and wetlands) that releases toxins into the water, lowers DO levels, kills fish and other aquatic organisms, and is harmful to the ecosystem (40).

4.6.2 Textile Industry

Due to the colourful wastewater that is produced during the textile dyeing process, it is one of the least ecologically friendly industrial processes due to the dyes, chemicals, and textile auxiliaries that are heavily disposed of. Additionally, the wastewater from textile finishing, particularly the effluent from dye houses, contains a variety of classes of organic dyes, chemicals, and additives. Due to their excessive pH, COD, and BOD values, colour, and presence of various salts, surfactants, heavy metals, mineral oils, and other substances, dye bath effluents must be treated before being released into the environment or municipal sewage treatment plants (41). Wet processing of fibres in the textile industry uses large amounts of water and chemicals. A variety of colourants are used to impart colour to the fibres, including dyes, tannins, lignin, and inorganic pigments. Textile wastewater contains various dyes, but dye waste predominates. Wastewater from the textile industry contains a huge amount of complex chemicals (42).

Small amounts of dye in wastewater are highly visible and undesirable. The distribution of untreated wastewater from the textile industry is the largest and most damaging global problem. Physico-chemical and biological approaches are used for conventional wastewater treatment. Coagulation, flocculation, and ozonation are examples of physicochemical processes, while biological approaches are used to remove organics, metallic residues, nitrogen, and phosphorus. During the past decades, many approaches or processes have been developed to clean wastewater from the textile industry in an adequate and cost-effective manner. The existing literature has several establishments for decolorizing textile effluents, including physicochemical, biological, and several newly developed techniques such as microbial fuel cells, biofilms, etc. It shows the efficiency of the strategy used (43).

Microbial electro-Fenton technology for the treatment of dye wastewater has emerged as a result of scientific inquiry and has been demonstrated to be an inventive, multidimensional approach that can be used in batch and continuous modes, increasing the possibility of treating domestic and dye wastewater concurrently. The system is superior to other approaches in comparison because it uses time and resources more effectively. Critical evaluations of dye characteristics, classification, environmental destiny, toxicity, and degradation mechanisms have been conducted. For the treatment of dye wastewater, many types and operational models of MEFS have also been evaluated. When treating textile wastewater, a variety of physicochemical and biological techniques are used.

4.6.2.1 Adsorption

Adsorption is thought to be one of the most effective physical–chemical techniques used in the treatment of textile effluent. Techniques for adsorption have attracted interest lately because of how well they remove contaminants. It is an economically viable procedure that yields a high-quality product. In this technique, the contaminants found in wastewater are adsorbed and removed using a porous medium or filter. While kaolin, silicon polymers, and activated carbon are all frequently used adsorbents with the ability to selectively absorb colours, activated carbon has recently acquired popularity. The activated carbon has a unique ability to adsorb colours; yet, it is unable to absorb suspended particles or insoluble dyes. Instead, it can effectively remove water-soluble pigments from wastewater, including reactive, basic, and azo dyes. The adsorption process is greatly influenced by a number of variables, including temperature, pH, dye–sorbent interaction, sorbent surface area, particle size, and contact duration. Chitin, an amino nitrogen-based adsorbent, has a sizable quantity of acid dye adsorption capability (44).

4.6.2.2 Membrane Separation Process

This technique primarily relies on the selective permeability of microporous membranes to separate certain chemicals found in wastewater. Currently, membrane pressure-based separation techniques used to treat dyeing wastewater include reverse osmosis, nanofiltration, ultrafiltration, and MF. All mineral salts, hydrolysed reactive dyes, and chemical compounds found in dye effluent may be decoloured and removed using reverse osmosis in a single process. Nanofiltration, a different membrane-based separation technique with a retention capacity of 80–1000 Da molecular weights, is used to treat coloured wastewater. The adsorption step is used prior to nanofiltration to boost process output. Recently, nanofiltration and adsorption were combined for the treatment of dye wastewater. Recent studies have demonstrated the effectiveness of using nanofiltration to clean dyeing effluent and complicated solutions with high concentrations (45).

4.6.2.3 Ion Exchange Process

Since it cannot be used to remove a variety of colours, the ion exchange approach is not frequently used to treat textile effluent. In this technique, coloured wastewater is allowed to run over ion exchange resin until the available exchange sites are saturated. With certain benefits, including no adsorbent loss, solvent recovery after use, and the removal of soluble dyes, this approach allows the removal of both anionic and cationic dyes from wastewater. This method's use for the treatment of wastewater from the textile industry is restricted by its high operating costs (46).

4.6.2.4 Photochemical

By applying UV therapy when there is water present, colour molecules are degraded using this technique. The only source of colour loss is from very concentrated hydroxyl radicals. To effectively remove dyes, H_2O_2 is activated by UV radiation to produce a hydroxyl radical. Other elements, including pH, dye structure, UV radiation intensity, and dye bath composition, also play a significant role in this method's treatment procedure. When textile wastewater is treated, certain other by-products are also created, including organic aldehydes, metals, halides, inorganic acids, and organic acids. The absence of sludge development and unpleasant smells during wastewater treatment is one advantage of photochemical treatment over other approaches (47).

4.6.2.5 Biological Treatment

Wastewater discharged from the textile industry is a complex mixture of various components. Dyes, especially azo dyes, are major contaminants in textile wastewater. Complex aromatic compounds containing one or more azo (–N=N–) groups are known as azo dyes. These dyes are resistant to deterioration and have stubborn properties. Therefore, it is important to remove these dyes before they are released into the ecosystem. The decolourization process of textile wastewater is inevitably related to the removal of dyes from the wastewater. Dyes are designed to be permanent, chemically stable dyes, so they often do not degrade quickly or efficiently. Both the problems of textile dye degradation and foreign contaminants and excess BOD/COD in wastewater from the textile industry can be solved by biologically based processes. Several taxonomic groups of microorganisms (bacteria, fungi, and algae) and plants are capable of removing dyes from textile wastewater. Moreover, the biological treatment of textile wastewater is cheaper and more environmentally friendly than other treatment methods (48).

4.6.2.5.1 Using Bacteria

It is common practise to use several trophic groups of bacteria to remove colours from textile effluent. Under the right circumstances, these bacteria may be easily cultured, multiply quickly, and work well to decolourize, degrade, and mineralize textile colours. Anaerobic, aerobic, or a mix of anaerobic and aerobic conditions may be used in the bacterial system's colour removal process. However, the physiology of bacteria that are grown in anaerobic and aerobic environments differs greatly (49).

4.6.2.5.2 Anaerobic Conditions

Bacterial removal of azo dyes from textile effluents under anaerobic conditions is a simple, non-specific, and more feasible technique. Many cytosolic azoreductases are used to reduce the azo bonds of dye molecules. These azoreductases have minimal substrate selectivity and are soluble. Reductive cleavage of the azo bond yields a colourless aromatic amine. It is toxic or carcinogenic to organisms and is often resistant to anaerobic mineralization. Under anaerobic conditions, dye removal requires a complex organic carbon or energy source. Both the dye structure and the additional supply of carbon affect the dye removal rate. Various bacterial species, such as *Citrobacter* species, *Pseudomonas putida*, *Clostridium bifermentans*, *Pseudomonas luteola*, and *Staphylococcus hominins*, have been used to remove azo pigments from anoxic/anaerobic environments. Despite the fact that many of these bacteria can grow aerobically, dye removal could only be performed in an anaerobic environment. It has been found to effectively remove numerous azo pigments from various bacterial species, including Bacillus, Pseudomonas, Micrococcus, Proteus, Aeromonas, and sulphur-free purple photosynthetic bacteria. Additionally, several research papers have explored the role of bacterial cultures in anoxic/anaerobic conditions via the colour change of azo dyes (49). Upflow Anaerobic Sludge Blanket (UASB) reactors are used to treat a variety of industrial wastewater and are recognized as one of the most popular anaerobic systems. The system has low investment costs, high efficiency, constant performance, and good adaptability to different wastewater. UASB reactors have proven useful in the treatment of printing and dyeing effluents due to their high resistance to harmful substances in the medium (dyes, additives, aromatic amines, etc.). An anaerobic impactor reactor (ABR) usually has 3–8 chambers, and each chamber can be considered a separate UASB reactor. Liquid flow is regulated up and down between compartment dividers. Additionally, the unique design of the ABR enables phase separation of microorganisms within a single unit, including Acidogenesis and Methanogenesis. Phase separation can also protect microorganisms from harmful compounds and changes in environmental conditions (50).

4.6.2.5.3 Aerobic Conditions

Due to the fact that azo bond decomposition is often hindered in the presence of oxygen, the major dyes in textile effluents as major contaminants are not readily digested in an aerobic environment (51). However, under aerobic conditions, some bacteria are able to degrade azo dyes through a reduction process. These bacteria are highly sensitive to the structure of the pigment and use NADH as a cofactor for action. It produces oxygen-insensitive azoreductase. These bacteria are usually specific to their substrates. Under aerobic conditions, oxygen-insensitive azoreductases reductively cleave the azo bond of certain azo molecules to produce aromatic amines. The removal of azo dyes under aerobic conditions has been described using several species of bacteria and their strains (51).

4.6.3 Pharmaceutical Industry

Anaerobic techniques such as anaerobic membrane bioreactors (AnMBR), anaerobic sequenced batch reactors (AnSBR), moving bed biofilm reactors (MBBR), and other hybrid technologies are ideal for processing drugs in an efficient manner. The advantages of using MBBR for hospital wastewater has been demonstrated at pilot-scale. Pharmaceuticals are typically degraded simultaneously with the removal of COD and nitrogen. This suggests that co-metabolic processes play a major role. Further ozonation was more feasible due to the reduction in COD effluent achieved. Intermittent loading of biofilms enhances both biomass drug effective concentration and drug clearance. MBBR technology appears promising, but no mineralization from MBBR has been observed, so there may be permanent alteration products from the treatment process (52). To remove pharmaceuticals from effluents from conventional activated sludge treatment plants, biofilms grown on supports in reactors that occasionally feed BOD-rich (precipitated) effluents are used to purify the effluents. This biofilm also effectively removes ammonia. Biofilms of intermittently supplied MBBR have demonstrated the ability to degrade agents (such as diclofenac) previously considered refractory because activated sludge does not remove them from the treatment plant. In a study, it was discovered that the polishing biofilm, which was developed utilizing intermittent feeding, was able to remove diclofenac and atenolol from the WWTP effluent by >50% (53).

Chen et al. explained the treatment of ampicillin-rich pharmaceutical wastewater at various temperatures, and the results are summarized as follows. The advantages of Mt-ALMBR, such as improved gas–liquid distribution, high flow velocity, and regular circulation flow structure, promote microbial activity in the reactor, resulting in excellent COD and ampicillin removal performance. Chen et al.'s two main methods for removing ampicillin were membrane rejection (7–16°C) and bioreactor methods (16–48°C). The optimal temperature for removing COD was 36 °C (mainly with bioreactors), with an average removal rate of 98.2 ± 0.2%, while the optimal temperature for removing ampicillin was 7°C (mainly with bioreactors), with an average removal rate of 63.2 ± 5.6%. Adsorption and membrane repulsion were the dominant removal mechanisms in this reactor at low temperatures. Hydrolysis and biodegradation gradually took over as the main removal mechanisms as the temperature increased. According to impact rate studies, there are two levels of impact rates for membrane removal and biodegradation, each with a limit of 5%. Adsorption is followed by hydrolysis, followed by biodegradation and membrane rejection. As the temperature increased from 7°C to 48°C, the half-life of ampicillin was dramatically shortened from 330.0 days to 2.1 days. Also, the maximum adsorption capacities Kf were 0.1284, 0.2169, 0.0645, 0.1502, and 0.1524 mg/g at adsorption temperatures of 7, 16, 24, 36, and 48°C, respectively. The fact that all Delta G° results were negative indicates that ampicillin adsorption is the result of spontaneous physisorption (54).

It has been demonstrated that the use of catalysed ozonation pre-treatment in a semi-batch packed bed reactor is successful at improving the biodegradability index (BI), removing COD, changing the colour of the effluent, and removing toxicity. The O3/n ZVI pre-treatment procedure produced the best removal efficiency for COD, colour, and toxicity of 62.3%, 93%, and 98%, respectively, with a 3.5-fold increase in the biodegradability index of complex pharmaceutical wastewater. In a study by Malik et al., the pre-treated wastewater sample underwent GC-MS analysis, which revealed that stubborn organic molecules found in the untreated wastewater sample had vanished. It also revealed the emergence of novel intermediates via O_3, $O_3/Fe^{2}+$, and O_3/n ZVI during the pre-treatment process (55).

4.6.4 Paper and Pulp Industry

A photoelectrocatalysis (PEC) system has the advantage of improved wastewater treatment effectiveness, thanks to upgraded electrode materials, particularly photoanode materials. The creation of novel and useful electrodes with efficient electrochemical and catalytic capabilities at a modest cost would be appreciated by the industry, as the use of commercially available electrodes is constrained by their high costs. Operating factors like wastewater characteristics (pollutant type and concentration, turbidity/colour, conductivity, and solution pH), photoanode synthesis technique, applied current potentials, and irradiation source (irradiance and photon energy) can all have an impact on the overall performance of PEC. Reactor design in terms of high mass transfer efficiency, simplicity of operation and fabrication, and its technological viability at the pilot scale is another significant aspect that has contributed to the increased activity of PEC systems. Instead of integrated PEC reactors, slurry and packed-bed reactors have been suggested as efficient designs for the straightforward superposition of EC/PC (56).

The effectiveness of activated sludge in removing nutrients and organic matter during the co-treatment of pulp and paper effluent in a bench-scale EAAS system was examined in a study by Jagaba et al. The performance of the activated sludge system is evaluated by analysing the removal efficiency of phosphorus, nitrates, ammoniacal nitrogen, COD, and TSS. For ammonia, a mean influent concentration of 13.8 mg/L decreased to a mean effluent concentration of 5.2 mg/L, with a removal efficiency of 62.3%. Phosphorus, on the other hand, decreased from a mean influent concentration of 46.2 mg/L to a mean effluent concentration of 30.5 mg/L, with a removal rate of 34%. Results showed that ammonia effectively met the established limits. As a result, the removal rates of COD and TSS were 83% and 90%, respectively. However, most effluent COD values are below the required standards and require additional testing. The average BOD of the effluent was found to be 4.54 mg/L (57).

4.7 SCOPE OF ENERGY GENERATION FROM WASTE USING VARIOUS BIOREACTORS

Many reactor configurations for anaerobic BioH2 manufacturing processes have been reported in the literature. Although the fully stirred tank reactor (CSTR) is the most common, membrane bioreactor (MBR) technology has many advantages over other systems. For example, it can handle higher solids levels, has a smaller footprint, prevents cell washout, provides excellent separation of solids and metabolites, ensures better hydraulic retention time (HRT) and sludge age can be controlled. On the other hand, the main challenge of membrane-based systems is membrane fouling, which is a common problem. In addition to reactor type, appropriate operating conditions must also be selected for highly stable BioH2 production. There are a number of variables reported to influence anaerobic BioH2 production, including loading rate, HRT, pH, mixing, nutrient availability, and temperature. From an energy recovery perspective, a two-step fermentation process that converts his BioH2 effluent from the production process to CH4 is said to be beneficial. It has also been proposed to combine dark and light fermentation (58).

The use of high-rate UASB reactors to produce biogas, a sustainable energy source, from a variety of wastewater streams, including both industrial and municipal substrates, is expanding globally. It was shown that using improved UASB systems, a multi-stage anaerobic process, and appropriate substrate pre-treatment, the UASB reactor can efficiently handle both highly biodegradable substrates and resistant or diluted fractions. Municipal wastewater treatment using low-temperature decentralized UASB can combine power with end users, resulting in efficient and long-lasting renewable energy clusters (59).

4.8 CHALLENGES AND FUTURE SCOPE

Treatment of refractory organic compounds and inhibition of degradation intermediates in the anaerobic reactor, synergistic metabolic effects between refractory organics and other easily biodegradable organics, the impact of high salinity, the effect of sulphate concentration, and microbial activity are limitations for anaerobic processes used for textile wastewater treatment. Finding the essential salinity for high-performance textile printing and dyeing wastewater treatments will require a lot of work. To find the right sulphate content, more thorough research is required to understand the detrimental effects of sulphate on anaerobic processes. Future research should concentrate on: (1) creating quick start-up methods for anaerobic reactors and sludge granulation, (2) figuring out boundary conditions for anaerobic reactors to operate at high performance, (3) creating combined processes to further increase treatment efficiency, (4) revealing the working mechanisms, including pollutant degradation mechanisms, granular sludge characteristics, and microbial evolution mechanisms, and (5) identifying the microbial evolution mechanisms. Although a few pilot-scale reactors have previously been developed, the PEC reactor concept is still in the laboratory-scale development stage. There are many things that have prevented its commercialization, like the high investment costs and difficult operation processes. More efforts should be made to lower the cost of the PEC system so that it may be scaled up from the laboratory to the industrial level. Maintenance and capital investments include PV technology integration, solar-activated photoanode design, and reactor design optimization for high capacity. For AnMBRs, issues such as membrane fouling management, methane in the effluent, low COD/SO_4 2-S ratio, and insufficient alkalinity must be addressed. Membrane fouling control and dissolved methane recovery from wastewater are of particular importance, as both have a significant impact on energy savings and require further research. Reactor control, system development, and energy recovery for waste-activated sludge treatment all present significant hurdles. The most difficult issue in a real-world scenario is how to increase the system's size while simultaneously ensuring each person's performance. It is anticipated that these issues will be resolved by further developing microbial electrolytic cell reactors, creating novel materials, and fully comprehending and mastering microbiology.

4.9 CONCLUSION

This chapter focuses on the various methods of treating wastewater. Major sources of wastewater focusing more on food, pharmaceutical, paper and pulp, and textile industries, their treatments, and recent advances for the treatment of the effluents from these industries are discussed in this chapter. Vast varieties of effluents are released from different industries. The composition of wastewater from various industries has also been discussed. General methods of wastewater treatment and recent advances in reactors for wastewater treatment are also discussed. Apart from the treatment, scope of energy generation from wastewater is also discussed. An overview of major challenges and future scope of various techniques is given in this chapter.

REFERENCES

1. Sonune, A., & Ghate, R. (2004). Developments in wastewater treatment methods. Desalination, 167, 55–63. https://doi.org/10.1016/j.desal.2004.06.113.
2. Crini, G., & Lichtfouse, E. (2018). Advantages and disadvantages of techniques used for wastewater treatment. Environmental Chemistry Letters, 17, 145–155.
3. Gothandam, K. M., Ranjan, S., Dasgupta, N., & Lichtfouse, E. (Eds.). (2020). Environmental Biotechnology Vol. 2. Environmental Chemistry for a Sustainable World. Springer.
4. Valta, K., Kosanovic, T., Malamis, D., Moustakas, K., & Loizidou, M. (2015). Overview of water usage and wastewater management in the food and beverage industry. Desalination and Water Treatment, 53(12), 3335–3347.
5. Gulzar, T., Farooq, T., Kiran, S., Ahmad, I., & Hameed, A. (2019). Green chemistry in the wet processing of textiles. In the impact and prospects of green chemistry for textile technology (pp. 1–20). Woodhead Publishing.
6. Gadipelly, C., Pérez-González, A., Yadav, G. D., Ortiz, I., Ibáñez, R., Rathod, V. K., & Marathe, K. V. (2014). Pharmaceutical industry wastewater: Review of the technologies for water treatment and reuse. Industrial & Engineering Chemistry Research, 53(29), 11571–11592.
7. Kumari, V., & Tripathi, A. K. (2019). Characterization of pharmaceuticals industrial effluent using GC–MS and FT-IR analyses and defining its toxicity. Applied Water Science, 9, 185.
8. Metcalf, L., Eddy, H. P., & Tchobanoglous, G. (1991). *Wastewater Engineering: Treatment, Disposal, and Reuse* (Vol. 4). McGraw-Hill.
9. Batstone, D. J., & Virdis, B. (2014). The role of anaerobic digestion in the emerging energy economy. Current Opinion in Biotechnology, 27, 142–149.
10. Chisti, Y. (2019). *Biorefinery: Integrated Sustainable Processes for Biomass Conversion to Biomaterials, Biofuels, and Fertilizers*. J. -R. Bastidas-Oyanedel, J. E. Schmidt (Eds.), Springer.
11. Tabatabaei, M., & Ghanavati, H. (Eds.). (2018). Biogas: Fundamentals, Process, and Operation (Vol. 6). Springer.
12. Wainaina, S., Awasthi, M. K., Sarsaiya, S., Chen, H., Singh, E., Kumar, A., & Taherzadeh, M. J. (2020). Resource recovery and circular economy from organic solid waste using aerobic and anaerobic digestion technologies. Bioresource Technology, 301, 122778.

13. Arthurson, V. (2008). Proper sanitization of sewage sludge: A critical issue for a sustainable society. Applied and Environmental Microbiology, 74(17), 5267–5275. https://doi.org/10.1128/AEM.00438-08.
14. Luukkonen, T., Prokkola, H., & Pehkonen, S. O. (2020). Peracetic acid for conditioning of municipal wastewater sludge: Hygienization, odor control, and fertilizing properties. *Waste Management*, *102*, 371–379. https://doi.org/10.1016/j.wasman.2019.11.004.
15. Saravanan, A., Kumar, P. S., Jeevanantham, S., Karishma, S., Tajsabreen, B., Yaashikaa, P. R., & Reshma, B. (2021). Effective water/wastewater treatment methodologies for toxic pollutants removal: Processes and applications towards sustainable development. Chemosphere, 280, 130595.
16. Samal, S. (2020). Effect of shape and size of filler particle on the aggregation and sedimentation behavior of the polymer composite. Powder Technology, 366, 43–51.
17. Nyström, F., Nordqvist, K., Herrmann, I., Hedström, A., & Viklander, M. (2020). Removal of metals and hydrocarbons from stormwater using coagulation and flocculation. Water Research, 182, 115919.
18. Ahmad, A., Rutten, S., de Waal, L., Vollaard, P., van Genuchten, C., Bruning, H., & van der Wal, A. (2020). Mechanisms of arsenate removal and membrane fouling in ferric based coprecipitation–low pressure membrane filtration systems. Separation and Purification Technology, 241, 116644.
19. Medeiros, R. C., de MN Fava, N., Freitas, B. L. S., Sabogal-Paz, L. P., Hoffmann, M. T., Davis, J., & Byrne, J. A. (2020). Drinking water treatment by multistage filtration on a household scale: Efficiency and challenges. Water Research, 178, 115816.
20. Viccione, G., Evangelista, S., Armenante, A., & Ricciardi, V. (2020). Clogging process and related pressure drops in wire-wound filters: Laboratory evidence. Environmental Science and Pollution Research, 27(19), 23464–23476.
21. Mallakpour, S., & Rashidimoghadam, S. (2021). Utilization of starch and starch/carbonaceous nanocomposites for removal of pollutants from wastewater. In Handbook of Polymer Nanocomposites for Industrial Applications (pp. 477–502). Elsevier.
22. Tufail, A., Price, W. E., Mohseni, M., Pramanik, B. K., & Hai, F. I. (2021). A critical review of advanced oxidation processes for emerging trace organic contaminant degradation: Mechanisms, factors, degradation products, and effluent toxicity. Journal of Water Process Engineering, 40, 101778.
23. Khazaie, A., Mazarji, M., Samali, B., Osborne, D., Minkina, T., Sushkova, S., & Soldatov, A. (2022). A review on Coagulation/Flocculation in dewatering of coal slurry. Water, 14(6), 918.
24. Kumar, V., & Dwivedi, S. K. (2021). A review on accessible techniques for removal of hexavalent chromium and divalent nickel from industrial wastewater: Recent research and future outlook. Journal of Cleaner Production, 295, 126229.
25. Types of Reactors in Wastewater Treatment. (2018, June 26). AOS Treatment Solutions. https://aosts.com/types-of-reactors-in-wastewater-treatment/
26. Lebrero, R., Rodríguez, E., Martin, M., García-Encina, P. A., & Muñoz, R. (2010). H_2S and VOCs abatement robustness in biofilters and air diffusion bioreactors: A comparative study. Water Research, 44(13), 3905–3914.
27. Moussavi, G., Naddafi, K., Mesdaghinia, A., & Deshusses, M. A. (2007). The removal of H_2S from process air by diffusion into activated sludge. Environmental Technology, 28(9), 987–993.
28. Barbosa, T. M., Fernandes, R. J., Morouco, P., & Vilas-Boas, J. P. (2008). Predicting the intra-cyclic variation of the velocity of the centre of mass from segmental velocities in butterfly stroke: A pilot study. Journal of Sports Science & Medicine, 7(2), 201.
29. Kiesewetter, F., Arai, A., & Schell, H. (1993). Sex hormones and antiandrogens influence in vitro growth of dermal papilla cells and outer root sheath keratinocytes of human hair follicles. Journal of Investigative Dermatology, 101(1), S98–S105.

30. Vertz, J., Van Durme, G. P., & McKnight, M. D. (2006). Activated sludge diffusion provides cost effective biological odor control at two Texas wastewater treatment facilities. Proceedings of the Water Environment Federation, 2006(8), 4440–4447.
31. Fan, F., Xu, R., Wang, D., & Meng, F. (2020). Application of activated sludge for odor control in wastewater treatment plants: Approaches, advances, and outlooks. Water Research, 181, 115915.
32. Waqas, S., Bilad, M. R., Man, Z., Wibisono, Y., Jaafar, J., Mahlia, T. M. I., & Aslam, M. (2020). Recent progress in integrated fixed film activated sludge process for wastewater treatment: A review. Journal of Environmental Management, 268, 110718.
33. Mahat, S. B., Omar, R., Man, H. C., Idris, A. M., Kamal, S. M., Idris, A., & Abdullah, L. C. (2021). Performance of dynamic anaerobic membrane bioreactor (DAnMBR) with phase separation in treating high strength food processing wastewater. Journal of Environmental Chemical Engineering, 9(3), 105245.
34. Al-Khafaji, S. S., & Al-Rekabi, W. S. (2022). Apply membrane biological reactor (MBR) in industrial wastewater treatment: A mini review. Eurasian Journal of Engineering and Technology, 7, 98–106.
35. Asgharnejad, H., Khorshidi Nazloo, E., Madani Larijani, M., Hajinajaf, N., & Rashidi, H. (2021). Comprehensive review of water management and wastewater treatment in food processing industries in the framework of water-food-environment nexus. Comprehensive Reviews in Food Science and Food Safety, 20(5), 4779–4815.
36. Javadinejad, S., Dara, R., Hamed, M. H., Saeed, M. A. H., & Jafary, F. (2020). Analysis of gray water recycling by reuse of industrial waste water for agricultural and irrigation purposes. Journal of Geographical Research, 3(2).
37. Cassano, A., Conidi, C., Ruby-Figueroa, R., & Castro-Muñoz, R. (2018). Nanofiltration and tight ultrafiltration membranes for the recovery of polyphenols from agro-food by-products. International Journal of Molecular Sciences, 19(2), 351.
38. Pichardo-Romero, D., Garcia-Arce, Z. P., Zavala-Ramírez, A., & Castro-Muñoz, R. (2020). Current advances in biofouling mitigation in membranes for water treatment: An overview. Processes, 8(2), 182.
39. Castro-Muñoz, R. (2020). Breakthroughs on tailoring pervaporation membranes for water desalination: A review. Water Research, 187, 116428.
40. Fallahi, A., Rezvani, F., Asgharnejad, H., Nazloo, E. K., Hajinajaf, N., & Higgins, B. (2021). Interactions of microalgae-bacteria consortia for nutrient removal from wastewater: A review. Chemosphere, 272, 129878.
41. Hassan, M. A., Li, T. P., & Noor, Z. Z. (2009). Coagulation and flocculation treatment of wastewater in textile industry using chitosan. Journal of Chemical and Natural Resources Engineering, 4(1), 43–53.
42. Nigam, P., Armour, G., Banat, I. M., Singh, D., & Marchant, R. (2000). Physical removal of textile dyes from effluents and solid-state fermentation of dye-adsorbed agricultural residues. Bioresource Technology, 72(3), 219–226.
43. Robinson, T., McMullan, G., Marchant, R., & Nigam, P. (2001). Remediation of dyes in textile effluent: A critical review on current treatment technologies with a proposed alternative. Bioresource Technology, 77(3), 247–255.
44. Ravi Kumar, M., Rajakala Sridhari, T., Durga Bhavani, K. and Dutta, P.K. (1998) Trends in Color Removal from Textile Mill Effluents. Colourage, 45, 25.
45. Babu, B. R., Parande, A. K., Raghu, S., & Kumar, T. P. (2007). Cotton textile processing: Waste generation and effluent treatment. Journal of Cotton Science, 11(3).
46. Slokar, Y. M., & Le Marechal, A. M. (1998). Methods of decoloration of textile wastewaters. Dyes and Pigments, 37(4), 335–356.
47. Singh, R. P., Singh, P. K., Gupta, R., & Singh, R. L. (2019). Treatment and recycling of wastewater from textile industry. In *Advances in Biological Treatment of Industrial Waste Water and Their Recycling for a Sustainable Future* (pp. 225–266). Springer.

48. Singh, R. L., Singh, P. K., & Singh, R. P. (2015). Enzymatic decolorization and degradation of azo dyes–A review. International Biodeterioration & Biodegradation, 104, 21–31.
49. Singh, R. P., Singh, P. K., & Singh, R. L. (2017). Present status of biodegradation of textile dyes. Current Trends in Biomedical Engineering & Biosciences, 3(4), 66–68.
50. Xu, H., Yang, B., Liu, Y., Li, F., Shen, C., Ma, C., & Sand, W. (2018). Recent advances in anaerobic biological processes for textile printing and dyeing wastewater treatment: A mini-review. World Journal of Microbiology and Biotechnology, 34(11), 1–9.
51. Ola, I. O., Akintokun, A. K., Akpan, I., Omomowo, I. O., & Areo, V. O. (2010). Aerobic decolourization of two reactive azo dyes under varying carbon and nitrogen source by *Bacillus cereus*. African Journal of Biotechnology, 9(5).
52. Tang, K., Ooi, G. T., Litty, K., Sundmark, K., Kaarsholm, K. M., Sund, C., & Andersen, H. R. (2017). Removal of pharmaceuticals in conventionally treated wastewater by a polishing moving bed biofilm reactor (MBBR) with intermittent feeding. Bioresource Technology, 236, 77–86.
53. Ooi, G. T., Tang, K., Chhetri, R. K., Kaarsholm, K. M., Sundmark, K., Kragelund, C., & Andersen, H. R. (2018). Biological removal of pharmaceuticals from hospital wastewater in a pilot-scale staged moving bed biofilm reactor (MBBR) utilising nitrifying and denitrifying processes. Bioresource Technology, 267, 677–687.
54. Chen, Z., Min, H., Hu, D., Wang, H., Zhao, Y., Cui, Y., & Liu, W. (2020). Performance of a novel multiple draft tubes airlift loop membrane bioreactor to treat ampicillin pharmaceutical wastewater under different temperatures. Chemical Engineering Journal, 380, 122521.
55. Malik, S. N., Khan, S. M., Ghosh, P. C., Vaidya, A. N., Kanade, G., & Mudliar, S. N. (2019). Treatment of pharmaceutical industrial wastewater by nano-catalyzed ozonation in a semi-batch reactor for improved biodegradability. Science of the Total Environment, 678, 114–122.
56. Rajput, H., Kwon, E. E., Younis, S. A., Weon, S., Jeon, T. H., Choi, W., & Kim, K. H. (2021). Photoelectrocatalysis as a high-efficiency platform for pulping wastewater treatment and energy production. Chemical Engineering Journal, 412, 128612.
57. Jagaba, A. H., Kutty, S. R. M., Fauzi, M. A. H. M., Razali, M. A., Hafiz, M. F. U. M., & Noor, A. (2021, August). Organic and nutrient removal from pulp and paper industry wastewater by extended aeration activated sludge system. In IOP Conference Series: Earth and Environmental Science (Vol. 842, No. 1, p. 012021). IOP Publishing.
58. Akca, M. S., Bostancı, O., Aydin, A. K., Koyuncu, I., & Altinbas, M. (2021). BioH2 production from food waste by anaerobic membrane bioreactor. International Journal of Hydrogen Energy, 46(55), 27941–27955.
59. Mainardis, M., Buttazzoni, M., & Goi, D. (2020). Up-flow anaerobic sludge blanket (UASB) technology for energy recovery: A review on state-of-the-art and recent technological advances. Bioengineering, 7(2), 43.

5 Aerobic and Anaerobic Digestion of Textile Industry Wastewater

C. Nagendranatha Reddy, Divyamshu Surabhi, Matta Chenna Keshava Charan, Reena Pravallika Balla, Hamsini Katla, Kavya Pasirika Pathipaka, Rajasri Yadavalli, Bishwambhar Mishra, Sanjeeb Kumar Mandal, and Suresh Sundaramurthy

5.1 INTRODUCTION

The 7 billion mark was just reached by the world's population. Food and clothes are the two basic needs of these people; thus, the textile industry, which directly employs people, plays a huge part in the global economy, and in the textiles we encounter every day in our homes, automobiles, workplaces, public areas, and even outdoors. One of the most highly advanced and historically significant sectors is textile processing (Shishoo, 2012). Because of its breakthroughs in spinning and weaving, England served as the foundation for the growth of the global textile industry. The global increase in wool, cotton, and silk production in recent years has boosted the development of the textile sector. Despite the fact that the textile industry originated in the United Kingdom, textile manufacture expanded to Europe and North America in the 19th century as a result of these regions' modernization processes (Sivaram et al., 2019). An interesting area for the examination of scheduling problems is the textile production system and structure. The sector has grown as a result of both vertical and horizontal integration, particularly between spinning and weaving businesses. The latter was facilitated by the potential requirement for a whole line of textile products for successful marketing. The textile sector is important because it supports employment, industrial output, and foreign trade via exporting goods (Gera, 2012). Global exchange in textiles and clothing has an enormous scope. Textile and clothing producers were responsible for 93% of world exports in 1993.

There are various steps in the manufacturing of textiles, including sizing, desizing, scouring, bleaching, mercerizing, dyeing, printing, and finishing activities. In addition to using a lot of water and energy, these production techniques also have a number of negative side effects. As the demand for textile products increased, weaving factories and their wastewater overflowed, creating a serious global contamination problem. Numerous chemicals used in textile wet processing, such as dyes and auxiliary chemicals, pose risks to the environment and public health. The usage of toxic chemicals during processing and water contamination from the discharge

DOI: 10.1201/9781003381327-5

of untreated wastewater are the two inherent issues that the textile industry often encounters on a global basis. Textile effluent is a basic ecological issue because of its hydrosulphide content, the fact that it blocks light from reaching water bodies, and lowers oxygen levels, all of which are hazardous to the aquatic environment. Therefore, the methodologies for treating textile wastewater and the physicochemical treatment boundaries taken into account during primary, secondary, and tertiary treatment processes are the main topics of this survey. Also covered are pH, total dissolved solids (TDS), total suspended solids (TSS), turbidity, and the effluent of biological and chemical oxygen demand (BOD and COD). Future limits are expected to be more stringent, so control measures should be taken to reduce effluent pollution. To minimize the effects of textile process pollution, practical dyeing, the adoption of cutting-edge, less harmful technologies, successful effluent treatment, and waste recycling methods should be modified. This review study discusses the use of several natural approaches and their benefits for treating wastewater from the textile sector (Azanaw et al., 2022).

5.1.1 Characteristics of Textile Industry Effluents

The various processes of the textile industry use a variety of chemicals, fabrics, inks, colours, binders, etc. which end up being discarded when not efficiently used. Some of the processes and their characteristics are given below.

5.1.1.1 Sizing

Prior to twisting or winding, cotton and a few other man-made yarns are measured to gain strength and reduce strand breaking. The size is pressed into the fibre while the strings move between rollers, and the strings are then dried. Glucose or glucose derivatives are used in 75% of the jobs involving estimation. Thus, starch becomes the primary substance for estimation. Various materials, such as polyvinyl alcohol (PVA), polyacrylates, and carboxymethyl cellulose, are used for measurement. There is an excess of estimating fluid that is wasted, typically in little amounts but with exceptionally high levels of suspended particles, BOD, and COD. A reduced BOD results from synthetic sizes. When recyclable materials like PVA can be securely recovered, the overall organic burden can be reduced by 90% (Bisschops and Spanjers, 2003). As a result, this stage of textile processing generates little to no waste, with the main wastes being fibre lint, yarn waste, and starch-based sizing used in the process (Sarayu and Sandhya, 2012).

5.1.1.2 Desizing

Prior to the fabric's further processing, the size should be removed after weaving. It can serve as a barrier to colours and other chemicals because it covers the yarn (Sarayu and Sandhya, 2012). Antacids, acids, compounds, or surfactants are utilized based on the size that is being used. While starch is normally removed using enzymes, washing with detergents may be sufficient for some sizes. The texture is cleaned up after being desized. Because wastewater comprises both the used sizes and the professionals who were used for desizing, its qualities change depending on the sizes that were used. A material wastewater can have a very high commitment to its overall BOD and total solids load. When starch is used as size, the desizing stage can represent up to 50% of the overall BOD in the handling of woven fabrics (Bisschops and Spanjers, 2003). Water-soluble materials, synthetic materials, lubricants, biocides, and antistatic solvents, all of which are heavily used in textile processing, are substantially concentrated in the wastewater produced during this phase.

5.1.1.3 Scouring

The contaminants are eliminated through scouring. Water scouring is typically preferred over dissolvable scouring since it is non-combustible, non-toxic, abundant, and less expensive. However, it can be done with either water or solvents. In comparison to cotton or wool, synthetic fibres require less scouring. Alkalis, wetting agents, and lubricants are examples of scouring agents. After scrubbing, the items are carefully rinsed (or washed) to get rid of extra agents. Scouring can be done in a batch or continuous process, adding a considerable organic load to textile effluents that have accumulated from the use of NaOH, disinfectants, pesticide residues, detergents, fats, oils, pectin, wax, knitting lubricants, spin finishes, and spent solvents (Sarayu and Sandhya, 2012).

5.1.1.4 Bleaching

Bleaching is usually used to eliminate natural colouring from cotton, mixed fabrics, or yarn, and is sometimes expected on wool and some synthetic fibres. Utilized synthetics incorporate sodium hypochlorite and hydrogen peroxide, too, as optical brighteners. Also, some auxiliary compounds are utilized and delivered to the wastewater. BOD levels are low, yet the solid content of the wastewater can be high. Denim handling results in a very high suspended solid content in the bleaching step because of the utilization of pumice stone. High volumes of chlorides or peroxide could cause inhibition issues, and this contributes to no or zero residual waste with high pH-containing wastewater (Sarayu and Sandhya, 2012).

5.1.1.5 Mercerizing

The process of mercerization enhances the durability, gloss, and dye affinity of cotton fabrics. The application of a cold NaOH liquid causes the fibres to swell and take on a circular cross-section. Through an acid wash, the solution is removed. To help reduce waste, the majority of mercerization machines feature their own caustic recovery systems. Although the wastewater produced by mercerizing typically has minimal BOD, it also includes natural oils, sodium hydroxide, and cotton waxes (Bisschops and Spanjers, 2003).

5.1.1.6 Dyeing

Industry-wide dyes are often synthetic and made from intermediates based on petroleum and coal tar. They are made up of molecules responsible for the colour, known as chromophores, as well as auxochromes, which are electron-withdrawing or -donating substituents that try to increase or target the colour of the chromophores. Azole (–N=N–), carbonyl (–COO), methine (–CHO), nitro (–NO_2), and quinoid groups make up the chromophores. Auxochromes with the symbols amine (NH_3), carboxyl (COOH), sulphonate (SO_3H), and hydroxyl (OH) are the most important auxochromes.

It is important to note that the dyes have a high degree of aqueous solubility thanks to the sulphonate groups. The classes of auxochromes include sulphur, solvent, direct acid, basic, mordant, dispersion, pigment, vat, anionic, and ingrain. The most popular method of colouring textiles is dyeing, which frequently uses large amounts of water both during the washing process and in the dye bath. Various synthetic materials, including metals, salts, surfactants, organic processing aids, sulphide, and formaldehyde, may be added during the dyeing process to enhance colour adsorption onto the fibres, which are the major pollutants in the effluent. Dyes must be more resistant to washing with chemicals including chlorine, ozone, nitrogen peroxide, and light hydrolysis. Although these dyes can be broken down, they are not alarming, but the hue they give the effluent is highly dubious (Sarayu and Sandhya, 2012).

5.1.1.7 Printing

The dyes and chemicals used in printing are similar to those used in fabric dyeing, with the exception that the colour is only applied to specified areas of the fabric. The print pastes, which are made up of water, thickeners, dyes, urea, and various man-made substances, including surfactants and solvents, play a vital role in textile printing. In reactive dye printing, urea is the most often used chemical. To improve the water-solvent dyes' solvency and fix them to the texture, urea is used while printing cotton, wool, and silk. The paste arrangement and equipment cleaning processes produce a lot of leftover pastes. Remaining print pastes are mostly the result of cleaning the printing frameworks, tanks for paste arrangement, and protective tissues for the framework.

The cleaning of textile products and dye spills are two different sources. Weakening the remaining glues and removing them with the remaining wastewater streams, where they significantly increase COD, nitrogen, and colour burdens, is a usual removal strategy. The printing method determines the characteristics of the effluent. Despite being tiny in volume, printing wastewater is challenging to clean. Textile printing wastewaters resemble material dyeing wastewaters without the particles and chemicals from the print glue arrangement. Additionally, printing wastewater has a larger concentration of contaminants than dyeing wastewater (Bisschops and Spanjers, 2003).

5.1.1.8 Finishing

The majority of printing is done using a lot or rotating screen, and after each printing lot, some leftover paste is left in the wastewater. By using fresh stock, you can use this again to print hues that are comparable. Ink-jet printing and electrostatic printing are examples of screen-free printing techniques that have recently been invented. These techniques use electrical control of colour distribution on fabric. The elimination of contamination makes screen-free printing systems interesting (Bisschops and Spanjers, 2003; Shah 2020).

5.1.2 Negative Impacts of Textile Industry Wastewaters

Nowadays, a large portion of the textile and dyeing industries are found in underdeveloped nations, frequently with subpar wastewater treatment. It is possible that India contributes the most textile wastewater to South Asia (Khan and Malik, 2014).

Numerous dyes contain substances that are known to cause cancer, like benzidine and other aromatic chemicals (Ali, 2010).

These chemicals, which are utilized in the textile industry, seriously harm the environment and human health. Dye chemicals are among the many compounds found in textile effluent and are significant pollutants. Water-related global environmental issues are frequently connected to the textile industry (Khan and Malik, 2014; Shah 2021b).

5.1.2.1 Water Pollution Due to the Release of Untreated Effluents

Among all industrial sectors, the textile industry ranks as one of the most polluting and severely worsens the quality of surface water (Odjegba and Bamgbose, 2012). Since high levels of textile dyes in water bodies interfere with biological activity in aquatic life and the photosynthesis process of aquatic plants or algae by obstructing sunlight and lowering the capacity of the receiving water to re-oxygenate, the presence of untreated dyes in water bodies poses the greatest threat to the environment (Zaharia et al., 2009; Shah 2021a; Rajasri et al., 2021). The aquatic ecology is seriously threatened by the presence of hydrosulphite in the effluents because it lowers oxygen levels and prevents light from entering the water body (Khan and Malik, 2014). Water cannot purify itself as a result of this. Concentrated dye effluents with high temperatures and an acidic pH are occasionally released after dyeing techniques. The method for transferring oxygen and the procedure by which environmental water bodies purify themselves will be hampered by this phenomenon. Due to their contamination of water sources and difficulty in using water, these effluents cause harm to the ecosystem when they are released into the environment after usage. The stench and soreness caused by dye effluents mixed with natural water sources are unpleasant. Aquatic and terrestrial life, starting with plants and animals, can be negatively impacted by textile effluents. When dye effluents are added to water sources, the turbidity of the water increases because dye effluents have a tendency to form a visible layer above the water surface due to their reduced density (Katheresan et al., 2018).

5.1.2.2 Land Pollution

The soil is a natural substance composed of both organic and mineral components. The foundation of agriculture is the soil. The soil is contaminated with textile waste. Root penetration is prohibited, and the soil's texture has been toughened. These dye effluents are likely to cause damage to soil production by blocking soil pores when they end up in fields and forests (Katheresan et al., 2018). Clogging pores cause low soil yield. Dyes also have an effect on the sapling growth of plants. These also have additional harmful effects on the soil and have an impact on the rates of plant germination, growth, and biomass. Dyes are having a direct impact on soil fertility (Pokharia and Ahluwalia, 2015).

5.1.2.3 Health Impacts

Depending on the length of exposure and the concentration of the dye, exposed organisms may experience acute or chronic effects. The textile industry's usage of dyes poses a risk to human health since anaerobic settings can transform them into dangerous or cancer-causing chemicals. Among the various components of a toxic effluent, a well-known carcinogen, organically bonded chlorine, is found in around 40% of colourants used worldwide (Khan and Malik, 2014). Numerous dyes have an effect on the flora and fauna, are harmful to fish and mammals, and prevent the growth of microorganisms. Additionally, it has been demonstrated that a variety of dyes, as well as the by-products of their degradation, are harmful to aquatic life, including fish, animals, plants, and microorganisms (Kim et al., 2004). They can also cause prenatal brain defects and intestinal cancer due to their carcinogenic properties (Doble and Kumar, 2005). Textile dyes have the potential to cause allergic reactions that result in disorders like contact dermatitis, respiratory infections, allergic eye reactions, skin rashes, and mucous membrane and upper respiratory tract irritation (Walsh et al., 1980). The long-term presence of dyes in the environment (i.e., their half-life of several years), their accumulation in sediments, but especially in fish or other aquatic life forms, the decomposition of pollutants into carcinogenic or mutagenic compounds, as well as their low aerobic biodegradability, can also have a toxic effect on aquatic environments. Due to their synthetic makeup and predominately aromatic makeup, the majority of dyes are not biodegradable, have cancer-causing effects, or cause allergies, dermatitis, skin irritation, or other tissue changes. Numerous azo dyes also show both acute and long-term toxicity, mostly aromatic compounds. The generation of haemoglobin adducts and interference with blood formation, as well as the absorption of azo dyes and their breakdown products (toxic amines) through the skin, lungs, and digestive tract, constitute a major threat to health. For aromatic azo dyes, the published values for the median lethal dose (LD50) range from 100 to 2,000 mg/kg of body weight. A number of azo dyes can damage DNA, which can lead to the growth of malignant tumours. Direct Black 38 azo dye, a precursor of benzidine, and azo disalicylate, a precursor of 4-phenylenediamine, are two of the most well-known azo dyes and their breakdown products that result in cancer in humans and animals (Carmen and Daniela, 2012). By consuming a large amount of potable water, the textile industry also contributes to environmental problems. Byssinosis is currently one of the biggest health problems in the textile industry as a whole. When dry procedures are involved, textile industry machines can produce noise levels that are beyond the legal limit, which can lead to hearing problems. There are several harmful health effects that might result from the use of dyes and pigments. When clothing is in close proximity to the skin for an extended period of time, toxic substances can be absorbed through the skin, especially when the body is warm and the pores on the skin have opened to enable perspiration. When heavy metals are eaten by humans, they frequently accumulate in the liver, kidney, bones, heart, and brain. Significantly detrimental effects on health may occur when significant levels of accumulation are established.

Children are especially at risk since exposure to toxic colouring and/or heavy metal deposits may adversely damage their development and potentially put their lives in peril. A few research on azo dyes found a link between them and numerous cancers of different organs, such as the bladder, spleen, and liver, as well as with usual aberrations in model organisms and chromosomal abnormalities in mammalian cells (Bhatia et al., 2017). Chemicals can also cause harm to children even before they are born since they can vaporize into the air we breathe or absorb through our skin. It can cause allergic reactions, alter animal cell/biological pathways, and obstruct essential functions including breathing, osmoregulation, reproduction, and even mortality.

Additionally, because heavy metals in wastewater from the textile industry cannot be biodegraded, they may build up in the body's major organs over time, start to fester, and eventually cause a variety of disease symptoms. Around 40% of the colourants used globally contain organically bound chlorine, a well-known carcinogen that is one of the many components of a hazardous effluent. As a result, untreated or improperly treated textile effluent can significantly influence the natural ecosystem, disrupt aquatic and terrestrial life, and have long-lasting detrimental effects on human health. A class of organic compounds known as textile dyes are primarily discharged into wastewater during chemical textile finishing processes. More than 7105 tonnes of synthetic dyes are produced annually worldwide, and it is estimated that 10,000 different dyes and pigments are used in the industry. Dye chemicals are the most important ones used in the textile industry to give yarn or fabric colour. Numerous harmful health effects and potential side effects have been noted for this group of compounds (Khan and Malik, 2014).

5.1.2.4 Environment Issues by Effluents

These textile firms' effluent adds caustic properties, disagreeable scents, less dissolved oxygen in water, and a rise in insoluble compounds to their physical qualities. The water's turbidity changes when colloidal components from effluent mingle with freshwater streams. The water becomes unfit for irrigation and human consumption as a result of the increase in salt content. Additionally, marine life and the microbes required to clean the water streams are seriously threatened (Bhatia et al., 2017). Multiple chemicals used in the creation of textiles cause the pH of the water to fluctuate, which has an adverse effect on marine life. If effluent from the textile sector is used to irrigate plants, the concentration of chlorophyll in the plants also decreases. One of the most important problems facing developing countries is the careless handling of the massive volumes of rubbish produced by various anthropogenic activities. It is more challenging to release these contaminants hazardously into the environment. The three main environmental issues associated with the textile business are water use, aqueous effluent treatment, and disposal (Objegba and Bamgbose, 2012).

The mechanical and chemical processes used in the textile industry are numerous, and each one has a different effect on the environment. The presence of heavy metals like copper, arsenic, lead, cadmium, mercury, nickel, and cobalt, as well as sulphur, naphtha, vat dyes, nitrates, acetic acid, soaps, and chromium compounds, makes the effluent incredibly hazardous. The ecosystem of the water is harmed by the pigment deposit that stops sunlight from penetrating. Additionally, when this effluent is allowed to flow through the fields, it blocks the pores of the soil, lowering the land's productivity. The soil's texture hardens, which prevents root penetration. The effluent that flows down the drains causes the sewerage pipes to corrode and encrusted. If wastewater is permitted to flow into drains and rivers, it degrades the quality of drinking water in hand pumps, rendering it unsafe for human use. It is understood that the colouring of watercourses is more of an aesthetic than an environmental risk. Most procedures in textile mills produce emissions into the atmosphere (Carmen and Daniela, 2012).

5.2 BIOREMEDIATION OF TEXTILE INDUSTRY EFFLUENTS

The salt load, BOD/COD, and colour content in textile effluents are all increased. Due to the presence of numerous reactive dyes that are difficult to remove, the textile industry's effluent is severely contaminated (Holkar et al., 2016). Dye-containing effluents damaging rivers and soil are among the environmental issues. Because more people are becoming aware of and concerned about the discharge of synthetic dyes into the environment and their persistence there, a lot of emphasis has been paid to cleaning up these pollutants. Among the current pollution control strategies in use, the degradation of synthetic dyes using diverse processes is emerging as a successful and promising strategy (Ali, 2010). Treatment of harmful dye effluents is crucial to preventing their negative effects on receiving waters, animals, and people (Katheresan et al., 2018). In order to successfully and economically clean textile wastewater before it is discharged into rivers, numerous treatment techniques have been developed. These methods have shown to be highly successful at treating wastewater from the textile industry (Holkar et al., 2016). Textile wastewater is primarily treated using three technologies. These methods' underlying operating principles have been separated into distinct categories. Chemical treatment employs a chemical method to remove colours, whereas biological treatment involves the biodegradation of dye. Physical treatment is the physical removal of dye from textile effluent. However, in order to effectively treat the effluents from the textile sector, biological approaches are preferable due to a number of drawbacks associated with physical, chemical, and electrochemical procedures (Gosavi and Sharma, 2014; Holkar et al., 2016).

5.2.1 Biological Methods

Physical and chemical techniques are not very suitable for removing the colours from textile effluent due to their high cost, poor efficiency, and highly specialized nature. Biological processes are an environmentally friendly method for biologically removing colour from textile effluent that costs the least amount of money and takes the least amount of time to operate. However, decolourization of textile effluent might be challenging. Biological techniques can be used to decrease turbidity and COD. Anaerobic treatment followed by aerobic treatment is the most effective biological treatment sequence for the decolourization of textile effluent (Gosavi and Sharma, 2014). Biologically based methods have been used for the effective degradation of the effluent from the textile industry. Biological degradation, sometimes referred to as bioremediation, is more economically feasible, less wasteful overall, and ecologically beneficial when compared to alternative treatments. Bond breakdown (i.e., chromophoric group) results in the breakdown of synthetic dyes into a more benign inorganic chemical, which ultimately helps with colour removal. Biologically based methods have been used for the effective degradation of the effluent from the textile industry. Biological degradation, sometimes referred to as bioremediation, is more economically feasible, less wasteful overall, and ecologically beneficial when compared to alternative treatments. Bond breakdown (i.e., chromophoric group) results in the breakdown of synthetic dyes into a more benign inorganic chemical, which ultimately helps with colour removal (Bhatia et al., 2017). In textile wastewater, the biological process only gets rid of the dissolved material. The amount of organic load/dye, the system temperature, and the oxygen content all have an effect on the removal efficiency. Biological processes can be classified as aerobic, anaerobic, anoxic, facultative, or a combination of these based on how much oxygen they require. The wastewater is treated using bacteria in aerobic methods as opposed to anaerobic procedures, which treat textile effluent when there is no oxygen present. The biological methods for completely metabolizing textile effluent include benefits such as being environmentally benign, economical, creating less sludge, giving non-hazardous metabolites, or full mineralization, and consuming less water with respect to physical and chemical methods (Holkar et al., 2016).

5.2.1.1 Bacteria

Although many different microorganisms have been studied, only a few numbers of bacteria that degrade colour have been discovered. The ability of bacteria to catabolize organic pollutants in the context of bioremediation has been the subject of numerous studies. The bacteria used as pollution indicators for various toxins in wastewater and crucial in the elimination of organic contaminants need further investigation. Working with bacteria is fundamentally advantageous since they cultivate easily and multiply more quickly than other types of germs. Bacteria's capacity to break down dyes can be easily increased by molecular genetic manipulation. Organic pollutants comprised of chlorinated and aromatic hydrocarbons can be oxidized and catabolized by bacteria, which can then use this energy (carbon source) to break down the pollutants. Finding the bacteria that can more quickly degrade different azo-based dyes has been the subject of numerous studies, with encouraging findings. Under normal aerobic, anaerobic, and extremely low oxygen conditions, a separate bacterial group induces an azo dye reduction for decolourization. When azo dyes are reduced in an anaerobic environment, the azoreductase enzyme breaks azo bonds (–N=N–), resulting in a colourless solution of aromatic amines (Bhatia et al., 2017). The majority of the bacteria belonged to the *Comamonas* genus. To achieve optimal dye degradation efficiency during bacterial growth, nitrogen supplies and improved electron donors are two essential components (Deng et al., 2020).

5.2.1.2 Algae

Algae are widespread and are getting increased attention in the field of treating wastewater from the textile industry (Holkar et al., 2016). Both freshwater and saltwater algae are frequently seen, and they are currently the focus of significant investigation as a bio-sorbent. Due to their enormous surface area and binding capability, algae have the greatest electrostatic force of attraction and bio-sorption potential for contaminants in wastewater. Numerous studies have shown that a range of metabolites of pollutants contained in the wastewater, such as –OH, RCOO, $–NH_2$, and PO_4^3, are absorbed by the surface of the algae (Bhatia et al., 2017; Rajasri et al., 2020). Algal colour breakdown is brought on by three distinct processes, including dye consumption for growth, dye conversion to non-coloured intermediates such as CO_2 and H_2O, and chromophores adsorption on algae. Both biodegradation and biosorption involve quite different mechanisms. While biodegradation is the process by which enzymes dissociate the chemical bonds that comprise the dye's chemical structure in order to transform it into other chemical compounds, biosorption is the process by which the dye is transferred from the liquid phase to the solid phase (the bioadsorbent). Green macroalgae of the *Cladophora* species may predominantly break down azo dyes due to the presence of the azoreductase enzyme (Holkar et al., 2016). The treatment process is influenced by a variety of operational factors, including biosorbent dosage, dye concentration, pH, biosorbent type (ash or dry), temperature, and contact time. *Phormidium animale* algae had the highest dye biosorption rate at pH 2, removing 99% of the dye (Deng et al., 2020).

5.2.1.3 Fungi

A fungal culture's metabolism may change in response to changing environmental factors. Their survival depends on their capacity to do so. Here, both internal and external enzymes help with metabolism. These enzymes can degrade certain hues that already exist in effluent from textile production. Fungal cultures seem to be the best option for breaking down dyes in textile effluent since they contain these enzymes. These enzymes include lignin peroxidase (LiP), manganese peroxidase, and laccase. White-rot fungal cultures have been used to remove azo dyes in the majority of cases. Fungal strains such as *Coriolopsis sp. & Pleurotus eryngii* along with *Penicillium simplicissimum* has showed effective removal of COD and deterioration of azo dye. However, white-rot fungi that break down dyes in textile wastewater have several intrinsic limitations, such as a protracted growth phase that needs nitrogen-restrictive conditions, variable enzyme production, and a large reactor size since total degradation takes so long to occur. The main problem with employing only fungi is that the system is unstable; after 20–30 days, bacteria will start to proliferate, and the fungi will lose their ability to control the atmosphere and break down the dyes (Holkar et al., 2016). The results of batch absorption demonstrated that the fungi had a significant capacity for decolourization in addition to a high efficacy of 98% for eliminating acid dyes. The fungal variety *Aspergillus* is a potential biosorbent for wastewater treatment, and modifying the pH of the dye solution may further increase effectiveness (Deng et al., 2020).

5.2.1.4 Yeast

Similar to algae, yeast decolourizes colours through adsorption, enzymatic breakdown, or a combination of the two. Since yeast has advantages over bacteria and filamentous fungi, there has been a lot of research into how different yeast species can remove colour in recent years. They can survive in harsh environments, such as low pH, in addition to rapidly reproducing, much like bacteria. In liquid-aerated batch cultures, the yeast strain *Candida zeylanoides* destroyed three simple azo dyes after 7 days of treatment. A 44–90% colour loss range was observed. It was shown that enzyme-mediated biodegradation followed the decolourization of numerous azo dyes by different ascomycetes yeast species, including Candida tropicalis, *Debaryomyces polymorphus*, and *Issatchenkia occidentalis* (Singh and Singh, 2017).

5.2.1.5 Plants

The most effective technique for removing heavy metals and organic pollutants from soils and groundwater is phytoremediation, which provides a unique, cost-effective alternative way for biotreatment of wastewater. Recent studies propose employing plants to biodegrade colour in wastewater. *Petunia grandiflora Juss* was used by the researchers. Cleanup of wastewater with a variety of hues and dyes is possible with phytoremediation. *Phaseolus mungo*, *Sorghum vulgare*, and *Brassica juncea* have all been studied for their potential to remove azo hues from textile effluents. These plants (*B. juncea*, *P. mungo*, and *S. vulgare*) reduced the colour of textile effluent to a maximum of 79%, 53%, and 57%, respectively. Numerous additional plants, including *Typhonium flagelliforme* and *Blumea malcolmii*, have also been shown to decolourize dye. The removal of colour is controlled by the plant's inherent enzymatic system. It is uncommon to employ plants to remove toxins because of a lack of complete knowledge about the natural metabolic pathways that plants use to break down the poisons. The main advantages of utilizing plants to remove dye are that they have a large biomass, an autotrophic system, require minimal nutrient input, are simple to handle, and are well-liked by the general people due to their demand for both aesthetics and environmental sustainability. The number of pollutants that plants can tolerate, the bioavailable fraction of the contaminants, the evapotranspiration of volatile organic pollutants, and the need for large areas to establish phytoremediation on a large scale make it currently impractical, despite extensive research being done to develop effective and efficient phytoremediation techniques for the decolourization and degradation of azo dyes (Singh and Singh, 2017).

5.3 AEROBIC BIOREMEDIATION OF TEXTILE INDUSTRY EFFLUENTS

Aerobic treatment frameworks are powerful oxidizers of solvents and natural and nitrogenous chemicals. The removal of a variety of suspended particles and bacteria from wastewater industry material and the lowering of COD, biological oxygen demand, and other high-impact treatments of effluents are advanced by commercially available oxygen-consuming treatment reactors. According to Khelifi et al., high-impact techniques for the purification of material waste water are competent and shrewd. Since the influent is not consistent in small-scale material enterprises, the majority of oxygen-consuming reactors operate as constant-volume reactors with full mixing and an irregular stream. To advance and intensify the contact between disintegrating oxygen, microorganisms, and wastewater, complete blending of materials is assured in the air circulation chamber. A clarifier receives the water that is streaming from the air circulation chamber. As soon as the influent stream rate changes, the profluent rate of release changes accordingly. For the treatment of wastewater from material industries, common bioprocesses that will be explored in this section include the drawn-out air circulation process and suspended and linked development processes (Khelifi et al., 2008; Venkata Mohan et al., 2013).

5.3.1 Mechanism of Aerobic Biodegradation

For the treatment of various effluents from textile companies, biological degradation is the key area of focus. Different microbial strains are chosen and trained to grow and function in the presence of poisonous, complicated, and resistant effluents to the point where they can transform the contamination into less hazardous substances. They are practical and eco-friendly, and they don't waste a lot of water when compared to other physicochemical methods used to treat these effluents. They also don't remove a lot of water from the ecosystem. A wide assortment of microorganisms has been confined with the capacity to biodegrade various classes of colours generally utilized in the textile industry. Major mechanisms by which wastewater can be dealt with utilizing microorganisms can be ordered into biosorption and enzymatic corruption.

5.3.1.1 Biosorption

By the mechanism of biosorption, microorganisms are known to effectively remove harmful substances that are dissolved in water (whether they are organic or inorganic). The existence of many active sites on the surface of microorganisms causes biosorption. Ionic exchange, precipitation, chelation, and complexation are a few of the procedures. Azo colours, which are important contributors to the hue of the material produced, are removed through biosorption, particularly by yeasts and organisms. The primary site of biosorption is thought to be within an organism's cell mass. Peptidoglycans, or proteins found on the cell wall, as well as dynamic assemblages, including polysaccharides, lipids, and amino acids, take on a crucial role due to biosorption by yeasts. The pH, temperature, initial colour fixation, and dose affect biosorption. At pH 6.0, *Saccharomyces cerevisiae* displayed the most intense sorption. *Aspergillus niger* dead biomass has been effectively employed as a biosorbent, with a preferred pH of 5 (Mullai et al., 2007).

5.3.1.2 Enzymatic Degradation

Azo dyes provide a large range of types to be used by the material industry and, in contrast to regular colours, are steady, modest, and simple to organize. Azoreductases, laccases, and peroxidases are three important catalysts in the microbial framework that contribute to azo colour debasement (Nagendranatha Reddy et al., 2018). Azoreductases are flavoproteins that can either be extracellular in nature or may be limited to the cytoplasm of microorganisms. Microbial organisms degrade azo colours completely in two steps. The reductive breakage of the azo bond in the first step produces dry metabolites such as aromatic amines in anaerobic conditions. The intermediates break down into stable final products during the following phase under demanding conditions. According to the theory, these chemicals reduce colours by moving electrons from the catalyst's redox centre to a go-between (NADH, NADPH, FMN, etc.) and then exchanging those electrons with the colours (Arvind et al., 2022). When laccases tolerate an electron from an azo dye and transfer it to O_2 (via an intermediate), the colour separates because of the oxidation of the azo dye. Laccases may adhere to numerous substrates and can therefore be used to remediate wastewaters that include various azo colours. Haeme-containing oxidoreductases called lignin peroxidases are recognized for degrading various aromatic combinations, including phenyls and synthetic colours (Nagendranatha Reddy et al., 2018). Various biomass types are used for colour ejection via biosorption, including *Aspergillus parasiticus*, *Aspergillus fumigates*, *Trichoderma species*, and *Kluyveromyces marxianus IMB3* (Mullai et al., 2017).

5.3.2 Role of Various Biocatalysts in Aerobic Biodegradation

Aerobic microorganisms are significant degraders of various textile industry effluents rich in certain components. Based on the type of textile effluent bioremediation, the aerobic microbe's distribution appears to be patchy, and their activities seem to be variable accordingly. Aerobic microbes grow on various substrates, thereby having a broad substrate range. Therefore, aerobic biodegradation is a significant factor in the natural attenuation of textile industry effluents.

5.3.2.1 Bacteria

Since bacteria decolourize and mineralize colours more quickly than plants do, they have been preferred for the purification of certain textile sector wastewaters. In addition to this significant benefit, the use of microorganisms for effective and efficient treatment renders the entire process sustainable. Some of the species that are generally thought of as bacteria for the degradation of colours and other toxic effluents belong to the genera *Pseudomonas*, *Bacillus*, *Aeromonas*, and *Proteus*. Specifically, added oxygen is used by aerobic bacteria to remove the pollutants in wastewater and convert them to energy. This energy is used by aerobic bacteria to grow and reproduce (Mullai et al., 2007).

In new treatment facilities, aerobic bacteria are mostly used in an environment known as circulating air through climate. The wastewater pollutants are decoloured and degraded by this bacterium, which also uses the free oxygen present in the water to produce energy. This type of bacteria has to have the right amount of oxygen introduced in order to be effectively used. This will guarantee that the bacteria can carry out their duties effectively and continue to develop and multiply on their food source (Katuri et al., 2019).

5.3.2.2 Microalgae

In contrast to the energy-intensive and common natural treatment procedures that are currently used, microalgae, such as eukaryotic microalgae and cyanobacteria, have demonstrated to be a manageable and ecosystem-safe alternative. It also serves as a sustainable hotspot for biomass, with microalgae being used to clean wastewater and as a potential method for bio-obsession of CO_2. Mixotrophic microalgae are used to clean wastewater because of their ability to utilize both natural and synthetic carbon as well as synthetic N and P for growth, which reduces the concentration of these elements in the water.

The main advantage of using tiny microalgae for wastewater treatment is that they release oxygen through photosynthesis, which is necessary for heterotrophic bacteria to decompose carbonaceous pollutants. While it can be challenging to evaluate the role that algal culture plays in the treatment of wastewater, numerous studies have demonstrated how algal arrangements can support the removal of excess nutrients from wastewater. For the removal of phosphorus and its recovery and reuse from the obtained P-rich algal biomass, the use of algae granules in produced wastewater has been shown to be incredibly effective. Microalgae have furthermore been shown to be a potential source of energy, in addition to being efficient at removing CO_2 and other supplements from wastewater.

Microalgae may utilize both organic and inorganic nitrogen sources, including nitrite and nitrate, as well as natural nitrogen sources like urea and ammonium. N-evacuation is still occurring due to the ecological conditions, which lead to the release of N_2O during the wastewater treatment cycle. A reduced number of ozone-depleting substances are released during microalgae treatment operations. For instance, most nitrogen is absorbed by microalgae rather than converted entirely to oxides of nitrogen.

Numerous investigations have revealed information about the insignificant N_2O emission caused by microalgae connected to connected microorganisms in wastewater treatment. 0.0047% g N_2O-N g/1 N-input is estimated to be the outflow component of the wastewater treatment process based on the analysis of microalgae. Supplying wastewater with broken-up O_2 through small-scale green growth photosynthesis is a sure bet for significant energy interest savings and reductions in linked ozone-depleting material discharges (Mohsenpour et al., 2021).

5.3.2.3 Filamentous Fungi

On relatively expensive substrates like starch or molasses, filamentous fungi are frequently grown in the food industry as a source of by-products like protein and biochemicals, among others. It makes sense to utilize filamentous fungi to clean wastewater with high strength. Furthermore, in addition to producing highly dewaterable fungal biomass that can be used as a source of animal feed and possibly even human diets, fungi treatment also transforms wastewater organics into high-value fungal proteins and useful biochemicals. In comparison to bacteria, fungi are better at metabolizing complex carbohydrates like starch because they can manufacture a large variety of fine biochemicals and enzymes. Fungi are frequently farmed in industry in addition to being used as food supplements to create a number of useful products such as amino acids, enzymes, pigments, organic acids, organic alcohols, and others. It is frequently rare for more nutrients to be needed throughout the healing phase for fungus development. Some have looked into the potential for wastewater purification utilizing yeasts and moulds to produce microbial biomass proteins (MBPs). In addition to producing useful by-products like amylase, chitin, and lactic acids, fungi have a number of significant benefits over bacteria in the biological treatment of wastewater. First, through an unintended oxidation reaction, fungi are endowed with a collection of extracellular enzymes that aid in the biodegradation of refractory substances, including phenolic compounds, pigments, and polyaromatic hydrocarbons (PAHs), among others. By contrast, bacterial cells are able to create enzymes that are specific to a given target and may degrade these resistant pollutants. Additionally, compared to bacterial species, fungi have a higher resistance to inhibitory substances. For their delicate organelles, fungi that grow in hyphal growth offer better protection. Fungi are protected from inhibiting substances through adsorption by the extra-polysaccharide matrix that makes up their cell walls. In addition, fungi are eukaryotes with a higher gene count than bacteria, which gives them a wider range of inhibitory chemical tolerance. Greater reproductive selectivity is a result of the increased gene count in fungi, which may lead to superior environmental adaptations. Bacterial biomass's well-understood growth dynamics are one of the main justifications for employing it in biological textile wastewater treatment. To further understand the dynamics of fungal cultivation and the techniques used to clean wastewater from textile manufacturing, more study is required. In order for this technology to be accepted as a wastewater treatment method, a deeper understanding of fungal systems is required (Sankaran et al., 2010).

5.3.2.4 Yeast

Yeast, a very essential microbial resource, can adapt to a variety of unusual situations and has a good protein structure in the body. This makes it an important component of wastewater's organic treatment. Different artificial substances and compounds can be found in the effluent from different businesses. For many years, yeast has been used in the natural treatment of contemporary wastewater. It has amazing advantages like a large ooze load, great effectiveness, excess muck that may be used as feed protein, and more. There may be nitrogen and carbon that need to be removed from wastewater with high natural strength. Anaerobic, high-impact, nitrification, and denitrification cycles are the typical types of evacuation. More and more yeasts are becoming familiar with the cycles these days. The carbonaceous substrates and natural nitrogen are transformed by the fermentative bacteria into smelling salts and VFA, which can be broken down to provide a substrate for the growth of yeast. The ability of yeast biomass to consume, collect, and degrade hazardous chromophores into further clear combinations is directly related to the interest in using yeasts to remediate material effluent. According to an analysis of a previous fragment, dead yeast biomass has been used as a biosorbent for the biosorption of different species. Additionally, the protein structure of yeasts allows them to breakdown colours found in wastewater from materials. In yeast, the chemical structures known as peroxidases, reductases, and laccases are part of the process that degrades different kinds of dyes. The biodegradation of various kinds of dyes has been predominantly linked to *Candida krusei, Trichosporon beigelii, Galactomyces geotrichum* and *S. cerevisiae* (Wang et al., 2018). However, additional species and their dye removal efficiencies have been listed in Table 5.1.

5.3.3 Various Aerobic Processes to Treat Textile Industry Effluents

5.3.3.1 Aerobic Granular Sludge–Membrane Bioreactor (AGS-MBR)

A promising breakthrough for treating urban and modern wastewater is the MBR, a crossover structure with natural treatment and filtration. However, the use of MBR has been significantly constrained by biofouling, which refers to the association and

TABLE 5.1
Dye Decolourization Using Different Microorganisms

Biocatalyst	Dye Tested	Initial Dye Concentration (mg/L)	pH	Temperature (°C)	Time (h)	Decolourization (%)	References
BACTERIUM							
Micrococcus glutamicus NCIM 2168	Reactive Green 19 A	50	6.8	37	42	100	Saratale et al. (2010)
Bacillus sp. *VUS*	Navy Blue 2 GL	50	7	40	18	94	Ayed et al. (2021)
Acinetobacter calcoaceticus NCIM-2890	Direct Brown MR	50	7	30	48	91.3	Mohd Rosli and Ahmad (2018)
Enterococcus gallinarum	Direct Black 38	100	–	–	480	100	Birmole et al. (2019)
Pseudomonas species SU-EBT	Congo red	1,000	8	40	12	97	Joe et al. (2011)
Rhizobium radiobacter MTCC 8161	Reactive Red 141	50	7	30	48	90	Rau et al. (2002)
Comamonas sp. *UVS*	Direct Red 5B	1,100	6.5	40	13	100	Liu et al. (2015)
Exiguobacterium sp. *RD3*	Navy Blue HE2R	50	7	30	48	91	Tan et al. (2009)
Proteus mirabilis	Red RBN	1,000	6.5–7.5	30–35	20	95	Saratale et al. (2010)
Aeromonas hydrophila	Red RBN	3,000	5.5–10.0	20–35	8	90	Hsueh et al. (2009)
Citrobacter sp.	Azo and triphenylmethane	5 mM	7–9	35–40	1	100	Bhuktar et al. (2010)
Paenibacillus azoreducens sp. *Nov*	Remazol Black B Amaranth,	100	–	37	24	98	Meehan et al. (2001)
Bacteroides fragilis	Orange II Disperse Blue 79	0.1 mM	8	35	–	95	Vymazal et al. (2021)
Bacillus fusiformis KMK5	Acid Orange 10	1,500	9	37	48	100	Kamaruddin et al. (2013)
FUNGI							
Aspergillus ochraceus	Reactive Blue 25	100	–	30	7	100	Gosavi and Sharma (2014)
Ganoderma austral	Poly R-478	40	6.7	25	18	93.4	Alsalami et al. (2017)

(Continued)

TABLE 5.1 *(Continued)*
Dye Decolourization Using Different Microorganisms

Biocatalyst	Dye Tested	Initial Dye Concentration (mg/L)	pH	Temperature (°C)	Time (h)	Decolourization (%)	References
Pleurotus ostreatus sp. *3*	Disperse Orange 3	100	5	30	5	66.25	Seker et al. (2014)
Trametes versicolor	Remazol Black B	100	4	30	12	88.4	Blánquez et al. (2004)
Cunninghamella elegans	Orange II	100–2,000	5.8	28	4	88	Ikram et al. (2022)
Lentinula edodes	Poly R-478	1,500	6.5	30	11	72	Alsalami et al. (2017)
Ganoderma sp. *WR-1*	Amaranth	50	–	28	8	96	Mohd Hanafiah et al. (2019)
Pycnoporus sanguineus MUCL 41582	Acid Blue 62	20–60	–	30	11	99	Cui et al. (2009)

growth of microorganisms as well as the adsorption of exopolysaccharides (EPS) and dissolvable microbial items onto the film surface or potentially inside layer pores. The granulation cycle can be advanced by the biomass maintenance provided by MBR because the EPS can be taken on by the AGS for its improvement rather than making the cake layer on film. This is why many researchers have developed powerful granular film bioreactors (AGMBR) to further develop treatment execution by moderating biofouling.

In the interim, microorganisms are more inclined to adhere to the AGS surface rather than obstructing pores, which simultaneously contributes to the moderation of biofouling and biomass granulation. Studies on AGMBR are currently at an important stage as the majority of analyses used engineered or urban wastewater as influent to focus on the reduction of biofouling. Even though it is thought to be a viable choice for modern emanation treatment, there are currently few studies that use modern effluents as influent to judge the viability of this invention. Effluents such as high salinity wastewater was effectively treated by AGMBR.

The results showed that incredible execution in removing contamination and reducing biofouling was displayed continuously. However, as the majority of stubborn organics are kept far from the environment and in bioreactors, it is also proposed that the accumulation of toxicity and recall of citrange are particularly likely to occur in MBRs. The existence and accumulation of these dangerous and resistant compounds may negatively affect microbial activities and treatment execution if they are not biodegraded or transferred into safe, non-harmful substances. Prolonged stability becomes a critical concern when it comes to the treatment of refractory contemporary effluents by AGMBR.

According to research, the AGS-MBR blend might produce significantly more permanent resistance than normal MBR in terms of treating high-intensity contemporary lighting, which could shorten the film administration term. It is obvious that this framework's long-term security in treating current effluents cannot be guaranteed, and more research is needed to solve this conundrum (Cai et al., 2021). An up-stream fixed-bed bioreactor is the reactor configuration that is used the most frequently to decolourize material wastes (Mullai et al., 2017).

5.3.3.2 Extended Aeration

Most high-influence reactors, in actuality, function as extended air course units. The term enlarged refers to the air course's range. Long support periods, low food-to-microorganism (F/M) extent, long stretch air distribution, and low biomass accumulation are indicative of it. Microorganisms will quickly exhaust the bioavailable normal carbon, including biomass, if excess separated oxygen and insignificant dissolvable regular matter are provided to ensure their endogenous improvement time. The goal is to strike the proper balance between the number of newly transmitted cells and the number of endogenously corrupted cells each day.

5.3.3.3 Attached Growth Bioreactors

The scaled back implemented sludge plants are termed suspended-improvement high-influence units. Started slop is a mixed microbial culture that includes parasites, rotifers, protozoa, and microbes. Microorganisms are necessary for the processing of ordinary materials. Protozoa and rotifers act as trackers and enable the isolation of dispersed microorganisms from the bio-coursed air through unit treated spouting. The biomass is totally incorporated with the biodegradable regular component, and various living things congregate to form an ever-expanding mass of microbes that is commonly referred to as natural flocs. Mixed liquor is the name given to this mixture of common flocs and wastewater. The term "fixed-film reactors" is also used in the writing to refer to connected development frameworks. The carrier is an inactive medium that is provided for the association of microorganisms. The digestion of the fine, suspended, colloidal, and separated normal solids by the natural film is how the wastewater is moved over the medium.

5.3.3.4 Suspended-Growth Bioreactors

The miniaturized actuated slime plants are the name given to the aerobic units in suspended development. Enacted slop is conceptualized as a heterogeneous microbiological culture made up of bacteria, protozoa, rotifers, and parasites. It is possible for natural stuff to be absorbed when bacteria are present. Protozoa and rotifers, which serve as predators, help to isolate dispersed bacteria from the bio-circulated air through unit-treated profluent. Individual organic entities aggregate to form a constantly changing mass of microorganisms, also referred to as organic flocs, after the biomass is thoroughly combined with the biodegradable natural division. Blended alcohol is the name given to this mixture of wastewater and organic flocs.

5.3.4 Factors Affecting Aerobic Bioremediation and Its Mechanism

Microorganisms (fungus, bacteria, algae, etc.) for the breakdown of pollutants can be treated in bioremediation procedures for the removal of dyes. Each microbe has its own unique degrading capabilities, restrictions, and operating environments (temperature, oxygen, pH, dye concentration, dye structure, and availability of nutrient sources).

5.3.4.1 Oxygen

The essential factor to take into account is how oxygen affects dye fading and cell growth. If the extracellular environment is aerobic during the dye drop stage, oxygen, a high-redox-potential electron acceptor, may suppress the colour decrease component. This is due to the fact that the cells use the electrons they liberate from the oxidation of electron donors specifically to decrease oxygen rather than the azo dye, and that the decrease item, water, is not a reducing agent. The dye's hypothesized intermediates are also reduced (Pearce, 2003).

5.3.4.2 Temperature

In many systems, the rate of variety ejection increases with rising temperature within a defined range that varies according to the system. The optimal temperature for cell culture development is between 35°C and 45°C, and this temperature is believed to give the fastest rate of variety evacuation. The lack of cell functionality or the denaturation of the azo reductase molecule may be to blame for the decrease in variety evacuation activity at higher temperatures. However, it has been demonstrated that the azo reductase protein is generally heat-stable and may remain active up to temperatures of 60°C for very brief periods of time in specific whole bacterial cell configurations.

The ideal variety expulsion temperature changes towards high quality when a cell culture is immobilized in a support medium because the milieu inside the support provides protection for the cells.

5.3.4.3 pH

The best pH for variety expulsion is generally neutral or slightly basic, and the rate of variety expulsion will typically slow down at strongly corrosive or unambiguously soluble pH values. As a result, the darkened wastewater is typically cradled to improve the cell culture's variety expulsion execution. Because more fundamental amine metabolites are produced as the azo bond organically decreases than the original azo molecule, the pH can rise as a result. A change in pH within the range of 7.0–9.5 has an impact on the dye oxidation process. It was discovered that as the pH increased from 5.0 to 7.0, the dye degradation rate virtually doubled, but between 7.0 and 9.0, the rate became chilly (Pearce, 2003).

5.3.4.4 Dye Concentration

The poisonousness of the dye at higher fixations, the ability of the protein to effectively perceive the substrate at the extremely low focus that may be present in some wastewaters, and a variety of other factors can all have an impact on the productivity of dye evacuation. The motor constants that control cycle efficiency may undoubtedly be represented using Michaelis–Menten energy, much like other catalyst-catalysed processes.

The use of Michaelis–Menten energy, for example, can allow predictions to be made about the productivity of a process, such as the amount of biomass stacking or the functional temperature required to maintain dye expulsion at a given proficiency within the constraints set by the reactor volume available, the creation of the foundation arrangement, and stream rates (Naresh Kumar et al., 2014).

5.3.4.5 Dye Structure

Some azo dyes are less likely to be ejected by bacterial cells. Variety ejection is more difficult with extraordinarily intricate, high sub-atomic weight colours, whereas dyes with simple patterns and low atomic loads show better rates of variety evacuation. Azo mixes with hydroxyl bunches or amino groups are more likely to degrade than those with methyl, methoxy, sulpho, or nitro groups. The number of azo bonds present in the dye particle has a direct impact on variety evacuation. The colour of mono azo dyes fades faster than the colour of diazo or tri azo dyes (Pearce, 2003).

5.3.4.6 Nutrient Source Availability

The growth of microorganisms in the aerobic oxygen-consuming reactor is improved by nutrients like C, N, and P, which increases the effectiveness of the therapy. The concept behind supplement types is that they contain microorganisms, specifically microscopic creatures like parasites, yeasts, and so forth. The most urgent problem is the perfect supplement requirement for the biggest material profusion decolourization (Mullai et al., 2017). Consequently, changes in a few key areas, such as pH, temperature, the convergence of dye and supplement sources, and aerobic/anaerobic conditions, have a significant impact on the bacterial biodegradation of different colours. These limitations have a fundamental impact on the bacterial decolourization tool, which in turn influences how the treatment interacts moving forward. Additionally, the process of biodegrading textile dyes occurs in two phases in both aerobic and anaerobic conditions (Nagendranatha Reddy et al., 2019).

Under aerobic conditions, dye degradation is often catalysed by catalysts, the majority of which are azo reductases. Aerobic microbes may degrade azo dyes in the presence of azo reductase chemicals and supply intermediate metabolites, such as sweet-smelling amines, which can then be further degraded by microorganisms under high-impact circumstances to achieve complete degradation. The azo bond (-N=N-) is broken in anaerobic conditions with the aid of azo reductase catalysts, which are dependent on redox intermediates like NADH/FADH for their function.

Decolourization of the dye is achieved by exchanging four electrons (lessening reciprocals) at the azo linkage over the course of two steps, with two electrons being transferred to the azo dye acting as an electron acceptor in each step. The reduced rate of azo dye degradation under anaerobic conditions can potentially be improved by utilizing cofactors and redox mediators to enhance dye degradation efficiency. Because of its low redox potential, it has been explained that anaerobic interactions are thought to be effective for the decolourization of azo dyes.

As a result, the cleavage of the –N=N– bond has been used to explain why organic decolourization of azo dyes is possible only in anoxic/anaerobic conditions. Aerobes are actually thought to be more responsible for degrading dye metabolite intermediates in the interim. As a result, the consecutive anaerobic-oxygen-consuming interaction is the best option for treating material effluent because it performed significantly better than a single aerobic or anaerobic reactor (Arvind et al., 2022; Srivastava et al., 2022).

5.4 ANAEROBIC BIOREMEDIATION OF TEXTILE INDUSTRY EFFLUENTS

Wastewater contaminated with various chemicals and dyes has become a significant environmental concern across many parts of the world. Human health and the photosynthetic activity of aquatic plants and animals are significantly impacted by the presence of heavy metals and chlorine in man-made dyes. As a result, wastewater from textile factories needs to be treated before being dumped into a body of water (Adane et al., 2021). In many nations, the discharge of untreated wastewater is a significant problem as well. One of the main xenobiotic substances that affect terrestrial and aquatic ecosystems and have a considerable impact on microbial populations is the use of azo dyes in the textile industry (Krishnamoorthy et al., 2021). According to the research done by Krishnamoorthy et al. (2021), prolonged exposure to wastewater from the textile sector has changed the bacterial diversity and enzymatic capability of soil. Due to the azo-dye contamination that resulted from this, the population of *Saccharibacteria* multiplied higher than that of other bacterial phyla (Krishnamoorthy et al., 2021). By using biological treatment of the textile effluent, it is possible to accomplish the decolourization of azo dyes and prevent such catastrophes. Adsorption to cell biomass and its subsequent destruction, caused by azo-bond reduction during anaerobic digestion with the aid of non-specific enzymes, are both involved in the therapy. Their role as an electron acceptor for starting the ETC, which calls for a labile carbon, is important to the reduction process (O'Neill et al., 2000). The fermentation process known as anaerobic digestion converts organic material into biogas, which is largely composed of methane and carbon dioxide. Wherever there is an abundance of organic material and a low redox potential (no oxygen), anaerobic digestion takes place often. This is typical in the stomachs of ruminants, marshes, lake sediments, ditch sediments, municipal landfills, and even municipal sewers (van Lier et al., 2020).

The use of aerobic treatment for textile wastewater or effluent results in good or even excellent organic load removal. However, this is not complemented by an equally effective colour removal. Traditional or advanced chemical decolourization methods are expensive and cannot reproduce accuracy in their results efficiently, making microbial decolourization more appealing. Anaerobic methods appear to be the most promising of the various processes (Stern et al., 2003). Biodegradable organic chemicals are effectively removed by anaerobic treatment, although mineralized compounds such NH_4^+, PO_4^{3-}, and S_2^- remain in the solution. Anaerobic treatment can be applied at any scale and practically everywhere because it is carried out using technically straightforward systems. Additionally, the surplus granular anaerobic sludge produced in the bioreactor is extremely tiny, well stabilized, and even has a market value (van Lier et al., 2020). Nevertheless, the breakdown of azo dyes produces aromatic amines, which have the potential to be poisonous and resistant to anaerobic treatment. Therefore, they get accumulated in the treated effluent, and such compounds can be degraded aerobically. Unlike the azo dyes, which do not inherently project carcinogenic, cytotoxicity, or mutagenic characteristics, the products of the anaerobic degradation of these dyes have been shown to project such characteristics in some cases. For example, some anaerobes can synthesize aromatic amines such as benzidine and 4-aminoaniline, which carry carcinogenic properties. These aromatic amines can be treated via aerobic processes that involve hydroxylation and the opening of the ring present within the aromatic compounds using nonspecific enzymes (O'Neill et al., 2000).

5.4.1 Mechanism of Anaerobic Treatment of Textile Wastewater

5.4.1.1 Anaerobic Degradation of Organic Polymers

The organic matter anaerobic degradation route comprises a number of activities that happen simultaneously and in sequential order. There are four phases to the decomposition of organic matter: acidogenesis, hydrolysis, methanogenesis, and acetogenesis. Anaerobic digestion is the process by which organic matter is serially digested by a variety of bacteria. It involves a complicated food web. Methane (CH_4), carbon dioxide (CO_2), ammonium (NH_3), hydrogen sulphide (H_2S), and water (H_2O) are produced because of the cooperative efforts of the engaged microbial consortiums. The anaerobic environment results from intricate interactions between different species of bacteria. The primary bacterial groups and the reactions they mediate include fermentation, hydrogen-producing, hydrogen-consuming, carbon dioxide-reducing, and aceticlastic methanogenesis (Nagendranatha Reddy et al., 2022; van Lier et al., 2020).

5.4.1.1.1 Hydrolysis

Exoenzymes, which fermentative bacteria expel, break down complicated substances into simpler ones so they may get past the cell walls and membranes of the fermentative bacteria. To put it another way, hydrolysis is a process that converts complicated polymeric substrates into monomeric and dimeric molecules that are easily utilized by acidogenic bacteria. The first phase in the anaerobic digestion of complex substrates is frequently hydrolysis, although occasionally a physicochemical pre-treatment is necessary to speed up hydrolysis. Due to the particle/substrate structure's free surface area, this process has a rate-limiting property (van Lier et al., 2020).

5.4.1.1.2 Acidogenesis

The dissolved cell suspensions of the fermenting bacteria are converted here into a variety of simple external chemicals. These substances include lactic acid, volatile fatty acids (VFAs), alcohols, CO_2, hydrogen, ammonia, and H_2S. A wide variety of hydrolytic and non-hydrolytic bacteria engage in acidogenesis. It is a rapid conversion phase in the anaerobic food chain where the G of acidifying processes produces ten to twenty times the bacterial growth rates, five times the bacterial yields, and better conversion rates (van Lier et al., 2020).

5.4.1.1.3 Acetogenesis

The breakdown of the digestive contents into acetate, hydrogen, and carbon dioxide leads to the synthesis of short-chain fatty acids (SCFA), such as acetate, in an intermediate acid generation stage that is comparable to acidogenesis. Propionate, butyrate, lactate, ethanol, methanol, H_2, and CO_2 all serve as crucial intermediates in the anaerobic digestion process and are hence essential acetogenic substrates. Since methanogenic bacteria use molecular hydrogen quickly, a partial pressure drop of around 10–4 atm results, which starts the hydrogen-producing acetogenic reactions. The two processes are interdependent. Such a link suggests that methanogenic bacteria, which actively scavenge electrons, are crucial for the breakdown of higher fatty acids and alcohols (van Lier et al., 2020).

5.4.1.1.4 Methanogenesis

During this process, CH_3COOH, CH_3OH, and H_2 and CO_2 are converted into CH_4. Such conversion is conducted by methanogenic archaea, which are obligate anaerobes that interact with only a narrow spectrum of substrates. They are classified into aceticlastic methanogens and hydrogenotrophic methanogens. Aceticlastic methanogens grow slowly and take a long time to double in size. However, hydrogenotrophic bacteria grow more quickly, with a doubling time of 4–12 hours.

In order to achieve large sludge concentrations, anaerobic reactors need a long start-up period when using unsuitable seed material. *Methanosaeta sp.* which are found in anaerobic filters with long solids retention durations and sludge bed systems, are the most common acetotrophic methanogens. The cause of this phenomenon could be attributed to wastewater treatment facilities' constant pursuit of the lowest possible effluent concentrations, while in anaerobic systems, substrate concentrations inside biofilms and sludge granules approach "zero" when bulk liquid concentrations are low. In these circumstances, *Methanosaeta sp.* clearly possesses a kinetic advantage over *Methanosarcina sp.* (van Lier et al., 2020)."

5.4.1.2 Anaerobic Degradation of Azo Dyes

Azo dyes are among the earliest synthetic compounds and are still widely used in the printing of textiles and in the culinary sector. Some colours are poisonous and/or cancer-causing, and they are therefore considered to be major environmental pollutants. Some of their N-substituted aromatic biotransformation products, in particular, bear this out. During standard aerobic wastewater treatment, azo dyes do not degrade, but in anaerobic conditions, azo linkages are quickly broken. Redox mediators and azo dyes act as electron acceptors and electron donors, respectively, during the reduction of flavin nucleotides. The production of aromatic amines is accelerated by azo dye degradation. Most of these aromatic amines that are gathered after the reduction process cannot be released back into the environment in their purest or least hazardous form when they occur in anaerobic environments. However, the majority of aromatic amines that accumulate following azo breakdown are not mineralized anaerobically, with the significant exception of some aromatic amines that have replacements for the hydroxyl and carboxyl groups present in them; these can be destroyed under methanogen conditions. On the other hand, aromatic amines are quickly broken down aerobically. The process of biodegradation of azo dyes, which is fuelled by bacteria, typically occurs in two stages. In the first stage, the dye's azo links are reductively severed, resulting in the generation of typically colourless but potentially dangerous aromatic amines. The second stage involves the degradation of the aromatic amines. Azo dye reduction frequently requires anaerobic (without oxygen) conditions, whereas the majority of aromatic amines degrade under aerobic circumstances. Beginning with the breakage of the -N=N- link in a reducing state, bacteria break down azo dyes. Azo dyes are decoloured by several bacterial species in methanogenic, anoxic, and aerobic environments (Alabdraba and Albayati, 2014).

5.4.1.2.1 Decolourization of Azo Dye under Anaerobic Conditions by Bacteria

The reductive breakage of the azo links results in the formation of aromatic amines, which also entails the transfer of reducing equivalents. Because aromatic amines are typically colourless and have a chromatic character, azo dye reduction is also known as azo dye decolourization. The possibility of hazardous aromatic amines building up in humans has been a major source of worry related to their synthesis. Traditional studies on bacterial azo dye reduction have primarily concentrated on facultative) anaerobes and their activity behaviour, which were isolated from the intestines of mammals (Alabdraba and Albayati, 2014).

Anaerobic, static culture conditions are known to facilitate the initial reductive phase of azo dye decolourization (Khalid et al., 2008, 2010; Khehra et al., 2005). Many dyes exhibit polarity, making it extremely improbable that they would cross the cell membrane and enter the interior of the cell, where they can be utilized by non-specific reductase enzymes. As a result, it is believed that bacteria's use of dye reduction is predominantly an extracellular activity (Pearce, 2003; Robinson et al., 2001; Khalid et al., 2010). According to Brigé et al., a reduction process is used to decolourize the dye in the extracellular space. This procedure or technique requires an extracellular component, cytoplasmic and periplasmic membranes, and an electron transfer channel (Brigé et al., 2007). Khalid et al.'s (2010) work also demonstrated the effectiveness of the bacterial strain in removing the dye's colour from the solid agar medium. Such a discovery hinted at the accumulation of active redox enzymes that were released into the media throughout the bacterial growth phase (Khalid et al., 2008, 2010). Therefore, it is stated that dye is reduced extracellularly via non-enzymatic methods (Khalid et al., 2010) (Figure 5.1).

5.4.2 Biocatalysts for Anaerobic Degradation of Azo Dyes

5.4.2.1 Biodegradation of Azo Dyes under Anaerobic Conditions

In addition, Khalid et al.'s (2010) study found that the effectiveness of the bacterial strain in removing the dye's colour from the phenazopyridine is biodegraded through several phases or processes, including hydrolysis, acidogenesis, acetogenesis, and

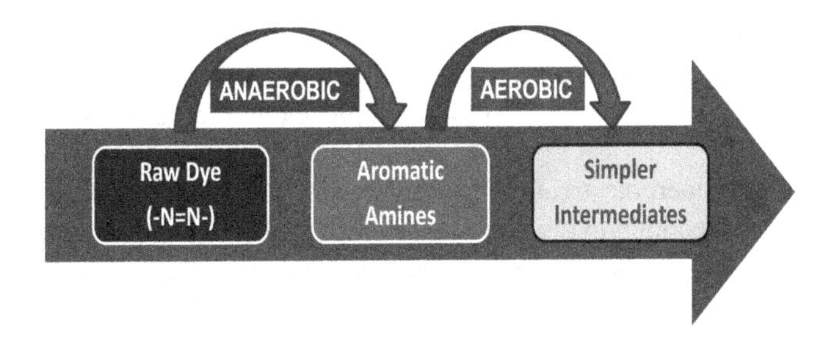

FIGURE 5.1 Bacterial removal of azo dyes: Potential mechanisms.

methanogenesis. Finally, the azo dye in question is decoloured through a reductive process. Although most dyes made of phenazopyridine cannot be degraded in the absence of oxygen (i.e., anaerobic conditions), aromatic amines are synthesized as a by-product of the biodegradation process as a result of these sequential procedures and show resistance to anaerobic degradation. For dye decolourization during methanogenesis, a source of organic carbon and energy are required. Simple substrates have been used as dye-decolourizing substrates, including sugar, starch, protein (whey), acetate, and ethanol (Sandhya, 2010). According to Chinrelkitvanich et al. (2000) and Brás et al. (2001), in the solid agar medium, both acidogenic and methanogenic bacteria contribute to dye decolourization. Such a finding suggested the accumulation of active redox enzymes that were released into the media throughout the bacterial growth phase (Sandhya, 2010; Brás et al., 2001).

Sulphate-reducing bacteria (SRB) and proteobacteria were found to be significant constituents of the mixed bacterial community, which was described using genetic approaches. This information was obtained via monitoring industrial dye pollution (Sandhya, 2010; Brás et al., 2001). According to Yoo et al.'s (2001) research, the addition of 2-bromo-ethane sulphonic acid, which acts as an inhibitor with a preference for methanogens, significantly affects the dye (Orange 96)'s capacity to be decoloured. Such a discovery suggested that methanogens, while taking part in the biodegradation process, do not take part in the decolourization process, while sulphate, molybdate, and acetate present in combination slow the rate of decolourization (Yoo et al., 2001).

5.4.3 Types of Reactors Utilized in Anaerobic Degradation of Wastewater

The first anaerobic reactor was developed by Karl Imhoff in 1905, which stabilizes solid sediments in a single tank. Germany's Ruhrverband, at Essen-Relinghausen, developed a method for precisely controlling the digestion of solids that have been encapsulated. Later, Buswell started applying the same method to the treatment of industrial wastewater and liquid waste. The majority of these systems were classified as low-rate systems since none of the designs included any unique elements that would have increased anaerobic catabolic capability. The anaerobic consortia's pace of growth has a big impact on how viable these systems' processes are.

Reactors grew increasingly large and hazardous to operate in the late 19th century; as a result, early studies on reactors with upward flow-fixed film systems were conducted, but the designs remained too underdeveloped to be practical (McCarty, 2001).

A low-loaded anaerobic treatment system is another way to conceptualize the anaerobic pond. Accompanying facultative and maturation ponds are frequently erected as anaerobic ponds. Anaerobic ponds have pond depths of 4 m, and the loading rate utilized ranges from 0.025 to 0.5 $KgCOD/m^3/day$. The drawbacks of anaerobic ponds include odour problems since these systems can readily get overloaded and the release of energy-rich CH_4 into the atmosphere, which is also considered a drawback. The anaerobic reactors used can be divided into three categories: high-rate anaerobic reactors, single-stage anaerobic reactors (such as anaerobic sludge bed reactors, anaerobic filters, anaerobic expanded, anaerobic contact processes (ACP), and fluidized bed systems), and up-flow anaerobic sludge blanket reactors (van Lier et al., 2020).

5.4.3.1 High-rate Anaerobic Systems

Contrary to aerobic processes, the percentage of viable anaerobic biocatalysts or anaerobic bacteria in full interaction with the wastewater elements in either an anaerobic or anoxic (denitrification) process determines the maximum permissible load rather than the highest rate within which a required reactant can be provided (e.g., oxygen during aerobic processes). In anaerobic high-rate systems, sludge concentrations are high due to the physical immobilization or holding of anaerobic sludge. High biomass concentrations make it possible to use high COD loading rates while still sustaining lengthy SRTs at relatively short HRTs. In anaerobic high-rate systems, sludge concentrations are high due to the physical immobilization or retention of anaerobic sludge. High biomass concentrations make it possible to sustain lengthy SRTs at relatively fast HRTs while still using high COD loading rates. When treating a given effluent in a reactor system under anaerobic conditions, the following requirements must be satisfied (van Lier et al., 2020):

- Under operational conditions, there is a greater retention of credible sludge in the reactor.
- An adequate amount of contact between surviving biomass (bacterial) and wastewater.
- A rapid rate of reaction and no significant transport constraints.
- The viable biomass needs to be well acclimated and/or acclimatized.
- All necessary organisms are present within the reactor under all mandated operating conditions, with a focus on the rate-limiting steps.

5.4.3.2 Single-Stage Anaerobic Reactors

5.4.3.2.1 Anaerobic Contact Processes

Sludge return and external settlers are both used in ACP (van Lier et al., 2020). An example of an anaerobic activated sludge process is anaerobic contact. A fully mixed reactor is accompanied by a settling tank. The reactor is used to recycle the biomass that has settled. Therefore, regardless of HRT, ACP may maintain a high biomass content inside the reactor and a high SRT.

Biogas bubbles (CO_2, CH_4) clinging to sludge that would otherwise circulate to the surface are removed by de-gasifiers. ACP was initially developed to treat diluted wastewater from locations like meat packing industries, where settleable flocs were prone to forming. For the processing of wastewater containing suspended particles, which microorganisms grab onto and create settleable flocs, ACP is appropriate. The biomass concentration can be as high as 25–30 g/L, depending on how easily the sludge settles. The daily loading rate ranges from 0.5 to 10 kg COD/m^3. The appropriate SRT could now be observed by maintaining the recycle rate, just like the activated sludge procedure.

5.4.3.2.2 Anaerobic Filters

To handle potential diluted, soluble organic waste, McCarty and Young developed the first anaerobic filter in the late 1960s. The filter contained rocks, much like the trickling filter did. The entire filter was submerged in wastewater that was dispersed throughout the bottom and flowing upward through the rock bed. As a result, the anaerobic microbes would flocculate into the media's void spaces, preserving the filter's biomass. Unattached biomass helps with waste treatment by forming large flocs with granular-shaped particles. This structure results from the interaction between rising gas bubbles and liquid, which helps shape the flocs. Sedimentation, followed by the anaerobic digestion of suspended or dissolved debris in the filter and fermentation of the sludge that settles on the bottom constitute the basic processes of this anaerobic filter treatment concept (Nagendranatha Reddy et al., 2016).

5.4.3.2.3 Anaerobic Sludge Bed Reactors (ASBR)

The AnWT systems that have so far acquired the most popularity are without a doubt the anaerobic sludge bed reactors (ASBRs). Sludge retention in such a reactor (GLSS device) is based on the development of readily settling sludge aggregates and the use of an internal gas–liquid–solid separation process. The upward-flow anaerobic sludge bed reactor, commonly known as UASB, was created in the Netherlands in the early 1970s and is the most well-known implementation of this principle (Lettinga et al., 1980). Wastewater passes through the reactor at an upward angle, similar to the UAF system. However, there doesn't appear to be any packing material inside the reactor vessel, compared to the AF system. The idea of a sludge bed reactor is supported by the following principles (van Lier et al., 2020):

- If the mechanical churning in the reactor is gentle and done properly, anaerobic sludge is anticipated to have acceptable sedimentation qualities.
- A proper link among the sludge and wastewater within the UASB systems is required, and it is achieved by uniform feeding, or it occurs as a product of agitation initiated by biogas production.
- When dealing with low-strength wastewater, reactors with an increased height-diameter ratio are used, reaching elevations of 20–25 m. Results in aiding the feeding system, followed by building up of a turbulent flow,

alongside with enhanced up flow velocity, thus procuring improved contact between the pollutants and the sludge.
- To avoid the drainage of sludge aggregates, a gas collection dome is used at the reactor's summit to separate the created biogas.

5.4.3.2.4 Anaerobic, Expanded, and Fluidized Bed Systems

The second generation of sludge-based bed reactors, known as extended bed/EGSB and fluidized bed/FB systems, can attain excessive organic loading rates (higher than 30–40 kgCOD/m^3/day). The EGSB is a growing system that includes some suspended biomass. On biotransporters like GAV, pulverized polyvinyl chloride, etc., this biomass grows. Due to the up-flow velocity produced by the influent wastewater and the recirculated effluent, these carriers expand by themselves.

This expansion reduces the likelihood of a blockage issue and improves substrate diffusion inside the biofilm. The FB system's configuration is relatively similar to that of the EGSB system. Because of the strong up-flow velocity, which causes the biomass suspends to drain out, it is a real fixed-film reactor. The expansion requires an up-flow velocity of between 10 and 25 m/hr and ranges from 25% to 300% of the settled volume of the bed. The biocarriers are allowed to move around the bed in this area. The fluidized bed reactor, like the extended bed, has no clogging issues and exhibits good substrate diffusion efficiency in the biofilm.

5.4.3.3 Up-Flow Anaerobic Sludge Blanket Reactor

The most common and efficient high-rate anaerobic technique for treating many types of wastewater is the UASB reactor. The UASB reactor's success can be due to its capacity to effectively separate the solids, liquids, and water phases while still retaining a significant amount of sludge. The UASB reactor consists of a circular or rectangular tank where sludge is pumped upward across an activated anaerobic sludge bed that occupies about fifty percent of the reactor's volume and is composed of effectively settleable flocs. The process happens during the passage as a result of particles getting trapped and organic material getting turned into sludge and biogas. An efficient GLSS is created as they ascend to the top under the influence of a gas–liquid interface. As the gases are trapped, the solid particles that had previously risen return to the top of the sludge blanket. Water carries a few solid suspends as it flows through the openings located between the baffles. The water removes the settlers as these solids settle on the blanket of sludge. The following aspects need to be considered when designing UASB: hydraulic surface loading capacity at its maximum, ensuring appropriate organic loading capacity is maintained, and the reactor internals such as inlet feed distribution, outlet effluent, and the GLSS need to be kept in check (van Lier et al., 2020).

5.4.4 Factors that Affect the Anaerobic Degradation of Azo Dyes

To deteriorate correctly, the dye needs to be in the appropriate environment. The efficiency of dye degradation can be affected by a number of physicochemical factors, including dye concentration, ionic strength (pH), temperature changes, glucose concentration, time, NaCl levels, urea levels, and redox mediators, among others (Ikram et al., 2022; Aruna et al., 2022). The studies conducted by Bardi and Marzona in 2010 indicated that the following parameters project an influence, to some extent, on the mineralizing process of Azo dyes. The studied parameters are the availability of oxygen, bioavailability, the availability of nutrients, the occurrence of cometabolic induction, pH levels, changes in temperature, etc. (Bardi and Marzona, 2010).

5.4.4.1 Availability of Oxygen

A larger or smaller extent of reduction of azo dyes occurs as a result of a reduction in the number of electrons surrounding the diazo link, which makes it simpler to decrease the azo group and liberate an aromatic amine. A charged functional group attached to the phenazopyridine group, or perhaps the presence of a second polar functional group, prevents the reaction from occurring, whereas the electron-withdrawing groups diminish the density of electrons. Electron-donor groups abridge diazo groups by creating hydrogen bonds with them while they are in ortho position. In water-soluble pigments with groups like $-SO_3Na$ or $-COOH$, a less complex diazo reduction process is seen. The azo bond breaking down, a chemical, unspecific process, either occurs within or outside the cell depending on the redox capacity of the intermediaries as well as of the azo compounds. Redox mediator reduction can also occur chemically or enzymatically. As a result, it serves as proof that environmental factors can affect the amount of azo dye breakdown both directly by affecting the environment's reductive or oxidative condition and indirectly by affecting microbial metabolism. A larger or smaller extent of reduction of azo dyes occurs as a result of a decrease in the number of electrons surrounding the diazo link, which makes it simpler to decrease the azo group and liberate an aromatic amine (Bardi and Marzona, 2010). Anaerobiosis is the most extensively studied environmental element driving this response. From a biological perspective, facultative anaerobes, fully pure anaerobes, and aerobes have all been researched for anaerobiosis (Adamson et al., 1965; Bragger et al., 1997; Dubin and Wright, 1975). These studies show that almost all azo compounds can undergo biological reduction in anaerobic environments (Stolz, 2001).

5.4.4.2 Bioavailability

The various chemical-physical characteristics of the dye molecules as well as their intermediates can have an impact on bioavailability in various ways, such as their initial concentration during the aqueous stage, when microbes or enzymes are active, or their capacity to pass through the cell membrane to be metabolized inside the cell. When azo dyes are exposed to biological systems, one of the main determinants of their fate is whether they are hydrophilic or hydrophobic. Due to their high polarity, water-soluble azo compounds such as sulphonated azo dyes demonstrate an inability to pass the plasma membrane barrier (Bardi and Marzona, 2010).

5.4.4.3 Availability of Nutrients and the Occurrence of Cometabolic Induction

Numerous bacteria that decolourize azo dyes by reductive pathways need an additional carbon source present, proving that they are not the only source of carbon and energy for these organisms (Stolz, 2001). However, under highly salinized conditions, a *S. putrefaciens* isolate demonstrated an inhibitory effect of glucose on the breakdown of azo dyes, which was explained by a preference for utilizing glucose for cell development (Khalid et al., 2008). By using cometabolic induction processes, certain bacteria strains are now capable of destroying azo dyes. Quite complex azo compounds, such as carboxy-Orange, were destroyed by a bacterial consortium that also degraded 4,4-dicarboxyazobenzene, demonstrating the ability of microorganisms to acquire novel metabolic properties. However, it wasn't really possible to adapt to Acid Orange 20 and Acid Orange 7, despite their structural similarities (Bardi and Marzona, 2010; Kulla et al., 1983).

5.4.4.4 Temperature and pH Effects

For industrial procedures that use azo dyes, which are reactive dyes that are often operated under alkaline conditions, tolerance to high pH is particularly crucial. The capacity of a *C. bifermentans* strain isolated from contaminated soils to break down Reactive Red 3B-A at various pH levels between 5 and 12 was examined. However, at pH 5, decolourization was not noticed. The dye, however, almost completely lost its colour after 48 hours of incubation when the pH ranged from 6 to 12 (Joe et al., 2008). The temperature at which each microbial species may thrive in cell culture is often associated with the maximum rate of colour removal, with an increase in decolourization proportionate to a rise in temperature within an ideal temperature range (Ozdemir et al., 2007; Adedayo et al., 2004).

5.5 ANOXIC/MICROAEROPHILIC BIOREMEDIATION OF TEXTILE INDUSTRY EFFLUENTS

Different azo dyes are anoxically decolourized by facultatively anaerobic and mixed aerobic microbial consortiums. Even though a number of these cells can grow aerobically, dye removal could only be achieved anaerobically. In anoxic conditions, strains of bacteria, including *Pseudomonas luteola*, *Bacillus subtilis*, *Aeromonas hydrophila*, *Pseudomonas* sp., and *Proteus mirabilis*, decolourized the azo dyes. Yeast extract, peptone, or a combination of a complex organic source and a carbohydrate were examples of organic ingredients that were frequently needed for azo dye breakdown by both mixed and pure cultures. Glucose is the optimum substrate for anaerobic dye degradation under methanogenic conditions. However, it may or may not be appropriate for non-obligatory anaerobes and fermenting bacteria to decolourize anoxic dye, depending on the bacterial culture. Glucose significantly improved *Sphingomonas xenophaga.* Strain BN6 exhibited a superior ability to decolorize the azo dye - Mordant Yellow 3, while P. luteola, Aeromonas sp., and a few other hybrid cultures showed significantly lower efficiency in azo dye decolorization. The negative impact of glucose on anoxic decolourization has been linked to both acid generation and catabolic inhibition. After *P. leuteola* decolourized reactive red 22 in the culture filtrates, HPLC and mass spectrometry results revealed that there are two aromatic amines present, as well as a product that has been partially reduced (Pandey et al., 2007). For the reduction of azo dye, we use combinations of different organisms (combination of facultatively anaerobic and mixed aerobic microbial) and combination of different reactors, such as UASB-CSAR, SBR anaerobic-aerobic, UASB-aerobic-SBR, UASB agitated tank, anaerobic-SBR-aerobic-MBR, AN-SBR MB-SBBR, Batch anaerobic-aerobic reactor, and anaerobic-anoxic-aerobic reactor (Popli and Patel, 2015; Nagendranatha Reddy et al., 2018).

5.5.1 Mechanisms behind the Degradation of Azo Dye

There are two ways to degrade azo dyes: directly via enzymatic means and indirectly through biological means.

5.5.1.1 Direct Enzymatic Reduction in Anaerobic Conditions

During the process of substrate or coenzyme oxidation to azo dyes, reducing equivalents are generated and subsequently transferred to the dyes via enzymes. These enzymes may be selective or generic, only catalysing the degradation of azo dyes. These latter biocatalysts conduct a variety of substrate reductions. These enzymes unnecessarily degrade azo dyes because of their general nature.

5.5.1.1.1 Direct Enzymatic Reduction in Anaerobic Conditions

Anaerobic bacteria with azoreductase decolourized sulphonated azo dyes while they were growing on solid or complicated media. The majority of these strains belonged to the genera Eubacterium and Clostridium. Oxygen-sensitive azoreductases that were released extracellularly were produced by their strains. Azo dye reduction is considered to be catalysed by a biocatalyst called flavin adenine dinucleotide dehydrogenase, which also reduces aromatic compounds containing nitrogen (Sreelatha et al., 2015; Venkata Mohan et al., 2013). Analyses using an immunoelectron microscope revealed that the enzyme was secreted concurrently with its synthesis. In *Escherichia coli*, the *C. perfringens* gene encoding this enzyme has been translated and expressed. Additionally, flavin reductases provide azo dyes with electrons via dissoluble flavins, which may have an impact on the decolourization of dyes. Most reported in vitro and of little consequence in vivo, cytosolic flavin-dependent azo reductases reduce sulphonated azo dyes, according to research on the recombinant *Sphingomonas strain BN6*. The introduction of flavins to strain BN6 resting cells revealed no appreciable upgrade in azo dye degradation, providing yet more proof that flavin-dependent azo reductases are almost laboratory artefacts. These findings led to the concept that the degradation of azo dyes in living cells with a complete cell membrane is mediated by extra biocatalyst systems and some redox regulators. Like facultatively anaerobic or aerobic bacteria like *Sphingomonas strain BN6*, microorganisms with membrane-based electron transportation systems may directly move electrons from the respiratory chain to the proper redox mediators. Assuming intracellular reductases play a role in the process, it is reasonable to consider that there must be other mediators, aside from flavin cofactors, which have a higher capability to move across the membranes (Pandey et al., 2007).

5.5.1.1.2 Direct Enzymatic Reduction in Aerobic Conditions

Initially, the identification of azo reductases among obligate aerobic bacteria was established through the discovery and characterization of two azo reductases.

These reductases were found in the carboxy-orange-degrading strains, specifically *Pseudomonas KF46* (now reclassified as *Xenophilus azovorans KF46F*) and *Pseudomonas K22* (now reclassified as *Pigmentiphaga kullae K24*). These intracellular azoreductases had exceptional dye structure selectivity and reductively degraded the structural analogues of the dyes that were carboxylated and sulphonated, whereas the azo reductases isolated from these two strains were peptides devoid of flavin attachments and the essential NADPH for activity, making them noticeably different from one another. Analyses of many orange dyes used as carboxy orange II azo reductase substrates revealed that the azo link requires an ortho positional hydroxy group. The *P. kullae KF24* enzyme, in contrast, needed to add a hydroxy group to the azo group's para position. However, the organisms were unable to utilize sulphonated colours as the source of carbon since the related azo reductases could not only decolourize but also act as inducers for AO7 and AO20. Reactive colours: the azo reductase from *Bacillus* sp. strain OYl-2 decolourized numerous proprietary dyes, including AR88, AO7, and many others. Additionally, a tiny single protein with a molecular weight of about 20 KDa was discovered to be this enzyme. These three distinct azo reductases' gene sequences are currently recognized. This azo reductase does not appear to be related to two other well-studied azo reductases from *X. azovorans KF 46* and *P. kullae*, according to amino acid sequence alignments. *E. coli*, *Bacillus* sp., *Staphylococcus aureus*, *Shigella dysenteriae*, and *P. aeruginosa* aerobically cultivated cultures have all produced non-specific enzymes performing azo bond reduction. These enzymes have been shown to be flavoproteins when they have been described (Arvind et al., 2022).

In addition to the presence of azo reductases, a specific transport mechanism is required to enable dye penetrating into the cells for sulphonated azo dye reduction to occur intracellularly. It appears that there is presently no information available on the ways in which different sulphonated substrates, such as taurine, p-toluene sulphonate, and sulphonated alkanes, are carried into bacterial cells. The capabilities of *Sphingomonas* sp. and strain ICX, as well as the ability of *Azovorans KF46F* to absorb and reduce Acid Orange 7 (AO 7), suggest that these organisms possess both azoreductase enzymes and effective transport mechanisms. Consequently, for the development of recombinant organisms capable of removing sulphonated dyes from colour, it could be required to introduce the aerobic azoreductase gene into bacterial strains that can thrive on sulphonated aromatics. In general, these organisms have limited substrate specificity. Studies on cultures that can only use 4-ABS instead of other benzenesulphonates have revealed that these are highly specialized 4-ABS-degrading strains. Two more aromatic sulphonates, toluene and sulphonate benzene, can be used by the 2-ABS-degrading *Alcaligenes* sp. strain O-1 for growth. This strain's cell extracts, however, can desulphonate at least six different substrates. This shows that these cultures have very specialized transport systems for absorbing aromatic sulphonates. As a result, the generated recombinants might still only have limited substrate specificity (Pandey et al., 2007).

5.5.1.2 Mediated Biological Azo Dye Reduction

Because of their higher molecular weight, it was suggested that these dyes could be reduced through methods unrelated to their entry into cells. The cell membrane is impermeable to polymeric azo dyes. In recent years, several studies have concentrated on how redox mediators help anaerobic bacteria break down azo linkages. When riboflavin is present in catalytic amounts, one of the azo dye's component amines, AO 7, 1-amino 2-napthol, substantially increases the breakdown of mordant yellow 10 through anaerobic granular sludge. This is likely because it makes it easier to transfer reduced equivalents. Adding artificial electron transporters like AQDS (anthraquinone-2,6-disulphonate) will significantly speed up the reduction of several azo dyes. The first instance of the aerobic breakdown of a xenobiotic chemical by mediators of redox led to the anaerobic cleavage of azo dyes. When grown aerobically with 2-naphthyl sulphonate (NS), the speed of amaranth decolourization was 10–20 times faster when grown with *Sphingomonas* sp. strain BN6 cell cultures. By adding the culture filtrates obtained from these cells, anaerobic dye removal via cell suspensions cultivated without NS may even be improved. These findings led to the suggestion of a mechanism for *S. xenophaga's* mediated reduction of azo dyes. Other bacterial cultures that form redox intermediates when aromatic compounds are broken down aerobically could also contribute to the increase in dye removal under anaerobic circumstances. *Escherichia coli* strain NO3. which discolours dye, and its metabolites were added to culture supernatants, which increased the degradation of azo dye. The surface of activated carbon, i.e., recognized to be comprised of quinone groups, improved the degradation of dye. One of the oldest known uses of activated carbon-mediated biocatalysis can be found here.

In bacterial (*Pseudomonas aeruginosa*, *Aeromonas hydrophila*, *Bacillus* sp., *Bacillus fusiformis*, etc.) azo dye degradation, two stages are involved. Anaerobic metabolism occurs first, followed by aerobic metabolism. Seventy percent of the degradation is observed in the 1st stage and 30% in the second stage. In the initial stage, degradation of the azo bond takes place by the addition of 4[H^+] ions, which produce aromatic amines that are more hazardous than pre-existing azo dyes. Production of these aromatic amines is a non-specific and extracellular process. This degradation may happen either chemically or biologically. The aromatic amines that are synthesized after the first degradation are oxidized, which degrades the aromatic amines to carbon dioxide, water, and ammonia. Some examples of aromatic amines are aminobenzene, aminonaphthalene, and amino-benzidine (Van der Zee and Villaverde, 2005).

5.5.2 Biocatalyst

Organisms that we use in the degradation of the dye are said to be biocatalysts, or the metabolites that are produced by them that help in the reduction of the dyes are known as catalysts. Various types of microorganisms, such as *Pseudomonas, Bacillus*, *Enterococcus*, *Xenophilus*, *Brevibacillus*, *Shewanella*, *Klebsiella*, and *Halomonas*, are utilized for the degradation process. Some specific examples of these organisms include *Pseudomonas fluorescens*, *P. nitroreductase*, *B. fusiformis*, *Halomonas* sp., *P. aeruginosa*, *Klebsiella* sp., *Bacillus* sp., *S. aureus*, *Pseudomonas luteola*, *Enterobacter*, *Morganella*, *Bacillus* sp., *X. azovorans*, *Enterococcus faecalis*, *E. gallinarum*, *Brevibacillus latrosporus*, and *Aeromonas hydrophila* (Table 5.2).

TABLE 5.2
Numerous Anoxic Microorganisms That Can Decolourize Dyes

S. No.	Organism Name	Azo Dye Degraded	Reference
1	*Pseudomonas fluorescens*	Acid Yellow	Pandey et al. (2007)
2	*P. nitroreductase*	Methyl Red Vibrio logei	Adedayo et al. (2004)
3	*B. fusiformis*	Disperse Blue 79 Acid Orange	Kolekar et al. (2008)
4	*Halomonas* sp.	Remazol Black B	Asad et al. (2007)
5	*P. aeruginosa*	Navitan Fast Blue	Nachiyar and Rajkumar (2003)
6	*Klebsiella* sp.	Basic Blue 41, Reactive Black 5	Elizalde-González et al. (2009)
7	*Bacillus* sp.	Disperse dye	Maier et al. (2004)
8	*S. aureus*	Congo Red	Chen et al. (2006)
9	*Pseudomonas luteola*	Reactive Red, Reactive Blue, Reactive Black	Chen et al. (2006)
10	*Enterobacter, Pseudomonas, Morganella*	Acid Orange 7	Barragán et al. (2007)
11	*Bacillus* sp.	Methyl Red	Suzuki et al. (2001)
12	*X. azovorans*	Orange II	Blümel et al. (2001)
13	*Enterococcus faecalis*	Methyl Red and sulphonated azo dyes	Chen et al. (2006)
14	*E. gallinarum*	Direct Black 38	Bafana et al. (2008)
15	*Brevibacillus latrosporus*	Remazol Orange	Sandhya et al. (2008)
16	*Aeromonas hydrophila*	Triphenylmethane, azo, and anthraquinone	Adedayo et al. (2004)
17	*Shewanella putrefaciens*	Acid Red 88, Reactive Black 5, Direct Red 81, and Disperse Orange 3	Khalid et al. (2008)

5.5.3 Factors Effecting the Degradation of Dye

5.5.3.1 Molecular Structure

The elimination of aromatic amines (AAs) is significantly influenced by molecular structure. This study has shown that the quantity of aromatic rings in amino acids has an impact on elimination efficiency. In comparison to amino acids with one aromatic ring, AAs with two aromatic rings often demonstrated poorer removal efficiency. For example, the removal effectiveness of 4-aminoazobenzene, 2-naphthylamine, o-aminoazotoluene, 3,3-dichlorobenzidine, and 2-chloroaniline varied between 37% and 76%, whereas amino acids with one aromatic ring have been shown to remove over 80% of their molecules. Despite having two aromatic rings in their structure, 4,4-diaminodiphenylmethan and 4,4′-Oxydianiline exhibit removal efficiencies that are more than 90%. This is related to the fact that the structure has a strong EDG (electron-donating group) but no strong EWG (electron-withdrawing group). AAs may behave stubbornly, depending on how many aromatic rings they possess. Instead of an inherent weakness in oxygenases' capacity for biotransformation, this behaviour may largely be the result of the decreased bioavailability of AAs with complex structures. Due to their hydrophobicity, thermal stability, and potent adsorption affinity, AAs with fewer aromatic rings are more easily degraded by native bacteria than AAs with more aromatic rings.

5.5.3.2 Substitution Group

The number of substitution groups significantly affects the removal effectiveness of AAs, along with EDG and EWG. For instance, compounds like 2,4,5-trimethylaniline, 2,4-dimethylaniline, 2,6-dimethylaniline, o-toluidine, and 2,4-diaminoanisole, which contain two or more EDG and simultaneously lack EWG in their structures, demonstrated excellent removal effectiveness, typically more than 80%. The reduction effectiveness of amino acids that possessed both EDG and EWG, however, ranged from 37% to 68%. Furthermore, despite having two aromatic rings in its structure, the removal effectiveness of 4,4-diaminodiphenylmethan was 99% because it had both amino groups, which are thought to be effective EDGs. Biological breakdown is inhibited by compounds with functional groups like azo aromatic amine, nitro, sulpho, iodide, and halogen, while biotransformation may be accelerated by chemicals with functional groups like esters, aromatic alcohol, and nitriles. After the first electrophilic assault of oxygenases by aerobic bacteria, most organic compounds are anticipated to break down most quickly. The linked EWG functional groups consequently result in an electron shortage and reduce the compound's capacity to oxidize. However, EDG functional groups enhance aerobic bacteria's electrophilic attack, hastening the biodegradation of organic compounds.

The backbone structures of 4-aminoazobenzene and o-aminoazotoluene are identical. Due to their hydrophobic nature and shared backbone structure, the perceivable similarity in removal effectiveness (65% and 67%) may be explained. Aromatic amines with many aromatic rings and electron-withdrawing groups in their structure have a decreased capacity to oxidize under aerobic conditions; nonetheless, the elimination of aromatic amines with only one aromatic ring and electron-drawing groups was successful. EWG functional groups in AAs like azo (–N–N–), nitro ($-NO_2$), and chloro (–Cl) raise the standard reduction potential and lower the ability of anilines to oxidize. However, the ($-NH_2$) amino EDG functional group improves the biodegradation of organic compounds by reducing the usual reduction potential. Although it is difficult to link the reduction efficiency of aromatic amines to the molecular structure, simple AAs and compounds with EDG, like ($-NH_2$) and ($-CH_3$), have typically been shown to have high removal efficiency, but complex AAs and compounds with EWG, like (–Cl) and azo linkages, have generally been found to have moderate removal efficiency.

5.5.3.3 pH or Acid Disassociation Constant

The pH of the effluent suggested that most molecules were in their deprotonated form. Consequently, these deprotonated species were more rapidly attracted to the negatively charged surface of microorganisms or sludge, hindering adsorption and reducing removal efficiency. In contrast to AAs with high pKa values like 2,4-diaminoanisolem (pKa=5.15), 2,4-dimethylaniline (pKa=4.89), and o-toluidine (pKa=4.49), which showed great removal efficacy (95–99%), 4-aminoazobenzene, o-aminoazotoluene, 3,3-dichlorobenzidine, and 4,4-methylene, o-toluidine, 4,4′-Oxydianiline, 2,4-diaminoanisole, 4,4-diaminodiphenylmethan, 2,4-dimethylaniline, 2,6-dimethylaniline, and 2,4,5-trimethylaniline all displayed high removal efficiencies when the average removal efficiencies of the investigated AAs were plotted against their physicochemical Log K_{ow}. The high solubility and frequent biodegradation of hydrophilic compounds can be used to explain this.

In general, two findings about the physical–chemical properties were made. First, because the hydrophobic AAs did not appreciably adsorb into sludge as anticipated, a lower removal efficacy was noted. Such a phenomenon can be explained using the tested AAs' activity at pH levels higher than pKa, which produced a strong tendency towards protonation and a high level of negative charge that hindered sorption into sludge. Due to their greater solubility and pKa pH of the effluent, hydrophilic AAs experienced remarkable removal effectiveness, typically above 80%. The sole method to eliminate neutral and negatively charged OMP molecules with low log K_{ow} values is by biotransformation. On the other hand, organic matter that is often somewhat negatively charged may adsorb to OMP molecules with high K_{ow} and positive or neutral charges. Second, it was found that complicated AAs and compounds with EWG, such as (–Cl) and (azo bond), might exhibit moderate removal effectiveness. Simple AAs and compounds with EDG, such as ($-NH_2$) and ($-CH_3$), demonstrate great removal effectiveness.

The elimination of organic micropollutants is closely related to their physicochemical characteristics and molecular structure. In general, it was discovered that less complicated AAs and compounds with electron EDG are easily biodegradable, whereas those with electron EWG may be more challenging. However, it is challenging to link the AAs' removal effectiveness and molecular structure. In addition, the hydrophobic AAs had a strong tendency to protonate (pKa pH), which prevented them from adhering to sludge, but the hydrophilic AAs had great removal efficiency because of their high solubility.

5.5.4 Azo Dye

This is the most important textile effluent that is released from the process of dying fabrics. There are many dyes that are used for the process. Azo dyes are divided into monoazo, diazo, and polyazo groups. Some examples of azo dyes are Acidic orange 7, Acidic yellow 36 acid dye, acid red 114, acid blue 225, direct red 7, direct blue 14, direct yellow 15, direct blue 15, azo dye salicylate, and many more. These azo dyes are chemical compounds that have an azo group (R–N=N–R′). These dyes do not show reactivity in aerobic conditions. For the degradation of these dyes, reduction of the azo bond should take place. This can be accomplished with powerful reducing agents such as sodium formaldehyde sulphoxylate, sodium hydrosulphite, thiourea dioxide, and sodium borohydride. As a result of this degradation, amines, which are colourless chemicals, are produced. The produced amines are furthermore resistant to degradation in anaerobic conditions; these produced amines undergo mineralization in the aerobic process, which produces a complete treatment of the dyes (Delee et al., 1998; Nagendranatha Reddy et al., 2018; Naresh Kumar et al., 2015).

5.5.4.1 Production of Aromatic Amines

Recent studies have clearly shown that the concentrations of the AAs increased in an anoxic tank. Surprisingly, the anoxic tank's 2,6-dimethylaniline concentration rose from approximately 40 ng/L to 180 ng/L. The azo bond's ring can break under reduced circumstances, releasing AAs in the process. The study discovered a substantial variation in the concentration of aromatic amines between those who had internal recycling and those who did not, which suggests that other factors resulting from the diversity of the microbial life in the anoxic tank affect, in addition to the redox situation, an anoxic bioreactor's reduction of the azo dye. Four key benefits of internal recycling an aerobic bioreactor into an anoxic one are increased mixing and mass, medium dilution, the transfer of some DO, and the supply of nitrate. These enable a special atmosphere in the anoxic tank that encourages aerobic decomposition. The growth of bacterial cultures may also be responsible for the anoxic tank's accelerated azo dye degradation. During the lengthy operation of the MBR, a highly functional bacterial population for pollution removal may evolve as a result of the sludge exchange between the aerobic and anoxic tanks.

In addition to all of these considerations, this study found significant evidence that nitrate is present in reducing conditions; the anoxic tank's increase in aromatic amine concentration was larger than what was observed in the anaerobic tank. The increased colour removal efficiency in anoxic environments compared to anaerobic settings shows that an increased amount of azo dye reductions took place in the anoxic tank and that more AAs were formed as a result, further supporting these findings. Significantly, the concentrations of 2,6-dimethylaniline, 2,4-diaminoanisole, 4,4′-diaminodiphenyl, and o-toluidine increased by 469%, 407%, 175%, and 108%, respectively, while they grew by 20%, 50%, 141%, and 40%, respectively, in the anaerobic tank. In the anoxic tank (IR=2.0), the AAs rose generally by 91% (from 205.1 to 393.7 ng/L). Due to the different bacterial varieties and the higher IR operation strategy, this is possible. On the other hand, in the anaerobic tank (IR=0.0), AAs rose by 6.1% (from 489.26 to 519.1 ng/L). Furthermore, the number of AAs present in the aerobic tank remained steady in both instances, hovering close to 50 ng/L. Anaerobic/aerobic and anaerobic/anoxic/aerobic sequences in a batch reactor are being studied to see what happens to and how the azo dye Reactive Black 5 changes. Like the way we looked at it, they discovered that azo dye in wastewater can be entirely degraded using the combination of anoxic/aerobic procedure. The azoreductase enzyme, produced by microbiomes, started the azo dye's biodegradation routes in anoxic/aerobic systems. Under reducing conditions, this enzyme uses various reduced metabolites or redox mediators to catalyse the breaking of the azo bond. Some examples of aromatic amines produced are 2,4,5-trimethylaniline, 4-chloro-2-methylaniline, 2,6-dimethylaniline, o-anisidine, 2,4-dimethylaniline, 2-methoxy-5-methylaniline, 2-naphthylamine, 2,4-diaminoanisole, 2,4-diaminotoluene, 4-aminoazobenzene, 4-aminobiphenyl, O-aminoazotoluene, benzidine, 4,4′-diaminodiphenylmethane, and 4,4′-diaminodiphenyl O-tolidine, 4,4-methylene-di (o-toluidine), 3,3-dichlorobenzidine, o-dianisidine, 4,4′-thiodianiline, and 4,4-methylene-bis(2-chloroaniline).

5.6 CONCLUSIONS AND FUTURE PERSPECTIVES

Biological processes are effective in the treatment of dye effluent. The removal of dyes and other pollutants from wastewater has been shown to have tremendous potential when using microbes and plants, either separately or in combination. Compared to physical and chemical procedures, these processes are more energy-efficient, cost-effective, and environmentally benign. Various biological processes, such as biodegradation, phytoremediation, and bioremediation, have been successfully implemented for treating dye effluents. The dye molecules can be efficiently broken down into less harmful chemicals by microorganisms like bacteria, fungus, and algae. Plants, on the other hand, can absorb the dyes and pollutants from the wastewater through their roots and either store them in their tissues or break them down into harmless substances. However, the success of these biological processes depends on various factors, such as the type and concentration of the dye, the availability of nutrients and oxygen, and the pH and temperature of the wastewater. Therefore, further research is necessary to optimize these processes and develop more efficient and effective strategies for the treatment of dye effluent.

The future for dye effluent treatment using biological processes is promising, as advancements in dye effluent treatment processes are growing exponentially. Such emerging prospects have created the need to develop more efficient and robust biological treatment systems that can handle a wide range of dye pollutants and can be applied to different industrial settings. Some potential areas of focus for future endeavours are bioremediation of emerging contaminants, bioaugmentation, integrating biological processes with advanced treatment technologies, optimization, and automation. Overall, the future of dye effluent treatment using biological processes looks promising as more advanced and efficient treatment systems are developed to address the growing environmental concerns associated with the textile industry.

ACKNOWLEDGEMENTS

The authors thank the Principal and Management of CBIT for their constant support and encouragement in carrying out this work.

REFERENCES

Adamson, R.H., Dixon, R.L., Francis, F.L. and Rall, D.P., 1965. Comparative biochemistry of drug metabolism by azo and nitro reductase. Proceedings of the National Academy of Sciences, 54(5), pp. 1386–1391.

Adane, T., Adugna, A.T. and Alemayehu, E. (2021). Textile industry effluent treatment techniques. Journal of Chemistry, 2021, pp. 1–14. https://doi.org/10.1155/2021/5314404.

Adedayo, O., Javadpour, S., Taylor, C., Anderson, W.A. and Moo-Young, M., 2004. Decolourization and detoxification of methyl red by aerobic bacteria from a wastewater treatment plant. World Journal of Microbiology and Biotechnology, 20, pp. 545–550.

Alabdraba, W.M.S. and Albayati, M.B.B.A. (2014). Biodegradation of azo dyes—A review. International Journal of Environmental Engineering and Natural Resources, Volume 1(Number 4).

Ali, H. (2010). Biodegradation of synthetic dyes—a review. Water, Air, & Soil Pollution, 213, pp. 251–273.

Alsalami, R.S., Alta'ee, A.H., Hadwan, M.H. and Sami, F., 2017. Removal of p-chlorophenols from aqueous solutions using tyrosinase extracted from mushroom Lentinula edodes. Journal of Chemical and Pharmaceutical Sciences. 10(2), pp. 863–866.

Aruna, V., Chada, N., Tejaswini Reddy, M., Geethikalal, V., Dornala, K. and Nagendranatha Reddy, C. (2022). Plant synthesized nanoparticles for dye degradation. In M.P. Shah, A. Roy (Eds.), Phytonanotechnology. Springer. https://doi.org/10.1007/978-981-19-4811-4_1

Asad, S., Amoozegar, M.A., Pourbabaee, A., Sara Bolouki, M.N. and Dastgheib, S.M.M. (2007). Decolorization of textile azo dyes by newly isolated halophilic and halotolerant bacteria. Bioresource Technology, 98(11), pp. 2082–2088.

Ayed, L., Ksibi, I.E., Charef, A. and Mzoughi, R.E. (2021). Hybrid coagulation-flocculation and anaerobic-aerobic biological treatment for industrial textile wastewater: Pilot case study. The Journal of the Textile Institute, 112(2), pp. 200–206.

Azanaw, A., Birlie, B., Teshome, B. and Jemberie, M. (2022). Textile effluent treatment methods and eco-friendly resolution of textile wastewater. Case Studies in Chemical and Environmental Engineering, 6, p. 100230.

Bafana, A., Krishnamurthi, K., Devi, S.S. and Chakrabarti, T. (2008). Biological decolourization of CI direct black 38 by E. Gallinarum. Journal of Hazardous Materials, 157(1), pp. 187–193.

Blánquez, P., Casas, N., Font, X., Gabarrell, X., Sarrà, M., Caminal, G. and Vicent, T., 2004. Mechanism of textile metal dye biotransformation by Trametes versicolor. Water research, 38(8), pp. 2166–2172.

Bangaru, A., Sree, K.A., Kruthiventi, C., Banala, M., Shreya, V., Vineetha, Y., Shalini, A., Mishra, B., Yadavalli, R., Chandrasekhar, K. and Reddy, C.N. (2022). Role of enzymes in biofuel production: Recent developments and challenges. In P. Chowdhary, N. Khanna, S. Pandit, R. Kumar (Eds.), Bio-Clean Energy Technologies: Volume 1. Clean Energy Production Technologies. Springer. https://doi.org/10.1007/978-981-16-8090-8_4

Bardi, L. and Marzona, M. (2010). Factors affecting the complete mineralization of azo dyes. The Handbook of Environmental Chemistry, pp. 195–210. https://doi.org/10.1007/698_2009_50.

Barragán, B.E., Costa, C. and Marquez, M.C. (2007). Biodegradation of azo dyes by bacteria inoculated on solid media. Dyes and Pigments, 75(1), pp. 73–81.

Bhatia, D., Sharma, N.R., Singh, J. and Kanwar, R.S. (2017). Biological methods for textile dye removal from wastewater: A review. Critical Reviews in Environmental Science and Technology, 47(19), pp. 1836–1876.

Bhuktar, J.J., Kalia, U. and Manwar, A.V. (2010). Decolorization of textile sulfonated azo dyes by bacteria isolated from textile industry effluent, soil and sewage. Journal of Pure and Applied Microbiology, 4(2), pp. 617–622.

Birmole, R., Parkar, A. and Aruna, K. (2019). Biodegradation of reactive red 195 by a novel strain *Enterococcus casseliflavus* RDB_4 isolated from textile effluent. Nature Environment & Pollution Technology, 18(1), pp. 97–109.

Bisschops, I. and Spanjers, H. (2003). Literature review on textile wastewater characterisation. Environmental Technology, 24(11), pp. 1399–1411.

Blümel, S., Busse, H.J., Stolz, A. and Kämpfer, P. (2001). Xenophilus azovorans gen. nov., sp. nov., a soil bacterium that is able to degrade azo dyes of the Orange II type. International Journal of Systematic and Evolutionary Microbiology, 51(5), pp. 1831–1837.

Bragger, J.L., Lloyd, A.W., Soozandehfar, S.H., Bloomfield, S.F., Marriott, C. and Martin, G.P., 1997. Investigations into the azo reducing activity of a common colonic microorganism. International journal of pharmaceutics, 157(1), pp. 61–71.

Brás, R., Ferra, M.I.A., Pinheiro, H.M. and Gonçalves, I.C., 2001. Batch tests for assessing decolourisation of azo dyes by methanogenic and mixed cultures. Journal of Biotechnology, 89(2–3), pp. 155–162.

Brigé, A., Motte, B., Borloo, J., Buysschaert, G., Devreese, B. and Van Beeumen, J.J., 2008. Bacterial decolorization of textile dyes is an extracellular process requiring a multi-component electron transfer pathway. Microbial Biotechnology, 1(1), pp. 40–52.

Cai, F., Lei, L., Li, Y. and Chen, Y. (2021). A review of aerobic granular sludge (AGS) treating recalcitrant wastewater: Refractory organics removal mechanism, application and prospect. Science of the Total Environment, 782, p. 146852.

Carmen, Z. and Daniela, S. (2012). Textile Organic Dyes-Characteristics, Polluting Effects and separation/elimination Procedures from Industrial Effluents-a Critical Overview (Vol. 3, pp. 55–86). IntechOpen.

Chen, B.Y., Chen, S.Y., Lin, M.Y. and Chang, J.S. (2006). Exploring bioaugmentation strategies for azo-dye decolorization using a mixed consortium of *Pseudomonas luteola* and *Escherichia coli*. Process Biochemistry, 41(7), pp. 1574–1581.

Chinwetkitvanich, S., Tuntoolvest, M., Panswad, T., 2000. Anaerobic decolorization of reactive dyebath effluents by a two-stage UASB system with tapioca as a co-substrate. Water Research. 34, 8, 2223–2232. https://doi.org/10.1016/S0043-1354(99)00403-0.

Cui, T., 2009. Enhanced large-scale production of laccases from Coriolopsis polyzona for use in dye bioremediation. PhD thesis University of Westminster School of Biosciences https://doi.org/10.34737/90w97.

Delee, W., O'Neill, C., Hawkes, F.R. and Pinheiro, H.M. (1998). Anaerobic treatment of textile effluents: A review. Journal of Chemical Technology & Biotechnology: International Research in Process, Environmental AND Clean Technology, 73(4), pp. 323–335.

Deng, D., Lamssali, M., Aryal, N., Ofori-Boadu, A., Jha, M.K. and Samuel, R.E. (2020). Textiles wastewater treatment technology: A review. Water Environment Research, 92(10), pp. 1805–1810.

Doble, M. and Kumar, A. (2005). Biotreatment of Industrial Effluents. Elsevier.

Dubin, P. and Wright, K.L. (1975). Reduction of azo food dyes in cultures of *Proteus vulgaris*. Xenobiotica, 5(9), pp. 563–571. https://doi.org/10.3109/00498257509056126.

Elizalde-González, M.P., Fuentes-Ramirez, L.E. and Guevara-Villa, M.R.G. (2009). Degradation of immobilized azo dyes by *Klebsiella* sp. UAP-b5 isolated from maize adsorbent. Journal of Hazardous Materials, 161(2–3), pp. 769–774.

Gera, N. (2012). Significance and future prospects of textile exports in Indian economy. IARS' International Research Journal, 2(1).

Gosavi, V.D. and Sharma, S. (2014). A general review on various treatment methods for textile wastewater. Journal of Environmental Science, Computer Science and Engineering & Technology, 3(1), pp. 29–39.

Holkar, C.R., Jadhav, A.J., Pinjari, D.V., Mahamuni, N.M. and Pandit, A.B. (2016). A critical review on textile wastewater treatments: Possible approaches. Journal of Environmental Management, 182, pp. 351–366.

Hsueh, C.C., Chen, B.Y. and Yen, C.Y. (2009). Understanding effects of chemical structure on azo dye decolorization characteristics by *Aeromonas hydrophila*. Journal of Hazardous Materials, 167(1–3), pp. 995–1001.

Ikram, M., Naeem, M., Zahoor, M., Hanafiah, M.M., Oyekanmi, A.A., Ullah, R., Farraj, D.A.A., Elshikh, M.S., Zekker, I. and Gulfam, N., 2022. Biological degradation of the azo dye basic orange 2 by Escherichia coli: A sustainable and ecofriendly approach for the treatment of textile wastewater. Water, 14(13), p. 2063.

Joe, M.H., Lim, S.Y., Kim, D.H. and Lee, I.S., 2008. Decolorization of reactive dyes by Clostridium bifermentans SL186 isolated from contaminated soil. World Journal of Microbiology and Biotechnology, 24, pp. 2221–2226.

Joe, J., Kothari, R.K., Raval, C.M., Kothari, C.R., Akbari, V.G. and Singh, S.P. (2011). Decolorization of textile dye Remazol Black B by *Pseudomonas aeruginosa* CR-25 isolated from the common effluent treatment plant. Journal of Bioremediation & Biodegradation, 2(2), p. 2.

Kamaruddin, M.A., Yusoff, M.S., Aziz, H.A. and Nazmi Ismail, M., 2013. Current progress of textile waste water treatment by biological processes. Caspian Journal of Applied Sciences Research, 2(10). p. 7.

Katheresan, V., Kansedo, J. and Lau, S.Y. (2018). Efficiency of various recent wastewater dye removal methods: A review. Journal of Environmental Chemical Engineering, 6(4), pp. 4676–4697.

Katuri, K.P., Ali, M. and Saikaly, P.E. (2019). The role of microbial electrolysis cell in urban wastewater treatment: Integration options, challenges, and prospects. Current Opinion in Biotechnology, 57, pp. 101–110.

Khalid, A., Arshad, M. and Crowley, D.E. (2008). Decolorization of azo dyes by *Shewanella* sp. Under saline conditions. Applied Microbiology and Biotechnology, 79(6), pp. 1053–1059. https://doi.org/10.1007/s00253-008-1498-y.

Khalid, A., Arshad, M. and Crowley, D. (2010). Bioaugmentation of azo dyes. The Handbook of Environmental Chemistry, pp. 1–37. https://doi.org/10.1007/698_2009_42.

Khan, S. and Malik, A. (2014). Environmental and health effects of textile industry wastewater. In Environmental Deterioration and Human Health (pp. 55–71). Springer.

Khehra, M.S., Saini, H.S., Sharma, D.K., Chadha, B.S. and Chimni, S.S., 2005. Decolorization of various azo dyes by bacterial consortium. Dyes and pigments, 67(1), pp. 55–61.

Khelifi, E., Gannoun, H., Touhami, Y., Bouallagui, H. and Hamdi, M. (2008). Aerobic decolourization of the indigo dye-containing textile wastewater using continuous combined bioreactors. Journal of Hazardous Materials, 152(2), pp. 683–689.

Kim, T.H., Park, C., Yang, J. and Kim, S. (2004). Comparison of disperse and reactive dye removals by chemical coagulation and fenton oxidation. Journal of Hazardous Materials, 112(1–2), pp. 95–103.

Kolekar, Y.M., Pawar, S.P., Gawai, K.R., Lokhande, P.D., Shouche, Y.S. and Kodam, K.M. (2008). Decolorization and degradation of disperse blue 79 and acid Orange 10, by *Bacillus fusiformis* KMK5 isolated from the textile dye contaminated soil. Bioresource Technology, 99(18), pp. 8999–9003.

Krishnamoorthy, R. et al. (2021). Long-term exposure to azo dyes from textile wastewater causes the abundance of *Saccharibacteria* population. Applied Sciences, 11(1), p. 379. https://doi.org/10.3390/app11010379.

Lettinga, G.A.F.M., Van Velsen, A.F.M., Hobma, S.D., De Zeeuw, W. and Klapwijk, A., 1980. Use of the upflow sludge blanket (USB) reactor concept for biological wastewater treatment, especially for anaerobic treatment. Biotechnology and bioengineering, 22(4), pp. 699–734.

Liu, X., Chen, Y., Zhang, X., Jiang, X., Wu, S., Shen, J., Sun, X., Li, J., Lu, L. and Wang, L. (2015). Aerobic granulation strategy for bioaugmentation of a sequencing batch reactor (SBR) treating high strength pyridine wastewater. Journal of Hazardous Materials, 295, pp. 153–160.

Maier, J., Kandelbauer, A., Erlacher, A., Cavaco-Paulo, A. and Gübitz, G.M. (2004). A new alkali-thermostable azoreductase from *Bacillus* sp. Strain SF. Applied and Environmental Microbiology, 70(2), pp. 837–844.

McCarty, P.L. (2001). The development of anaerobic treatment and its future. Water Science and Technology, 44(8), pp. 149–156. https://doi.org/10.2166/wst.2001.0487.

Meehan, C., Bjourson, A.J. and McMullan, G. (2001). *Paenibacillus azoreducens* sp. nov., a synthetic azo dye decolorizing bacterium from industrial wastewater. International Journal of Systematic and Evolutionary Microbiology, 51(5), pp. 1681–1685.

Mohd Hanafiah, Z., Wan Mohtar, W.H.M., Abu Hasan, H., Jensen, H.S., Klaus, A. and Wan-Mohtar, W.A.A.Q.I., 2019. Performance of wild-Serbian Ganoderma lucidum mycelium in treating synthetic sewage loading using batch bioreactor. Scientific reports, 9(1), p. 16109.

Mohd Rosli, N.H. and Ahmad, W.A. (2018). Single cultures of *Acinetobacter* sp. and *Cellulosimicrobium* sp. Grown in pineapple waste: Adaption study And potential in reducing cod from real textile wastewater. Science Letters (ScL), 12(1), pp. 30–43.

Mohsenpour, S.F., Hennige, S., Willoughby, N., Adeloye, A. and Gutierrez, T. (2021). Integrating micro-algae into wastewater treatment: A review. Science of the Total Environment, 752, p. 142168.

Mullai, P., Sathian, S. and Sabarathinam, P.L. (2007). Kinetic modelling of distillery wastewater biodegradation using *Paecilomyces variotii*. Chemical Engineering World, 42(12).

Mullai, P., Yogeswari, M.K., Vishali, S., Namboodiri, M.T., Gebrewold, B.D., Rene, E.R. and Pakshirajan, K. (2017). Aerobic treatment of effluents from textile industry. In Current Developments in Biotechnology and Bioengineering (pp. 3–34). Elsevier.

Nachiyar, C.V. and Rajkumar, G.S. (2003). Degradation of a tannery and textile dye, navitan fast blue S5R by *Pseudomonas aeruginosa*. World Journal of Microbiology and Biotechnology, 19(6), pp. 609–614.

Nagendranatha Reddy, C., Kondaveeti, S., Krishna, G.M. and Min, B. (2022). Application of bioelectrochemical systems to regulate and accelerate the anaerobic digestion processes. Chemosphere, 287, p. 132299.

Nagendranatha Reddy, C., Kumar, A.N. and Venkata Mohan, S. (2018). Metabolic phasing of anoxic-PDBR for high rate treatment of azo dye wastewater. Journal of Hazardous Materials, 343, pp. 49–58.

Nagendranatha Reddy, C., Modestra, J.A., Kumar, A.N. and Venkata Mohan, S. (2016). Waste remediation integrating with value addition: Biorefinery approach towards sustainable bio-based technologies. In V.C. Kalia (Ed.), Microbial Factories, Biofuels, Waste Treatment (Volume 1, pp. 231–256). Springer. https://doi.org/10.1007/978-81-322-2598-0_14

Nagendranatha Reddy, C., Nguyen, H.T.H., Noori, M.T. and Min, B. (2019). Potential applications of algae in the cathode of microbial fuel cells for enhanced electricity generation with simultaneous nutrient removal and algae biorefinery: Current status and future perspectives. Bioresource Technology, 292, p. 122010.

Naresh Kumar, K., Nagendranatha Reddy, C., Prasad, R.H. and Venkata Mohan, S. (2014). Azo dye load-shock on relative behavior of biofilm and suspended growth configured periodic discontinuous batch mode operations: Critical evaluation with enzymatic and bio-electrocatalytic analysis. Water Research, 60, pp. 182–196.

Naresh Kumar, K., Nagendranatha Reddy, C. and Venkata Mohan, S. (2015). Biomineralization of azo dye bearing wastewater in periodic discontinuous batch reactor: Effect of microaerophilic conditions on treatment efficiency. Bioresource Technology, 188, pp. 56–64.

O'Neill, C., Lopez, A., Esteves, S., Hawkes, F.R., Hawkes, D.L. and Wilcox, S., 2000. Azo-dye degradation in an anaerobic-aerobic treatment system operating on simulated textile effluent. Applied microbiology and biotechnology, 53, pp. 249–254.

O'Neill, C., Hawkes, F.R., Hawkes, D.L., Lourenço, N.D., Pinheiro, H.M. and Delée, W. (1999). Colour in textile effluents–sources, measurement, discharge consents and simulation: A review. Journal of Chemical Technology & Biotechnology: International Research in Process, Environmental & Clean Technology, 74(11), pp. 1009–1018.

Odjegba, V.J. and Bamgbose, N.M. (2012). Toxicity assessment of treated effluents from a textile industry in Lagos, Nigeria. African Journal of Environmental Science and Technology, 6(11), pp. 438–445.

Ozdemir, G., Pazarbasi, B., Kocyigit, A., Omeroglu, E.E., Yasa, I. and Karaboz, I., 2008. Decolorization of Acid Black 210 by Vibrio harveyi TEMS1, a newly isolated bioluminescent bacterium from Izmir Bay, Turkey. World Journal of Microbiology and Biotechnology, 24, pp. 1375–1381.

Pandey, A., Singh, P. and Iyengar, L. (2007). Bacterial decolorization and degradation of azo dyes. International Biodeterioration & Biodegradation, 59(2), pp. 73–84.

Pearce, C. (2003). The removal of colour from textile wastewater using whole bacterial cells: A review. Dyes and Pigments, 58(3), pp. 179–196. https://doi.org/10.1016/s0143-7208(03)00064-0.

Pokharia, A. and Ahluwalia, S.S. (2015). Toxicological effect of textile dyes and their metabolites: A review. Current Trends in Biotechnology and Chemical Researchl January–June, 5(1).

Popli, S. and Patel, U. (2015). Destruction of azo dyes by anaerobic–aerobic sequential biological treatment: A review. International Journal of Environmental Science and Technology, 12(1), pp. 405–420.

Rajasri, Y., Hariprasad, R., Reddy, J., Nagendranatha Reddy, C. and Chandrasekhar, K. (2021). Simultaneous production of astaxanthin and lipids from *Chlorella sorokiniana* in the presence of reactive oxygen species: A biorefinery approach. Biomass Conversion and Biorefinery. https://doi.org/10.1007/s13399-021-01276-5

Rajasri, Y., Hariprasad, R., Snehasri, M., Nagendranatha Reddy, C., Ashok Kumar, V. and Chandrasekhar, K. (2020). Simultaneous production of flavonoids and lipids from the *Chlorella vulgaris* and *Chlorella pyrenoidosa*. Biomass Conversion and Biorefinery. https://doi.org/10.1007/s13399-020-01044-x

Rau, J., Knackmuss, H.J. and Stolz, A. (2002). Effects of different quinoid redox mediators on the anaerobic reduction of azo dyes by bacteria. Environmental Science & Technology, 36(7), pp. 1497–1504.

Robinson, T., McMullan, G., Marchant, R. and Nigam, P., 2001. Remediation of dyes in textile effluent: a critical review on current treatment technologies with a proposed alternative. Bioresource technology, 77(3), pp. 247–255.

Sandhya, S. (2010). Biodegradation of azo dyes under anaerobic condition: Role of azoreductase. The Handbook of Environmental Chemistry, pp. 39–57. https://doi.org/10.1007/698_2009_43.

Sandhya, S., Sarayu, K. and Swaminathan, K. (2008). Determination of kinetic constants of hybrid textile wastewater treatment system. Bioresource Technology, 99(13), pp. 5793–5797.

Sankaran, S., Khanal, S.K., Jasti, N., Jin, B., Pometto, I.I.I., A.L. and Van Leeuwen, J.H. (2010). Use of filamentous fungi for wastewater treatment and production of high value fungal byproducts: A review. Critical Reviews in Environmental Science and Technology, 40(5), pp. 400–449.

Saratale, R.G., Saratale, G.D., Chang, J.S. and Govindwar, S.P. (2010). Decolorization and biodegradation of reactive dyes and dye wastewater by a developed bacterial consortium. Biodegradation, 21, pp. 999–1015.

Sarayu, K. and Sandhya, S. (2012). Current technologies for biological treatment of textile wastewater–a review. Applied Biochemistry and Biotechnology, 167(3), pp. 645–661.

Seker, D., Kalyoncu, F., Kalmis, E. and Akata, I., 2014. Evaluation of Pleurotus ostreatus, P. sajor-caju and Ganoderma lucidum Isolated from Nature for their Ability to Decolorize Synazol Blue and Synazol Red Textile Dyes. J Pure Appl Microbiol. 2014;8(4):2687–2691.

Shah, M.P. (2020). Microbial Bioremediation & Biodegradation. Springer.

Shah, M.P. (2021a). Removal of Emerging Contaminants Through Microbial Processes. Springer.

Shah, M.P. (2021b). Removal of Refractory Pollutants From Wastewater Treatment Plants. CRC Press.

Shishoo, R. (Ed.). (2012). The Global Textile and Clothing Industry: Technological Advances and Future Challenges. Woodhead Publishing Ltd.

Sivaram, N.M., Gopal, P.M. and Barik, D. (2019). Toxic waste from textile industries. In Energy From Toxic Organic Waste for Heat and Power Generation (pp. 43–54). Woodhead Publishing.

Sreelatha, S., Nagendranatha Reddy, C., Velvizhi, G. and Venkata Mohan, S. (2015). Reductive behaviour of acid azo dye based wastewater: Biocatalyst activity in conjunction with enzymatic and bio-electro catalytic evaluation. Bioresource Technology, 188, pp. 2–8.

Srivastava, A., Rani, R.M., Patle, D.S. and Kumar, S. (2022). Emerging bioremediation technologies for the treatment of textile wastewater containing synthetic dyes: A comprehensive review. Journal of Chemical Technology & Biotechnology, 97(1), pp. 26–41.

Stern, S., Szpyrkowicz, L., Rodighiero, I., 2003. Anaerobic treatment of textile dyeing wastewater.. Water science and technology : a journal of the International Association on Water Pollution Research, 47, 10, pp. 55–9 . https://doi.org/10.2166/WST.2003.0537.

Stolz, A. (2001). Basic and applied aspects in the microbial degradation of azo dyes. Applied Microbiology and Biotechnology, 56(1–2), pp. 69–80. https://doi.org/10.1007/s002530100686.

Suzuki, Y., Yoda, T., Ruhul, A. and Sugiura, W. (2001). Molecular cloning and characterization of the gene coding for azoreductase from Bacillus sp. OY1-2 isolated from soil. Journal of Biological Chemistry, 276(12), pp. 9059–9065.

Tan, L., Qu, Y.Y., Zhou, J.T., Li, A. and Gou, M. (2009). Identification and characteristics of a novel salt-tolerant Exiguobacterium sp. for azo dyes decolorization. Applied Biochemistry and Biotechnology, 159, pp. 728–738.

Van der Zee, F.P. and Villaverde, S. (2005). Combined anaerobic–aerobic treatment of azo dyes—a short review of bioreactor studies. Water Research, 39(8), pp. 1425–1440.

van Lier, J.B., Zeeman, G. and Mahmoud, N. (2020). Anaerobic wastewater treatment. Biological Wastewater Treatment: Principles, Modeling and Design, pp. 701–756. https://doi.org/10.2166/9781789060362_0701.

Venkata Mohan, S., Nagendranatha Reddy, C., Kumar, A.N. and Modestra, J.A. (2013). Relative performance of biofilm configuration over suspended growth operation on azo dye based wastewater treatment in periodic discontinuous batch mode operation. Bioresource Technology, 147, pp. 424–433.

Vymazal, J., Zhao, Y. and Mander, Ü. (2021). Recent research challenges in constructed wetlands for wastewater treatment: A review. Ecological Engineering, 169, p. 106318.

Walsh, G.E., Bahner, L.H. and Horning, W.B. (1980). Toxicity of textile mill effluents to freshwater and estuarine algae, crustaceans and fishes. Environmental Pollution Series A, Ecological and Biological, 21(3), pp. 169–179.

Wang, Y., Qiu, L. and Hu, M. (2018). Application of yeast in the wastewater treatment. In E3S Web of Conferences (Vol. 53, p. 04025). EDP Sciences.

Yoo, E.S., Libra, J. and Adrian, L. (2001). Mechanism of decolorization of azo dyes in anaerobic mixed culture. Journal of Environmental Engineering, 127(9), pp. 844–849. https://doi.org/10.1061/(asce)0733-9372(2001)127:9(844).Table 5.1

6 Micropollutants Removal
Microbial Versus Physical-Chemical Treatment Process

Bianca Ramos Estevam, Daniele Moreira, Ana Flora Dalberto Vasconcelos, Rosane Freire Boina

6.1 INTRODUCTION

Micropollutants (MPs) are a large group of natural and synthetic chemicals that are difficult to degrade in nature, remain in ecosystems, and lead to the bioaccumulation phenomenon (Gutiérrez et al. 2021). These compounds are present in the environment at concentrations of micrograms or nanograms per liter. Even at low concentrations, MPs can cause damage to the environment and human health (Argun et al. 2020). They came from various sources, such as domestic wastewater; agricultural operations; foundry, textile, tanning, microelectronics, and metallurgical industries; landfills; and hospital wastewater (Bhatt et al. 2022). Among the MPs are metal ions, agrochemicals, dyes, pharmaceuticals, and polycyclic aromatic hydrocarbons (PAHs).

The removal of MP can be carried out through physical-chemical and biological processes. Biological or microbial treatment is divided into anaerobic and aerobic processes. In these systems, microorganisms (such as bacteria, fungi, and algae) are responsible for removing, degrading, or transforming organic matter and contaminants into other less toxic compounds through their metabolic routes and cellular characteristics (Saleh et al. 2020). Among the microbial wastewater treatments are biofilms, pond systems, activated sludge, and anaerobic sludge blankets. These processes were essentially developed for removing organic matter, but some techniques have presented good results in the management of MPs (Grandclément et al. 2017). The main advantage of biological processes is their low cost over physical-chemical treatments (Bhatt et al. 2022). The limitations of these systems are related to the long process time and the necessity of controlling the microorganism growth conditions, such as pH, temperature, and dissolved oxygen (Al-Maqdi et al. 2017).

Physical-chemical treatment aims to remove pollutants that are difficult to degrade in conventional microbial processes, such as inorganic pollutants, insoluble solids, metals, and recalcitrant organic compounds (Ahmed et al. 2021). These treatments are also applied to reduce the organic load before microbial treatment. The physical process aims to remove the pollutants using gravity, pressure, electricity, or surface interactions, while the chemical process consists of adding reagents that will degrade the pollutant or reduce its contaminant load (Saleh et al. 2020; Shah, 2020). The most commonly used physical-chemical treatments are coagulation-flocculation, electrochemical processes, adsorption,

DOI: 10.1201/9781003381327-6

membrane filtration, and advanced oxidation processes. Physical-chemical treatments are generally fast and have high MP removal efficiency (Paździor et al. 2019). However, they can be costly, with higher energy and reagent consumption, require greater process control, and generate large volumes of residual sludge (Dhangar and Kumar 2020).

Due to the complexity and persistence of MPs, a combination of treatment techniques has been analyzed to increase the removal efficiency of these compounds from wastewater (Prado et al. 2017). The hybrid system consists of a combination of biological and physical-chemical treatment techniques. This configuration minimizes the disadvantages and limitations of the processes involved once it combines the capabilities of each technique (Dhangar and Kumar 2020). Hybrid systems can also promote a reduction in cost, time, and space (Gutiérrez et al. 2021). To determine the more suitable system for treating wastewater, some parameters must be considered, such as operating cost, installation area, removal potential, type of pollutant, wastewater characteristics, and system limitations (Saidulu et al. 2021; Shah, 2021b).

Therefore, this chapter addresses the microbial and physical-chemical treatment systems most commonly used to remove MPs from wastewater, highlighting their main characteristics, process parameters, advantages, and limitations. The removal of a range of MPs using a single microbial or physical-chemical process or hybrid techniques is reviewed and compared. Thus, this chapter assists in the decision-making process for the most suitable treatment(s) to remove MPs.

6.2 OCCURRENCE AND IMPACTS OF MICROPOLLUTANTS IN WASTEWATER

MPs are toxic compounds made up of natural or anthropogenic substances. They are bioaccumulative and persistent once they are not completely degraded by nature. Thus, even at low concentrations, MPs can cause damage to the environment and human health (Bhatt et al. 2022). Bioaccumulation occurs due to the high liposolubility of the MPs, which increases their ability to accumulate in the tissues of the organism for years. These bioaccumulative pollutants move up the food chain, increasing their concentration as they are consumed and accumulated, increasing their toxicity in the environment (Grandclément et al. 2017). There are two classes of MPs: inorganic (metals and toxic anions) and organic (by-products of disinfection and oxidation, endocrine active/destructive compounds, dyes, PAHs, and pharmaceutical compounds) (Bodzek and Konieczny 2018). This classification takes into account the origin, use, and possible health effects on the ecosystem and organisms.

In general, these substances are present in various products consumed daily, including personal care products, batteries, pesticides, herbicides, fuels, and pharmaceuticals. Thus, MPs can come from various sources, such as hospitals, industries, domestic wastewater, and agricultural runoff. The city's Wastewater Treatment Plants (WWTPs) are the first (and sometimes only) barrier to the discharge of MPs into water. However, the traditional methods applied in the WWTPs are not adaptable to completely remove or degrade MPs, being more suitable for conventional macro-pollutants such as suspended solids, organic substances, and some microorganisms (Dhangar and Kumar 2020; Shah, 2021a).

Table 6.1 summarizes the occurrence and toxic effects of the most common categories of MPs. Besides, it presents the treatment process that is able to remove them

TABLE 6.1
Source, Toxic Effect, and Process Able to Remove the Main MPs Found in Ecosystems (Ahmed et al. 2021; Bhatt et al. 2022; Saidulu et al. 2021)

Micropollutant	Main Source	Toxic Effects		Treatment Process Able to Remove Them
		Human Health	Environmental	
Metals	Metallurgy and textile industries	• Affects the kidneys, liver, lungs, and other vital organs • Long-term exposure can lead to cancer or degenerative muscle and neurological disorders	• Pollution of soil, water, and air through the mechanism of bioaccumulation and biomagnification	• Activated sludge • Coagulation-flocculation • Electrochemical process • Adsorption • Membrane systems
Agrochemicals	Agriculture	• Neurological effects disorders • Intracellular oxidative stress and differentiation of brain cells • Autoimmune disorders • Exposure during pregnancy can cause mental disorders in children	• Ecosystem imbalance • Contamination of water and soil • Epithelial cell hyperplasia, aneurysms and capillary changes in fish • Decreased growth and replication of microorganisms	• Activated sludge • Coagulation-flocculation • Electrochemical process • Adsorption • Membrane systems • Advance oxidation process
Pharmaceuticals	Medicines	• Diplopia • Visual disturbances • Drowsiness • Paresthesia • Balance disorders • Leukopenia • Neutropenia • Skin eruptions	• Decreased hatching of bird, fish, and turtle eggs • Fish feminization • Problems in the reproductive system of fish, reptiles, birds, and mammals • Changes in the immune system of marine mammals	• Biofilms • Activated sludge • Anaerobic sludge blanket • Adsorption • Membrane systems • Advance oxidation process
Polycyclic aromatic hydrocarbons (PAH)	Production of dyes, drugs pesticides, plastics, and explosives. Result of incomplete combustion or pyrolysis of organic material	• Lung and skin cancer risks • Formation of specific DNA adducts and characteristic mutations in oncogenes and tumor suppressor genes • Eye and skin irritation	• Contamination and death of aquatic and terrestrial organisms • Tumors in marine organisms such as mollusk, bryozoans, and algae	• Biofilms • Pond systems • Activated sludge • Adsorption • Membrane systems • Advance oxidation process

(Continued)

TABLE 6.1 *(Continued)*
Source, Toxic Effect, and Process Able to Remove the Main MPs Found in Ecosystems (Ahmed et al. 2021; Bhatt et al. 2022; Saidulu et al. 2021)

		Toxic Effects		
Micropollutant	Main Source	Human Health	Environmental	Treatment Process Able to Remove Them
Dyes	Textile industries	• Carcinogenic potential	• Restrict the passage of solar radiation, reducing the natural photosynthetic activity, bringing changes to the aquatic biota, and acute and chronic toxicity to these ecosystems	• Biofilms • Activated sludge • Adsorption • Membrane systems • Advance oxidation process

from wastewater. It is worth mentioning that removal efficiency can vary with the experimental conditions applied in each process, and the toxicity is related to the MP concentration. Their concentrations vary according to the source and rate of pollution beyond the discharged volume, which hampers their analytical procedures and treatment methods. In addition, some of these MPs still do not have regulations that define their limited concentration in drinking water, natural water, or released wastewater into the environment (Bhatt et al. 2022). Thus, they are commonly discharged into surface water or groundwater, becoming a relevant problem for environmental conservation strategies. In surface water, MPs can also come from rainfall, surface runoff from landfills, and flow into soil drainage zones (Pisharody et al. 2022).

6.3 MICROBIAL BIOMASS IN WASTEWATER TREATMENT

Microbial wastewater treatments use the ability of microorganisms to degrade or accumulate pollutants. Thus, it depends on the metabolic activities of microorganisms, their resistance to the pollutant, and the cultivation and operational conditions applied (Grandclément et al. 2017). Bacteria, fungi, and algae are among the microorganisms most commonly used in wastewater treatment (Ahmed et al. 2021). These species can be used in active or inactive form. Bacteria are widely applied for the treatment of wastewater and soil contaminated by MPs due to their capability to degrade organic and inorganic compounds and tolerate extreme environmental conditions such as salinity, pH, and temperature (Pushkar et al. 2021). There are heterotrophic and autotrophic bacteria. In wastewater treatment, heterotrophic bacteria stand out as they can metabolize a range of organic compounds as sources of carbon and energy. Besides, the cell surface of the bacteria presents a range of functional groups that can interact with the medium, including toxic pollutants (Roccuzzo et al. 2021).

Fungi are eukaryote heterotrophs with absorptive nutrition and glycogen energy reserves. They can be unicellular, colonial, or multicellular, with yeast-like or filamentous structures (Madadi and Bester 2021). Fungi are widely distributed in nature, being important in the geochemical cycles and balance of aquatic and terrestrial ecosystems. They can grow in extreme environments and tolerate a range of pollutant concentrations (Morsi et al. 2020). The filamentous fungi have a complex macroscopic structure with various enzymes capable of forming surface interactions to retain or degrade the MPs (Ferreira et al. 2020). These characteristics highlight their use over unicellular microorganisms such as bacteria, yeasts, and microalgae. In addition, their biomass is easier to recover from the medium, and they have lower substrate specificity (Roccuzzo et al. 2021). The complex structure of fungi allows different ways to remove contaminants, such as Van der Waals forces, ionic bonds, extracellular precipitation, complexation and crystallization, the transformation of substances by enzymatic oxidation, methylation, dealkylation, energy-dependent intracellular accumulation, active uptake, and passive adsorption removal by living cells (Ferreira et al. 2020; Morsi et al. 2020). These mechanisms can operate individually or in combination. However, the studies with fungi in MP removal are still limited to lab-scale systems.

The taxonomic group of algae encompasses both prokaryotic and eukaryotic organisms with a wide genetic and metabolic range. Their structure is divided into unicellular, colonial, filamentous, or multicellular. Algae are commonly segmented into microalgae (with a size between 2 and 200 μm) and macroalgae – larger than 200 μm (Griffiths et al. 2016). These organisms have high photosynthetic efficiency (assimilating up to 50 times more CO_2 than terrestrial plants) and high metabolite productivity, such as lipids, proteins, carbohydrates, and pigments (Estevam et al. 2022b). In addition, most of the algae species are resistant to ammoniacal compounds, making them able to be grown in wastewater. In this case, the algae use the carbon, nitrogen, phosphorus, and sulfur from the wastewater as a source of macronutrients and metabolize metals or recalcitrant compounds in residual concentration (Vassalle et al. 2020). These species also oxygenate the water, reducing its biochemical oxygen demand (BOD). Thus, algae can simultaneously treat wastewater, sequester carbon, and produce metabolites with high added value (Abdelfattah et al. 2023). However, microalgae present a lower growth rate than bacteria and fungi, hindering the cultivation of a single species, especially in open systems (Estevam et al. 2022a). Thus, studies suggest a consortium of microalgae with heterotrophic microorganisms such as bacteria and fungi. With this approach, the limitations of one species are supplied by the other. This process involves multiple microbial interactions based on mutualism or commensalism (Abdelfattah et al. 2023). Although the cultivation of some algae species is quite consolidated on a commercial scale, their use in WWTP is still quite limited, especially with a focus on MP removal.

The microorganisms can bioremediate the wastewater pollutants through biosorption, bioaccumulation, or biodegradation routes (Figure 6.1). In biosorption, the pollutant remains adhered to the surface of the microorganism's cells. This process depends on the affinity of the biosorbent (living or dead microorganisms) to the target MP, which can interact through physical or chemical mechanisms (Tiquia-Arashiro 2018). If the MP crosses the cell membrane, the process changes from biosorption to

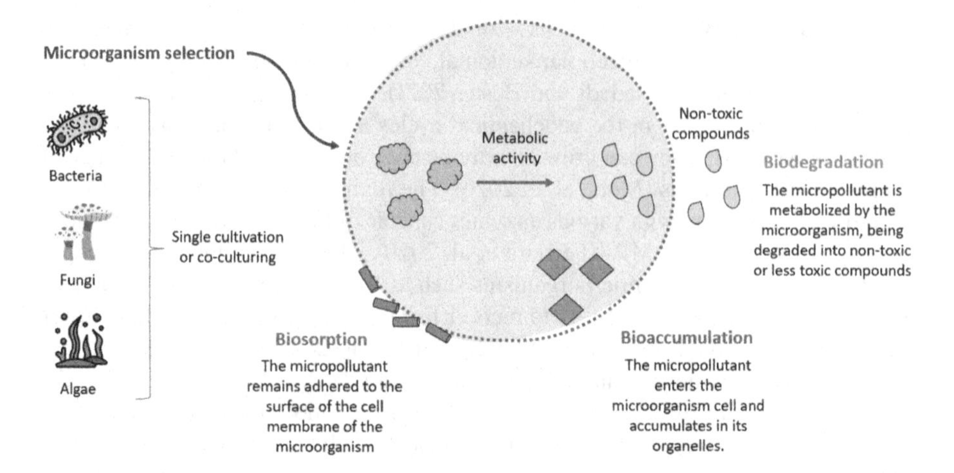

FIGURE 6.1 Microorganism bioremediation via biosorption, bioaccumulation, and biodegradation.

bioaccumulation (Abdelfattah et al. 2023). Bioaccumulation requires cellular energy and, therefore, occurs on a smaller scale than biosorption. Due to the hydrophobicity of the cell membrane, non-polar and liposoluble pollutants with low molecular weight can cross the cell membrane by passive diffusion (Rempel et al. 2021). Meanwhile, polar compounds with high molecular weight molecules are hardly diffused in microorganism cells. In the biodegradation process, the pollutant is used in the metabolic pathways of the microorganisms, degrading them into less or non-toxic substances (Abdelfattah et al. 2023). Microorganisms can mineralize pollutants (releasing only CO_2 and H_2O molecules) or biotransform them through a series of enzymatic reactions to produce various metabolic intermediates (Morsi et al. 2020). It is worth mentioning that biosorption, bioaccumulation, and biodegradation occur simultaneously, and their differentiation during the process is observed in lab-scale analysis.

6.4 CONVENTIONAL MICROBIAL WASTEWATER TREATMENT PROCESS

6.4.1 Biofilm Systems

Biofilm systems comprise a community of immobilized microorganisms and extracellular polymeric substances that grow attached to a support. This support can be composed of natural (stones, pebbles, and wood) or synthetic (metallic or plastic mesh) materials. Biofilm systems can operate in a stationary phase (fixed bed bioreactors) or with agitation (moving bed or fluidizing bed bioreactors). The complex consortium of microorganisms that grow in the support material leads to a high concentration of active biomass (Saidulu et al. 2021). The thickness of the biofilms can vary with the operational conditions and supporting medium. The different regions of the biofilms allow the occurrence of aerobic, anoxic, and anaerobic zones, enabling the simultaneous processes of nitrification, denitrification, and MP degradation (Paździor et al. 2019). Besides, immobilized cells are less sensitive to environmental variations, such as temperature, pH, and concentrations of toxic substances (Ahmed et al. 2021).

The most commonly used fixed biofilm is the trickling filter, in which the wastewater is distributed over a packed medium with microorganisms that metabolize the organic matter. They present simple operational requirements, low energy consumption, a self-cleaning mechanism, and better sludge thickening (Paździor et al. 2019). Among the biofilms with agitation systems are the Moving Bed Biofilm Reactors (MBBR), Aerobic Fluidized Bed Bioreactors (AFlBR), and Sequence Biofilm Bioreactors (SBBR). The agitation can be performed by aeration or mechanical stirring. This process allows a more homogeneous distribution of the biomass, with less dead space, less sludge generation, and a lower hydraulic retention time (HRT) (Jiang et al. 2018).

Although these systems were developed to remove biodegradable organic matter, some studies evaluate their capability to degrade recalcitrant compounds. To promote the removal of MPs, the operating conditions must be adapted, especially the HRT (Saidulu et al. 2021). Generally, the increase in HRT promotes a more efficient removal of MPs once the microorganism has more time to degrade the contaminant (Jiang et al. 2018). The HRT also impacts the mass transfer of MP from the wastewater to the biofilm. The supporting material and its interstitial retention spaces also play an important aspect in the treatment performance (Ahmed et al. 2021).

6.4.2 Pond Systems

Wastewater treatment ponds, also known as stabilization ponds, are one of the most commonly used methods of domestic wastewater treatment. Its application is associated with low installation and operational costs (Adhikari and Fedler 2020). The pond system is a tank in which the stabilization of the organic matter occurs through microbiological, hydrodynamic, biochemical, and photosynthetic processes. These processes occur through the interaction of several types of microorganisms, such as protozoa, fungi, bacteria, algae, and viruses (Li et al. 2018). Usually, the wastewater undergoes a preliminary treatment before pond systems to reduce the amount of solids or increase the biodegradability of the contaminants.

There are three types of pond systems: aerobic (which requires O_2), anaerobic (in O_2 absence), and facultative (with aerobic and anaerobic regions). Aerobic systems have a depth ranging from 1.0 to 1.90 m, where the oxidation process is carried out mainly by algae and bacteria. The anaerobic ponds have a depth of 2.0–5.0 m and a relatively short retention time. The facultative system is usually shallower than anaerobic ponds, reaching up to 2.0 m, with the upper part operating in aerobiosis and the lower part in anaerobiosis (Li et al. 2018). In addition to these systems, there are also the maturation ponds, designed to be used as the final stage of treatment to remove pathogenic organisms as an alternative to chlorination. These systems have a shallow depth (less than 1.0 m) and use ultraviolet radiation from sunlight to promote disinfection (Ali et al. 2020). The low concentration of biodegradable organic matter contributes to endogenous metabolism, promoting good sewage nitrification and enhancing BOD removal (Adhikari and Fedler 2020).

Although pond systems are widely used, they still have some limitations regarding the overload of organic matter, the generation of odors, sludge accumulation, the large implementation area, and the low removal of recalcitrant substances (Dhangar and Kumar 2020). Besides, the performance of the pond system can vary with the sludge accumulation, requiring a periodic desludging process (Adhikari et al. 2021). The efficiency of the stabilization ponds depends on climatic parameters, such as sunlight, temperature, wind, and rain. A combination of ponds with different characteristics has been proposed as an alternative to improve their performance, such as constructed wetlands, aeration tanks, Advanced Integrated Wastewater Pond Systems (AIWPS), and Wastewater Formation Ponds (WFP) (Paździor et al. 2019).

6.4.3 Activated Sludge

Activated sludge systems consist of an aerated tank with microorganisms able to degrade the nutrients, remove the nitrogen, and oxidize the carbonaceous biological matter (Manai et al. 2017). The acclimated microorganism culture is mainly composed of aerobic bacteria (around 90%) and protozoa able to form flocs (Paździor et al. 2019), which constitute the "activated" sludge of the process. The wastewater remains in the aerated tank for an adjusted time to ensure treatment. Then it is transferred to a gravimetric clarifier, where the sludge settles. The settled activated sludge is recycled into the aerated tank, with the process operating continuously (Grandclément et al. 2017). As the microorganisms multiply during the process, the volume of sludge increases in each treatment cycle. Thus, excessive sludge generation is one of the limiting aspects of the activated sludge treatment process (Martínez-Alcalá et al. 2017). It is worth mentioning that the wastewater must pass through a preliminary treatment to remove solids prior to activating sludge systems.

The pollutant removal in the activated sludge process is governed by three pathways: biodegradation (metabolic or co-metabolic), surface interactions in the sludge particles, and volatilization during the aeration (Saidulu et al. 2021). In biodegradation, the microorganisms assimilate the pollutants mainly through co-metabolic pathways since MP concentrations are commonly too low to serve as direct growth substrates. Besides, the ammonia-oxidizing bacteria in the sludge can co-metabolize oxidized MPs in the presence of an ammonia monooxygenase (Grandclément et al. 2017). Surface interaction pathways occur when the activated sludge flocs retain the pollutants, removing them from the wastewater. This process can occur with a range of hydrophobic and ionized MPs (Martínez-Alcalá et al. 2017). Manai et al. (2017), show that the use of fungi in the activated sludge can enhance the removal of recalcitrant compost. Thus, the composition of the microorganism consortium in the activated sludge is a determinant in removing MPs. To remove MPs, the volatilization pathway can be considered negligible (Saidulu et al. 2021).

Among the process parameters of activated sludge systems are the food-to-microorganisms (F/M) ratio, HRT, and sludge retention time (SRT). The F/M interferes with microbial growth and sludge settling. A reduction in the F/M ratio (more microorganisms than food) reduces the BOD in the aerated tank and can promote higher assimilation of MPs (Saidulu et al. 2021). HRT is the average time that the wastewater remains in the aerated tank. HRT also affects the F/M ratio and depends on the reactor volume (Martínez-Alcalá et al. 2017). Higher HRT provides more contact opportunities between the MPs and the microbes in the sludge, increasing the process efficiency. However, with higher HRT, a larger volume reactor is required, which can increase the expenses of the process (Saidulu et al. 2021). SRT is the period that the active sludge is in the tank, impacting the occurrence of the biodegradation and adsorption processes. SRT influences the adaptability of the microbe population and their potential to metabolize pollutants (Manai et al. 2017). Generally, the MPs' removal increases with a higher SRT. However, increased SRT leads to more generation of excess sludge and requires a larger operational area (Saidulu et al. 2021).

6.4.4 Anaerobic Sludge Blanket

The anaerobic sludge blanket is one of the most promising anaerobic treatment technologies in wastewater treatment due to its high efficiency associated with simple design, low installation and operational costs, and adaptability to a range of temperatures (from 15 to 55°C), pH (ranging from 5 to 9), and composition of the wastewater (Dutta et al. 2018). In these systems, the biodegradation process occurs through the metabolic activity of bacteria, which oxidize and reduce organic and inorganic compounds while producing biogas (Rani et al. 2022).

The Upflow Anaerobic Sludge Blanket (UASB) is the type of reactor most commonly used in this system. It is designed to degrade organic matter and separate solid, liquid, and gaseous materials. UASB has shown good results for removing some MPs such as EE2-ethinyl estradiol, E2-estradiol, ibuprofen, estrone, gemfibrozil, and bisphenol (Vassalle et al. 2020). However, this system does not eliminate pathogens, so it must be associated with other post-treatment technologies.

In the UASB, the affluent is pumped in an ascending flow and passes through three stages: the sludge bed (where there is a high concentration of active biomass), followed by a sludge blanket (with less dense biomass), and a three-phase separator, where the separation of the suspended solids (sludge), liquid, and biogas formed in the treatment occurs. The stabilization of the organic matter occurs in the reaction zones – bed or sludge blanket. During the biodegradation in the sludge bed, biogas is formed, which tends to remain attached to the biomass. Thus, this biomass becomes less dense due to the increase in volume (solids and biogas) and floats upon ascending. Then, it reaches the three-phase separator. Where the sludge touches the bottom of the decanter, releasing the biogas and allowing the solids (biomass) to decant (Dutta et al. 2018).

Among the parameters that influence the anaerobic digestion in UASB are the volumetric organic load, HRT, volumetric hydraulic load, biological load, ascending flow rate, surface application rate, temperature, and pH (Vassalle et al. 2020). The temperature is one of the most significant, as it interferes with microbiological activity and reaction speed. In the literature, three main temperature ranges are applied in UASB reactors: thermophilic (above 45°C), mesophilic (between 25 and 40°C), and psychrophilic (below 25°C). Mesophilic and thermophilic conditions are most commonly used since low temperatures hamper the growth of some methanogens (Dutta et al. 2018). Despite the wide range of pH (between 5 and 9) supported in UASB systems, the best performances are communally obtained at neutral pH – a condition that benefits the growth of the anaerobic bacteria used. The HRT in UABS usually varies from 6 to 48 hours but can be applied for longer periods, depending on the characteristics of the wastewater and the target pollutant to be removed (Rani et al. 2022).

6.5 CONVENTIONAL PHYSICAL-CHEMICAL WASTEWATER TREATMENT PROCESS

6.5.1 Coagulation-flocculation

The coagulation-flocculation process aims to reduce the insoluble suspended substances in water that do not settle with gravity. Thus, a coagulant agent is applied to destabilize the electrical charges on the surface of the particles, decreasing their repulsive potential and promoting their agglomeration. With this, the size of the particles increases, giving rise to flocs (Bodzek and Konieczny 2018). The coagulation-flocculation process occurs in stages of fast and slow mixing. The fast mixing aims to homogenize the aqueous medium with the coagulant, promote the reactions between the coagulant and water, and form electron clouds. The slow mixing promotes the connection of the particles electrically charged with the suspended solids, forming the aggregates. Then, the aggregates can be easily removed by sedimentation or flotation processes (Montaño-Medina et al. 2023).

The coagulation-flocculation can be applied as a pre- or post-treatment. This process has received great interest due to its simple operation requirements, high efficiency, and low cost (Zhang et al. 2021). Coagulation-flocculation can remove suspended solids, reducing phosphorus, COD, turbidity, and color. However, when applied alone, this method is inefficient for removing low concentrations of metallic ions, dyes, and other MPs (Zhang et al. 2021). Therefore, a post-treatment, such as sedimentation, adsorption, or filtration, is required. The efficiency of the conventional coagulation-flocculation process depends on the pH, coagulant type and concentration, mixing speed, and sedimentation time (Ayed et al. 2021).

6.5.2 Electrochemical Process

Electrochemical wastewater treatment is a group of processes in which electrical current is the driving force. It includes electroflotation, electrocoagulation, and electrochemical oxidation-reduction. The electrical current is applied to create a difference in potential and neutralize the pollutants, allowing efficient mineralization of non-biodegradable compounds. One of the greatest advantages of the electrochemical process is the non-consumption of chemical reactants. However, reducing or oxidizing chemical agents are commonly added to improve the efficiency of pollutant mineralization in the electrochemical cell (Song et al. 2020). Electrochemical treatment is mainly used to remove metallic ions, dyes, ionized organic matter, and salts. Besides, it is applied in gas desulfurization and desalination (Tran et al. 2017). Currently, this technology faces economic hurdles for organic compound removal since biological treatments are commonly less expensive (dos Santos et al. 2021).

The electrocoagulation process promotes water electrolysis and the attraction of the pollutant with ionized particles (Al-Shannag et al. 2015). This process is more efficient than chemical coagulation but has higher costs. Electrochemical reduction or oxidation can occur by direct or indirect electrolysis. Examples of direct electrolysis include the removal of metallic ions by cathodic deposition, in which the ions are reduced by electron addition at the cathode (Tran et al. 2017). Meanwhile, indirect electrolysis generates redox reagents, which convert the pollutants into less toxic products. The redox reagent acts as an intermediary to transport electrons between the polluting substrate and the electrode (Al-Shannag et al. 2015). An example is the generation of chlorine (Cl_2) from chloride (Cl^-) in the residual solution at the anode, which oxidizes pollutants. The effect of indirect oxidation depends on the material selected for the anode. In addition, it may contribute to pH reduction, leading to corrosive processes on the electrodes (Song et al. 2022).

Among the critical parameters of the electrochemical process is the current density, which impacts treatment efficiency and cost requirements (Tran et al. 2017). Besides, the mass transfer is limited by the electrode area, which can destabilize during the process by fouling (Song et al. 2022). Thus, the materials used as electrodes are also directly related to the efficiency and sustainability of electrochemical processes.

6.5.3 Adsorption

Adsorption is a separation process in which compounds in a fluid (liquid or gas) are transferred to a solid surface. The solid is the adsorbent, and the retained compounds are the adsorbates. This process gained prominence due to its simplicity, low cost, low energy consumption, and high efficiency in removing persistent contaminants (Ramos et al. 2021a). The contaminants can be adsorbed through chemical or physical interactions. In physical adsorption, the phenomenon occurs through Van der Waals interactions (dipole-dipole or dipole-induced), which are broad-spectrum intermolecular bonds (Lima et al. 2015). This process is reversible once physisorption does not involve chemical reactions. It can occur in multilayers without changing the chemical nature of the adsorbent. Meanwhile, chemical adsorption involves the exchange of electrons between contaminants and the adsorbent surface. In this process, specific areas of the adsorbent (activated sites) retain the pollutant through chemical bonds, forming covalent electrostatic interactions that are more difficult to reverse. Thus, only a single layer of adsorbed molecules is formed (Vieira et al. 2020).

The efficiency of the adsorption process is directly related to the surface area of the adsorbent and the chemical characteristics of the pollutant. Generally, the number of activated sites increases with the surface area. The surface charge, structure, and pore size are also important parameters, as they can allow or hinder the retention of the contaminant (Martins et al. 2022). Regarding the characteristics of the pollutant, its size and polarity have a great influence on its interactions with the adsorbent, being a crucial parameter to the efficiency of the process (Zonato et al. 2022).

The operational conditions of pH and temperature also impact the efficiency of the adsorption process. The temperature interferes with the kinetic energy of the molecules and their solubility, affecting intraparticle diffusion and the adsorption capacity. Generally, the kinetics of physical-chemical processes increase with temperatures (Vieira et al. 2020). The pH of the solution defines the chemical species distribution degree, their solubility, and the electrostatic interactions that can occur on the adsorbent surface (Ramos et al. 2022).

An ideal adsorbent for wastewater treatment should not cause environmental damage and have a high capacity and selectivity to adsorb contaminants at low concentrations (Martins et al. 2022). Besides, it must operate in multiple adsorption-desorption cycles. Otherwise, the process can become expensive and unfeasible (Vieira et al. 2020). Among the most commonly used adsorbents are activated carbons, clays, and clay minerals. To make the process even more economically feasible, a range of residual biomass has been applied as adsorbents, such as bamboo fiber, sawdust, sludge from water or wastewater treatment, and peels from passion fruit, coconut, cocoa, orange, palm, walnut, coffee, and rice (Ramos et al. 2021b).

6.5.4 Membranes Filtration System

Membranes are thin and semipermeable solid films that act as a selective barrier to the transport of pollutants. Membrane filtration can retain particles, microorganisms, ions, and dissolved salts (Dhangar and Kumar 2020). When the compounds cross the membrane, it is necessary to apply an external driving force, such as a concentration gradient, pressure, temperature, electricity, or a combination of these factors. Four membrane types operate by pressure gradient: microfiltration, ultrafiltration, nanofiltration, and reverse osmosis. The material used to manufacture the membranes (usually ceramics, polymers, or zeolites) affects their selectivity and efficiency (Bhattacharya et al. 2021). In addition, the electrodialysis technique, which uses electrical potential difference, is widely studied in membrane systems.

Each membrane filtration technique is suitable for removing contaminants with a specific size range. Microfiltration membranes can retain particles with a diameter between 0.1 and 10 μm, including protozoa cysts and oocysts, bacteria, algae, cyanobacteria, zooplankton, iron, and manganese oxides (Bodzek and Konieczny 2018). It is also used as a pre-treatment for reverse osmosis systems. The microfiltration process operates at atmospheric pressure, which reduces the cost of the other membrane systems (Vieira et al. 2020).

Ultrafiltration membranes are suitable for separating particles with dimensions between 0.001 and 0.02 μm, being able to remove substances with a molecular weight between 1,000 and 2,000 $g.mol^{-1}$, including colloids, soluble organic compounds, and viruses. Its applications are similar to microfiltration but with a greater capacity to separate dissolved organic substances with high molecular weights (Dhangar and Kumar 2020). Ultrafiltration is also used for water recovery in industrial facilities, enabling its reuse to wash operational components. Nanofiltration membranes can remove substances with a molecular weight between 200 and 1,000 $g.mol^{-1}$, including various chemical compounds and ions, such as Ca^{2+} and Mg^{2+}, responsible for water hardness. Besides, nanofiltration minimizes the need for disinfection by retaining viruses, bacteria, and organic and inorganic constituents (Bodzek and Konieczny 2018).

Reverse osmosis is based on the natural phenomenon in which the diluted solution crosses a semipermeable membrane towards the more concentrated solution. In this process, high pressure is applied to the concentrated solution (raw wastewater), forcing the flow in the opposite direction, which leads the pollutants to cross the membrane (Vieira et al. 2020). Reverse osmosis membranes can remove dissolved salts and organic matter with a molecular weight of less than 200 $g.mol^{-1}$. Its main use is desalination. In wastewater, reverse osmose is used to remove dissolved contaminants after preliminary filtration or microfiltration (Linares et al. 2016).

In the electrodialysis process, the most relevant factor is the ionic selectivity of the membranes, which are produced with ion exchange resins in the form of flat sheets (Giwa et al. 2019). The cationic and anionic membranes are alternately arranged between plastic spacers, an electrical potential difference is applied, and the ions are transported across membranes. Thus, the ions to be removed are attracted by the oppositely charged membranes and barred by the first membrane in the system that has the same charge as the pollutant.

The membrane filtration process is affected by the membrane and pollutant characteristics, such as pore size, functional groups, selectivity, hydrophilicity, and pKa (Vieira et al. 2020). Membrane fouling is one of the most relevant problems in these systems. Its occurrence depends on the physical, chemical, and biological characteristics of the wastewater, the membrane material, the configuration of the modules, and the operating conditions (Giwa et al. 2019). The occurrence of fouling defines the need for pretreatment, the cleaning frequency and method, the maintenance costs, and the membrane performance (Bodzek and Konieczny 2018). In addition, the management of the concentrate from membrane separation processes is a challenge of this technology, as it is mainly composed of inorganic compounds in alkaline pH material, which increases the probability of metal precipitation – if this pollutant were in the solution (Giwa et al. 2019).

6.5.5 Advanced Oxidation Processes

Chemical oxidation aims to transform the pollutants into molecules that are easier to assimilate by the environment. In complete oxidation, the products are water, carbon dioxide, and eventual ions that may constitute the starting pollutant molecule. Advanced oxidation processes (AOPs) arise from the necessity of guaranteeing the complete oxidation of contaminants (Paździor et al. 2019). AOPs can use a combination of oxidants (like hydrogen peroxide, persulfate, ferrate, permanganate, and ozone) and techniques – such as ultraviolet light, ultrasound, and inorganic catalysts – to increase the generation of highly oxidizing radicals, such as the hydroxyl radical (most used), sulfate, or chlorine (Vieira et al. 2021). These radicals can attack organic molecules by radical addition, hydrogen abstraction, electron transfer, and a combination of radicals (Lin et al. 2022). AOP processes generally reach high removal efficiency and increase the biodegradability of some constituents, potentially requiring a biological process as a second step to remove the residual biodegradable organic material (Pisharody et al. 2022).

Among the AOP techniques are Fenton, photocatalysis, ozonation, and electrochemical processes. These techniques can be applied individually or combined to enhance the formation of reactive species (Lin et al. 2022). The Fenton process uses iron-based or transition metal-based catalysts to produce OH^- from hydrogen peroxide (Pisharody et al. 2022). In the Fenton method, the pH must be controlled to avoid metal precipitation and the formation of excessive OH^- (Dhangar and Kumar 2020).

Anodic or electrochemical oxidation is the process in which pollutants are degraded due to anodic activity. These processes depend on the electrochemical cell voltage, the solution composition, and the electrode characteristics (dos Santos et al. 2021). In photocatalysis, semiconductors are photoactivated by irradiation. Titanium dioxide (TiO_2) and ozone/H_2O_2 are the most used semiconductors as TiO_2 has the properties of strong metal-support interaction and H_2O_2 directly photolyzes two hydroxyl radicals (Paździor et al. 2019).

Ozone is an oxidant known for its selectivity – unlike the OH^- radical. O_3 reacts with ionized or dissociated pollutants instead of their neutral form (Bodzek and Konieczny 2018). However, organic degradation with hydroxyl radicals is higher and faster than with ozone. Thus, it is interesting to potentiate the generation of hydroxyl radicals from ozone. For this, hydrogen peroxide, UV radiation, or heterogeneous catalysts can accelerate ozone decomposition and the production of hydroxyl radicals (Lin et al. 2022). Gamma radiation can also be used in AOPs to generate radicals, highly reactive electrons, ions, and neutral molecules. This process is more efficient than UV radiation due to the lower probability of the pollutant absorbing photons (Paździor et al. 2019). With the water absorbing more photons, there is a high generation of hydroxyl radicals.

6.6 HYBRID OR INTEGRATED SYSTEMS

As the wastewater presents a complex composition, no single technique can satisfactorily remove MPs (Dhangar and Kumar 2020). Thus, combining microbial with physical-chemical treatment processes – a hybrid system – has been proven to be more effective in removing MPs (Grandclément et al. 2017). In hybrid systems, the treatment methods are integrated, combining their advantages while reducing their challenges and drawbacks, which results in cost reduction and time savings (Prado et al. 2017). In addition, depending on the system configuration, space and energy requirements can also be reduced by combining different operations into only one reactor (Dhangar and Kumar 2020).

From the literature, it can be observed that combining the activated sludge process with membrane filtration allows the removal of a range of pollutants such as PAHs, metals, volatile organic compounds (VOCs), benzene, toluene, ethyl benzene and xylenes (BETEX), and organic composts (Argun et al. 2020). The addition of adsorbent material in activated sludge systems enhances the removal of pharmaceuticals (Mojiri et al. 2020). The use of a membrane bioreactor (MBR) followed by membrane filtrations (reverse osmosis/nanofiltration) can effectively remove a range of MPs, such as pharmaceuticals, beta-blockers, pesticides, and endocrine disruptors (Nguyen et al. 2013). Combining MBR with OAP systems also improves the degradation of pharmaceuticals (Prado et al. 2017). Coupling bioassimilation with the coagulation process allowed the degradation of phenol substances (Ayed et al. 2021). *In vivo* biosorption and biodegradation, followed by photoreduction, can remove dyes and metallic ions from wastewater (Tu et al. 2022). By coupling MBR with an electrochemical process, the treatment of metallic ions was enhanced (Giwa and Hasan 2015).

Argun et al. (2020) investigated the removal of several MPs in landfill leachate in a hybrid system that consists of a sequence of equalization ponds, anoxic and aerobic bioreactors, ultrafiltration, and nanofiltration. The authors state that the metal was removed by the coprecipitation of suspended solids with the microbial biopolymers in the bioreactor and separated in the membrane filtration. The biological process removed the higher fraction of VOC, BETEX, and PHAs, which may be related to the molecular weight of these composts, which is lower than the membranes used. Besides, the authors observed that the main removal mechanisms of VOCs were volatilization and biodegradation in the anoxic/aerobic biological process. Only 11–20% of color was reduced in the biological process, while filtration systems removed 59–93% of the color in ultrafiltration and 100% after nanofiltration. The bioreactor assimilates 48% of the ammonia and 54% of COD, which were further reduced to 96% and 88% using ultrafiltration and reached almost complete removal after the nanofiltration process.

Mojiri et al. (2020) added a powdered composite adsorbent (manufactured from a mixture of bentonite, zeolites, and biochar) in an activated sludge system to reduce the concentration of pharmaceuticals, ammonia, and COD. The authors show that the hybrid process increases the removal percentage of all the MPs analyzed. Without the activated carbon, 83% COD, 73% ammonia, 57% atenolol, 58% ciprofloxacin, and 58% diazepam were removed. The hybrid system allows the removal of 92% COD, 95% ammonia, 90% atenolol, 95% ciprofloxacin, and 96% diazepam. Besides, after adding the adsorbent material, the equilibrium time decreased from 24 h to 14 h. The authors also state that even in the reactor without the powdered composite adsorbent, pharmaceutical removal occurs mainly through adsorption in the activated sludge.

Prado et al. (2017) applied ozonation and ultrasound as pretreatments in a membrane bioreactor to remove pharmaceuticals from simulated domestic wastewater. Applying only the AOP process, 69–79% of the pharmaceuticals were removed. Using only the MBR, an efficiency of 74–77% was achieved. Combining the two processes, efficiencies reached 80–84%. Despite the good removal of the process individually, toxicity was reduced only by applying them together. The AOP system decreased toxicity by 20%, and the MBR system by 10%. When these systems were combined, toxicity reduction reached 80%. In addition, AOP before MRB reduces membrane fouling by about 50%.

Ayed et al. (2021) report up to an 80% reduction in the phytotoxicity of textile wastewater by combining coagulation-flocculation with bioassimilation. The authors added three bacteria (*Sphingomonas paucimobilis*, *Bacillus* sp., and filamentous bacteria) into a coagulation-flocculation tank and achieved 82% removal of the color, 76% of COD, and 36% of phenol. Tu et al. (2022) proposed an innovative solution to remove cadmium and dyes from an aqueous medium. The authors used in vivo precipitation of cadmium in a protozoan (*Tetrahymena thermophila*) to reduce toxicity and increase the bioremediation of the metal. Then, the cell containing cadmium nanoparticles triggers the photoreduction of Congo red under visible-light irradiation. With this approach, 94% of cadmium and 95% of Congo red were removed. Giwa and Hasan (2015) proposed a combination of biological decomposition, membrane filtration, and electrocoagulation in a hybrid electrically enhanced membrane bioreactor. With this system, the authors removed almost all the phosphorous and COD, 91% of nitrogen, and 79–89% of metallic ions.

Nguyen et al. (2013) evaluate the removal of 22 traces of organic pollutants communally found in domestic wastewater by combining MBR, UV oxidation, nanofiltration, and reverse osmosis processes. The authors observed that MP removal is governed by adsorption into the suspended solids and biological degradation. Hydrophilic composts containing chlorine or nitrogen groups are particularly resistant to biological degradation, demanding further treatment. The UV process was effective for degrading composts with a chlorinated and phenolic group but did not present a good performance in removing amide-containing pollutants. UV oxidation and membrane systems (nanofiltration and reverse osmosis) significantly complement MBR treatment to obtain high overall removal of hydrophilic and biologically persistent trace pollutants, reaching more than 90% removal of the contaminants.

6.7 HOW TO SELECT THE APPROPRIATE TREATMENT(S) PROCESS(ES)

To select the treatment(s) process(es) most appropriate to remove or degrade a target MP, it is necessary to understand the effectiveness and limitations of each process. Table 6.2 summarizes the results obtained in the literature to remove a range of MPs using microbial, physical-chemical, and hybrid treatment processes. It also presents the main advantages and disadvantages of each process, whether used individually or combined. From Table 6.2, it can be noted that the main limitations of biofilms and pond systems – the methods most used in WWTP – are related to the large area required and the long reaction time. Anaerobic sludge blankets and activated sludge processes optimize the use of space, requiring a smaller area and low dependence on climatic conditions but having greater energy expenditure. It is worth mentioning that microbial treatment was initially developed to remove organic matter from wastewater, and for this purpose, it is still the most cost-effective (Grandclément et al. 2017). However, the increased concentration of recalcitrant compounds and non-biodegradable material in the wastewater weakened the use of these technologies (Paździor et al. 2019). Although some research in the literature shows good results for the removal of persistent contaminants (especially recalcitrant organic matter) using biological treatment, the optimization of the process comes with the combined use of physical-chemical treatment (Table 6.2).

The physical-chemical treatment process presents fast degradation/removal of MPs as a common advantage (Paździor et al. 2019). However, they are associated with increased energy demand and installation and operation costs (Eniola et al. 2022). Even obtaining high MP removal efficiencies when used alone, the physical-chemical best performance is reached using them with microbial systems (Saidulu et al. 2021). AOPs, for example, can increase the biodegradability of MPs to enhance the effectiveness of biological treatment (Dhangar and Kumar 2020).

TABLE 6.2
Comparison of Individual and Combined Treatment Processes to Remove Micropollutants in Wastewater

Treatment Process						
Microbial	Physical-Chemical	Pollutant Removed	Removal Efficiency (%)	Advantages	Disadvantages	Reference
Biofilm	–	Ibuprofen Salicylic acid Triclosan Ketoprofen Primidone Carbamazepine Bisphenol A Atrazine	98 98 91 80 90 26 75 64–73	• Low operational cost • Low-energy consumption • Simple operational requirements • Can present aerobic and anaerobic regions in the same structure	• Low efficiency to some MPs • Demand long HRT to degrade recalcitrant composts • Necessity of large area	Jiang et al. (2018) and Saidulu et al. (2021)
Pond systems	–	BOD Ammonium Fecal coliforms Nitrate Phosphate PAHs	40–80 95 3-log 77 65 65–98	• Low installation and operation cost • Low-energy consumption • Efficiency for organic matter and pathogenic organisms	• Overload • Odors generation • Sludge build-up • Low removal efficiency for MPs • Large footprint	Adhikari et al. (2021), Badawy et al. (2010), and Ho and Goethals (2020)
Activated sludge	–	COD Diclofenac Ibuprofen Ketoprofen Naproxen Color	95 90 100 51 98 50–90	• Low area requirements • Reuse of part of the active biomass • Precise and flexible operational control • Low dependence on climatic and environmental conditions	• High-energy consumption • Limited by the COD • Excess sludge generation • Low removal of pathogens	Manai et al. (2017), Martínez-Alcalá et al. (2017), and Saidulu et al. (2021)

(Continued)

TABLE 6.2 *(Continued)*
Comparison of Individual and Combined Treatment Processes to Remove Micropollutants in Wastewater

Treatment Process						
Microbial	Physical-Chemical	Pollutant Removed	Removal Efficiency (%)	Advantages	Disadvantages	Reference
Anaerobic sludge blanket	–	Ibuprofen Carbamazepine Diclofenac 17α-ethinylestradiol Estrone Bisphenol Gemfibrozil COD BOD TSS	24 23 29 26 95 43 39 63 67 70	• Simple design • Low operating and installation costs • Resistance to fluctuations in temperature, pH, and affluent composition • Good efficiency in the removal of organic matter	• Low reaction rate • High HRT • Requires post-treatment	Mora-Cabrera et al. (2021), Stazi and Tomei (2018), and Vassalle et al. (2020)
–	Coagulation–flocculation	Turbidity COD Antibiotics Beta-blockers Diclofenac Acetaminophen Pesticides	82–97 58–79 2–20 5–15 2–7 2–7 2–5	• Increase the biodegradability of the wastewater • Low cost • Simple operational requirements	• Production of a large volume of sludge • Great dependence of the pH • Low efficiency when used alone	Choi et al. (2022) and Haddaji et al. (2022)

(Continued)

TABLE 6.2 *(Continued)*
Comparison of Individual and Combined Treatment Processes to Remove Micropollutants in Wastewater

Treatment Process						
Microbial	Physical-Chemical	Pollutant Removed	Removal Efficiency (%)	Advantages	Disadvantages	Reference
–	Electrochemical process	Cu^{2+}	49–99	• Control of the reaction • Selective process • Fast treatment • Small facilities • Decrease in the use of chemical reactant	• Electrode instability and encrustation • Mass transfer limitations • Financial obstacles for organic compounds removal • Energy consumption	Al-Shannag et al. (2015), Aoudj et al. (2015), and Gerek et al. (2019)
		Ni^{2+}	60–98			
		Zn^{2+}	55–99			
		Cr^{3+}	52–100			
		COD	89			
		Fluoride	90			
		Turbidity	85			
–	Adsorption	17β-Estradiol	94	• High efficiency • Simple operational requirements • Possibility of adsorbents regeneration and adsorbate recuperation	• Not indicated for high pollutant concentration • Affected by suspended particles • Generation of solid waste	Baharum et al. (2020), Hassan et al. (2018), Martins et al. (2022), Ramos et al. (2021b), and Zonato et al. (2022)
		17α-Ethinylestradiol	92			
		Reactive Blue 198	96			
		Direct Black 22	97			
		Cd^{2+}	99			
		Cr^{3+}	95			
		Diazinon	99			
		Pyrene	98			
		Benzo(a)pyrene	99			
		Zn^{2+}	93			
		Ni^{2+}	95			
		Cu^{2+}	99			

(Continued)

TABLE 6.2 *(Continued)*
Comparison of Individual and Combined Treatment Processes to Remove Micropollutants in Wastewater

Treatment Process						
Microbial	Physical-Chemical	Pollutant Removed	Removal Efficiency (%)	Advantages	Disadvantages	Reference
–	Membranes filtration	COD	78–98	• High selectivity • Adaptable to wastewater treatment facilities • Disinfectant capacity • Non-generation of by-products	• Membrane fouling • High-energy consumption • High cost requirement	Bodzek and Konieczny (2018) and Samaei et al. (2018)
		Turbidity	60–90			
		Microorganism	99			
		Estrone	63			
		17β-Estradiol	77			
		Estriol	71			
		Mestranol	100			
		17α-Ethynyl estradiol	90			
		Diethylstilbestrol	86			
		Benzo(b)fluoretane	98			
		Benzo(a)pyrene	96			
		Benzo(g, h, i) perylene	91			
–	Advanced oxidation processes	Bisphenol A	87	• Fast and effective degradation of recalcitrant components • No generation of secondary waste stream • High disinfectant potential	• Formation of excessive hydrogen peroxide • High installation and operational cost • Complex process	Antonelli et al. (2022), Pisharody et al. (2022), Ryu et al. (2021), and Singa et al. (2021)
		Atrazine herbicide	98			
		Sulfamethoxazole antibiotic	96			
		TOC	96			
		COD	85			
		PAH	92			
		Ciprofloxacin	100			

(Continued)

TABLE 6.2 *(Continued)*
Comparison of Individual and Combined Treatment Processes to Remove Micropollutants in Wastewater

Treatment Process						
Microbial	Physical-Chemical	Pollutant Removed	Removal Efficiency (%)	Advantages	Disadvantages	Reference
Equalization pond and activated sludge	Membranes filtration	COD	98	• High removal efficiency for various types of MPs • Disinfectant capacity • Non-generation of by-products	• High cost of the membranes system • Low removal of volatile organic composts • Membrane fouling	Argun et al. (2020)
		Ammonia	99			
		Suspended solids	95			
		Electrical conductivity	51			
		Color	99			
		Phosphorous	82			
		VOCs	81–100			
		BTEX	97			
		Copper	92			
		Nickel	91			
		Chromium	51			
		PAHs	62			
Activated sludge	Adsorption	COD	92	• Two treatment processes in a single reactor • Simple adaptation in the system • Low investment	• Generation of residual sludge	Mojiri et al. (2020)
		Ammonia	95			
		Atenolol	90			
		Ciprofloxacin	94			
		Diazepam	95			
Membrane bioreactor	Advanced oxidation processes	Diclofenac	83	• Fouling control • High disinfection • High efficiency for recalcitrant composts	• High operational costs • Possible generation of toxic by-product	Prado et al. (2017)
		Sulfamethoxazole	84			
		Carbamazepine	80			

(Continued)

TABLE 6.2 *(Continued)*
Comparison of Individual and Combined Treatment Processes to Remove Micropollutants in Wastewater

Treatment Process						
Microbial	Physical-Chemical	Pollutant Removed	Removal Efficiency (%)	Advantages	Disadvantages	Reference
Biodegradation	Coagulation–flocculation	Color	82	• Simple operation requirements	• Sludge generation	Ayed et al. (2021)
		COD	76	• Low cost	• Necessity of accurate process control	
		Phenol	36			
Membrane bioreactor	Advanced oxidation processes and membranes filtration	TOC	92	• Fast and effective degradation of MPs	• High cost	Nguyen et al. (2013)
		Endocrine disruptors	97–100	• Disinfectant capacity	• Possible generation of toxic by-product	
		Pharmaceuticals	85–100			
		Fenoprop	94			
		Pentachlorophenol	99			
		Triclosan	99			
		Salicylic acid	92			
In vivo biosorption and biodegradation	Advanced oxidation processes	Cadmium	94	• High degradation of the MPs	• Complex control of the operational conditions	Tu et al. (2022)
		Congo red	95	• Fast and effective process		
Membrane bioreactor	Electrocoagulation	Nickel	79	• Fast and effective process	• High installation cost	Giwa and Hasan (2015)
		Iron	89	• Single reactor with several treatment processes	• High-energy consumption	
		Chrome	80	• Control of membrane fouling		
		COD	99			
		Phosphorous	99			
		Ammonium	91			

COD, chemical oxygen demand; TOC, total organic carbon; PAH, polycyclic aromatic hydrocarbon; TSS, total suspended solids; HRT, hydraulic retention time; VOCs, volatile organic composts; BTEX, benzene, toluene, ethyl-benzene and the xylenes.

AOPs and membranes have a disinfectant capacity, which helps maintain the biota of microbial systems (Vieira et al. 2020). The main drawback of membrane systems is their fouling. Thus, their performance is enhanced when associated with a solid removal process. The adsorption process is also impacted by the presence of solids and is more suitable for low concentrations of MPs, being more efficient when applied at the final stages of treatment. It is worth mentioning that adsorption, membranes, and electrochemical treatment present high selectivity for MPs removal and are quite effective for removing contaminants that can be toxic to microorganisms in biological treatment (Paździor et al. 2019). The electrochemical processes are indicated to remove ionizable MPs, which are not economically viable for organic matter.

The economic feasibility of the process is a decisive parameter for decision-making and is commonly estimated with a life-cycle cost analysis. This analysis considers the capital expenditure (CAPEX) – mainly covered by the installation costs – and the operational expenditure (OPEX), which includes energy, reactants, and maintenance costs. These estimates may vary with local factors, such as the cost of energy and construction materials, property value, possible subsidies, and the composition of the raw wastewater (Linares et al. 2016). Aiming for water desalination and reuse, Linares et al. (2016) show that a hybrid system composed of membrane bioreactors–reverse osmosis–advanced oxidation process can reduce the OPEX by 10% in comparison with the physical-chemical desalination process performed by reverse osmosis. However, the authors claim that CAPEX can increase by 7% due to system adaptations. These values were mathematically estimated based on the treatment of 100,000 m^3 of sewage water per day.

Giammar et al. (2022) evaluated the domestic wastewater treatment, aiming for water reuse. They compared the reverse osmosis process with a hybrid system formed by ozonation, biological filtration, and a granular activated carbon system. The authors evaluate the treatment for seven months on a pilot scale (treating about 2.0 million gallons of wastewater per day) at four different WWTPs across the United States. They show that the cost to enable water portability using the hybrid system was 0.40 USD per cubic meter, and the electricity consumption was estimated at 0.30 kWh m^{-3}. Meanwhile, reverse osmosis treatment costs US$0.54 per cubic meter to achieve the same water quality and requires 0.84 kWh m^{-3} of electricity. Thus, the hybrid system requires less than half of the energy and presents a lower total cost than reverse osmosis alone.

Eniola et al. (2022) performed a literature review and showed the operational cost of biological processes is around 0.18–0.56 US$.$m^3$, while the physical-chemical cost is 0.9–11.0 US$.$m^3$, and the cost of hybrid systems stays between 0.17 and 0.75 US$.$m^3$. Thus, the hybrid treatment reduces the operating costs over physical-chemical systems without significant increases for biological treatment. It should be noted that the authors estimate only the OPEX and do not verify the CAPEX investment. Furthermore, the types of processes to be applied and the possibility of integrating them into the same reactor impact the economic evaluation of hybrid systems. Eniola et al. (2022) considered the following hybrid systems: adsorption and photocatalytic MBR; or biodegradation, adsorption, and membrane combinations. Therefore, other process associations may have different estimates of costs.

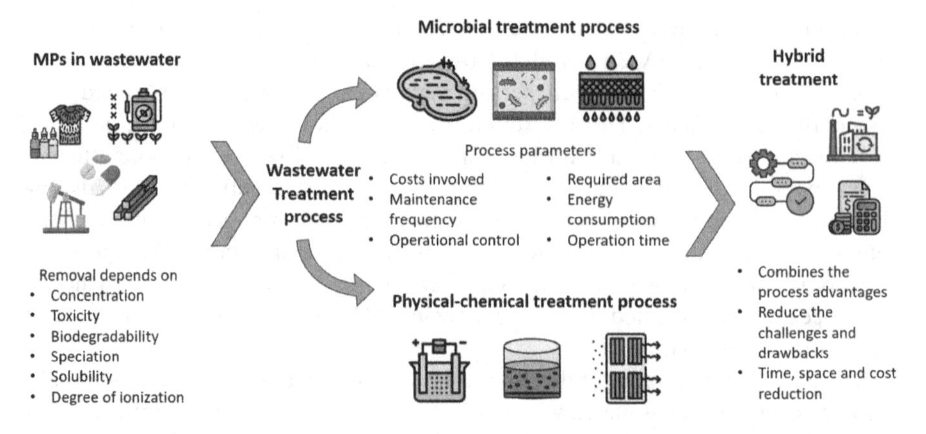

FIGURE 6.2 Overview of the parameters required to select the appropriate treatment process to remove MPs.

There is a lack of studies of the technical-economic assessment of the wastewater process in MP removal, essentially on a pilot or large scale. It may be due to the absence of regulations on the release of some MPs or the belated concern about the presence of these contaminants in water, whose removal is still not fully established in the WWTPs. Thus, studies in the field of MP removal from wastewater should focus both on identifying the processes and operational conditions that allow the removal of the contaminants, as well as the possibility of coupling these processes in an already established WTTP and verifying the technical-economic viability of the systems to allow the scale-up.

Considering all the topics discussed in this chapter, the selection of the processes for MPs removal must consider the characteristics of the pollutant, the possibility of integrating the process with the previous WWTP, the operational conditions, the financial requirements, the area available, and technological limitations (Saidulu et al. 2021). Figure 6.2 presents an overview of the parameters required to select the appropriate treatment to remove MPs and summarizes the advantages of hybrid systems.

6.8 CONCLUSION

A range of techniques are addressed in the literature aimed at MP removal from wastewater. The most commonly used microbial wastewater treatments are biofilms, pond systems, activated sludge, and anaerobic sludge blankets. The complex microbial interaction in some of these systems can enhance MP removal. However, these techniques were developed to remove organic load, and their process parameters need to be manipulated to allow efficient removal of recalcitrant composts. The removal of MPs – especially inorganic matter – in biological systems is commonly lower than in physical-chemical treatments. In addition, the bioremediation of MPs is mostly verified on a laboratory scale since biological processes are not applied alone on a large scale. Coagulation-flocculation, electrochemical processes, adsorption, membrane filtration, and AOP are widely used in MP removal. Despite the high efficiency of these techniques, they are more expensive than the biological process and require accurate process control, in addition to the possible generation of toxic by-products. This literature review shows that the removal of MPs increases by combining microbial and physical-chemical treatment processes. Physical-chemical systems can reduce the contaminant load and toxic composts in wastewater, increasing biological processes' efficiency. In addition, it can increase the biodegradability of the pollutant, allowing its assimilation in the microbial treatment. Biological treatment reduces the organic load of the wastewater, which can enhance the performance and selectivity of the physical-chemical treatments. Thus, integrating or combining these techniques can reduce the challenges and drawbacks of MP removal. Although the hybrid system presents a good perspective, its efficiency can vary with the selected process and its sequence, as well as with the pollutant characteristics. Besides, to reduce the costs and space required, the implementation of hybrid systems must consider the possibility of integrating the process into a single reactor and/or using a pre-established WWTP. There is a lack of studies regarding the technical and economic feasibility of hybrid systems and their scale-up, which is a demand to consolidate this technique's use.

ACKNOWLEDGMENT

The authors acknowledge the Coordination for the Improvement of Higher Education Personnel – CAPES (Grant Number 88887.679106/2022-00) for the financial support.

ABBREVIATIONS

AFBR Aerobic fluidized bed bioreactors
AOP Advanced oxidation process
BETEX Benzene, toluene, ethyl benzene, and xylenes
BOD Biochemical oxygen demand
CAPEX Capital expenditure
COD Chemical oxygen demand
F/M Food to microorganisms ratio
HRT Hydraulic retention time
MBR Membrane bioreactor
MBBR Moving bed biofilm reactors
MP Micropollutants
OPEX Operational expenditure
PHA Polycyclic aromatic hydrocarbon
SBBR Sequence biofilm bioreactors
SRT Sludge retention time
UASB Upflow anaerobic sludge blanket
UV Ultraviolet
VOC Volatile organic compound
WWTP Wastewater treatment plant

REFERENCES

Abdelfattah A, Ali SS, Ramadan H, El-Aswar EI, Eltawab R, Ho S-H, Elsamahy T, Li S, El-Sheekh MM, Schagerl M, Kornaros M, Sun J (2023) Microalgae-based wastewater treatment: Mechanisms, challenges, recent advances, and future prospects. Environ Sci Ecotechnology 13:100205. https://doi.org/10.1016/j.ese.2022.100205

Adhikari K, Fedler CB (2020) pond-in-pond: An alternative system for wastewater treatment for reuse. J Environ Chem Eng 8:103523. https://doi.org/10.1016/j.jece.2019.103523

Adhikari K, Fedler CB, Asadi A (2021) 2-D modeling to understand the design configuration and flow dynamics of pond-in-pond (PIP) wastewater treatment system for reuse. Process Saf Environ Prot 153:205–214. https://doi.org/10.1016/j.psep.2021.07.012

Ahmed SF, Mofijur M, Nuzhat S, Chowdhury AT, Rafa N, Uddin MA, Inayat A, Mahlia TMI, Ong HC, Chia WY, Show PL (2021) Recent developments in physical, biological, chemical, and hybrid treatment techniques for removing emerging contaminants from wastewater. J Hazard Mater 416:125912. https://doi.org/10.1016/j.jhazmat.2021.125912

Ali AE, Salem WM, Younes SM, Kaid M (2020) Modeling climatic effect on physiochemical parameters and microorganisms of stabilization pond performance. Heliyon 6:e04005. https://doi.org/10.1016/j.heliyon.2020.e04005

Al-Maqdi KA, Hisaindee SM, Rauf MA, Ashraf SS (2017) Comparative degradation of a thiazole pollutant by an advanced oxidation process and an enzymatic approach. Biomolecules 7:64. https://doi.org/10.3390/biom7030064

Al-Shannag M, Al-Qodah Z, Bani-Melhem K, Qtaishat MR, Alkasrawi M (2015) Heavy metal ions removal from metal plating wastewater using electrocoagulation: Kinetic study and process performance. Chem Eng J 260:749–756. https://doi.org/10.1016/j.cej.2014.09.035

Antonelli R, Malpass GRP, da Silva MGC, Vieira MGA (2022) Photo-assisted electrochemical degradation of ciprofloxacin using DSA® anode with NaCl electrolyte and simultaneous chlorine photolysis. J Water Process Eng 47:102698. https://doi.org/10.1016/j.jwpe.2022.102698

Aoudj S, Khelifa A, Drouiche N, Hecini M (2015) Development of an integrated electrocoagulation–flotation for semiconductor wastewater treatment. Desalin Water Treat 55:1422–1432. https://doi.org/10.1080/19443994.2014.926462

Argun ME, Akkuş M, Ateş H (2020) Investigation of micropollutants removal from landfill leachate in a full-scale advanced treatment plant in Istanbul city, Turkey. Sci Total Environ 748:141423. https://doi.org/10.1016/j.scitotenv.2020.141423

Ayed L, Ksibi IE, Charef A, Mzoughi RE (2021) Hybrid coagulation-flocculation and anaerobic-aerobic biological treatment for industrial textile wastewater: Pilot case study. J Text Inst 112:200–206. https://doi.org/10.1080/00405000.2020.1731273

Badawy MI, El-Wahaab RA, Moawad A, Ali MEM (2010) Assessment of the performance of aerated oxidation ponds in the removal of persistent organic pollutants (POPs): A case study. Desalination 251:29–33. https://doi.org/10.1016/j.desal.2009.10.001

Baharum NA, Nasir HM, Ishak MY, Isa NM, Hassan MA, Aris AZ (2020) Highly efficient removal of diazinon pesticide from aqueous solutions by using coconut shell-modified biochar. Arab J Chem 13:6106–6121. https://doi.org/10.1016/j.arabjc.2020.05.011

Bhatt P, Bhandari G, Bilal M (2022) Occurrence, toxicity impacts and mitigation of emerging micropollutants in the aquatic environments: Recent tendencies and perspectives. J Environ Chem Eng 10:107598. https://doi.org/10.1016/j.jece.2022.107598

Bhattacharya P, Mukherjee D, Deb N, Swarnakar S, Banerjee S (2021) Indigenously developed CuO/TiO_2 coated ceramic ultrafiltration membrane for removal of emerging contaminants like phthalates and parabens: Toxicity evaluation in PA-1 cell line. Mater Chem Phys 258:123920. https://doi.org/10.1016/j.matchemphys.2020.123920

Bodzek M, Konieczny K (2018) Membranes in organic micropollutants removal. Curr Org Chem 22:1070–1102. https://doi.org/10.2174/1385272822666180419160920

Choi S, Son H, Kim YM, Lee Y (2022) Abatement efficiencies of organic matter and micropollutants during combined coagulation and powdered activated carbon processes as an alternative primary wastewater treatment option. J Environ Chem Eng 10:107619. https://doi.org/10.1016/j.jece.2022.107619

Dhangar K, Kumar M (2020) Tricks and tracks in removal of emerging contaminants from the wastewater through hybrid treatment systems: A review. Sci Total Environ 738:140320. https://doi.org/10.1016/J.SCITOTENV.2020.140320

dos Santos AJ, Kronka MS, Fortunato GV, Lanza MRV (2021) Recent advances in electrochemical water technologies for the treatment of antibiotics: A short review. Curr Opin Electrochem 26:100674. https://doi.org/10.1016/j.coelec.2020.100674

Dutta A, Davies C, Ikumi DS (2018) Performance of upflow anaerobic sludge blanket (UASB) reactor and other anaerobic reactor configurations for wastewater treatment: A comparative review and critical updates. J Water Supply Res Technol 67:858–884. https://doi.org/10.2166/aqua.2018.090

Eniola JO, Kumar R, Barakat MA, Rashid J (2022) A review on conventional and advanced hybrid technologies for pharmaceutical wastewater treatment. J Clean Prod 356:131826. https://doi.org/10.1016/j.jclepro.2022.131826

Estevam BR, Pinto LFR, Filho RM, Fregolente LV (2022a) Growth and metabolite production in chlorella sp.: Analysis of cultivation system and nutrient reduction. BioEnergy Res. https://doi.org/10.1007/s12155-022-10532-z

Estevam BR, Ríos Pinto LF, Filho RM, Fregolente LV (2022b) Potential applications of *Botryococcus terribilis*: A review. Biomass Bioenergy 165:106582. https://doi.org/10.1016/J.BIOMBIOE.2022.106582

Ferreira JA, Varjani S, Taherzadeh MJ (2020) A critical review on the ubiquitous role of filamentous fungi in pollution mitigation. Curr Pollut Reports 6:295–309. https://doi.org/10.1007/s40726-020-00156-2

Gerek EE, Yılmaz S, Koparal AS, Gerek ÖN (2019) Combined energy and removal efficiency of electrochemical wastewater treatment for leather industry. J Water Process Eng 30:100382. https://doi.org/10.1016/j.jwpe.2017.03.007

Giammar DE, Greene DM, Mishrra A, Rao N, Sperling JB, Talmadge M, Miara A, Sitterley KA, Wilson A, Akar S, Kurup P, Stokes-Draut JR, Coughlin K (2022) Cost and energy metrics for municipal water reuse. ACS ES&T Eng 2:489–507. https://doi.org/10.1021/acsestengg.1c00351

Giwa A, Dindi A, Kujawa J (2019) Membrane bioreactors and electrochemical processes for treatment of wastewaters containing heavy metal ions, organics, micropollutants and dyes: Recent developments. J Hazard Mater 370:172–195. https://doi.org/10.1016/j.jhazmat.2018.06.025

Giwa A, Hasan SW (2015) Numerical modeling of an electrically enhanced membrane bioreactor (MBER) treating medium-strength wastewater. J Environ Manage 164:1–9. https://doi.org/10.1016/j.jenvman.2015.08.031

Grandclément C, Seyssiecq I, Piram A, Wong-Wah-Chung P, Vanot G, Tiliacos N, Roche N, Doumenq P (2017) From the conventional biological wastewater treatment to hybrid processes, the evaluation of organic micropollutant removal: A review. Water Res 111:297–317. https://doi.org/10.1016/j.watres.2017.01.005

Griffiths M, Harrison STL, Smit M, Maharajh D (2016) Major commercial products from micro- and macroalgae. In: F Bux, Y Chisti (Eds) Algae Biotechnology. Green Energy and Technology; Springer, pp. 269–300.

Gutiérrez M, Grillini V, Mutavdžić Pavlović D, Verlicchi P (2021) Activated carbon coupled with advanced biological wastewater treatment: A review of the enhancement in micropollutant removal. Sci Total Environ 790:148050. https://doi.org/10.1016/j.scitotenv.2021.148050

Haddaji C, Ennaciri K, Driouich A, Digua K, Souabi S (2022) Optimization of the coagulation-flocculation process for vegetable oil refinery wastewater using a full factorial design. Process Saf Environ Prot 160:803–816. https://doi.org/10.1016/j.psep.2022.02.068

Hassan SSM, Abdel-Shafy HI, Mansour MSM (2018) Removal of pyrene and benzo(a)pyrene micropollutant from water via adsorption by green synthesized iron oxide nanoparticles. Adv Nat Sci Nanosci Nanotechnol 9:015006. https://doi.org/10.1088/2043-6254/aaa6f0

Ho L, Goethals PLM (2020) Municipal wastewater treatment with pond technology: Historical review and future outlook. Ecol Eng 148:105791. https://doi.org/10.1016/j.ecoleng.2020.105791

Jiang Q, Ngo HH, Nghiem LD, Hai FI, Price WE, Zhang J, Liang S, Deng L, Guo W (2018) Effect of hydraulic retention time on the performance of a hybrid moving bed biofilm reactor-membrane bioreactor system for micropollutants removal from municipal wastewater. Bioresour Technol 247:1228–1232. https://doi.org/10.1016/j.biortech.2017.09.114

Lima ÉC, Adebayo MA, Machado FM (2015) Kinetic and equilibrium models of adsorption. In: Carbon Nanostructures. Springer International Publishing, pp. 33–69.

Li M, Zhang H, Lemckert C, Roiko A, Stratton H (2018) On the hydrodynamics and treatment efficiency of waste stabilisation ponds: From a literature review to a strategic evaluation framework. J Clean Prod 183:495–514. https://doi.org/10.1016/j.jclepro.2018.01.199

Lin W, Liu X, Ding A, Ngo HH, Zhang R, Nan J, Ma J, Li G (2022) Advanced oxidation processes (AOPs)-based sludge conditioning for enhanced sludge dewatering and micropollutants removal: A critical review. J Water Process Eng 45:102468. https://doi.org/10.1016/j.jwpe.2021.102468

Linares R, Li ZV, Yangali-Quintanilla V, Ghaffour N, Amy G, Leiknes T, Vrouwenvelder JS (2016) Life cycle cost of a hybrid forward osmosis – Low pressure reverse osmosis system for seawater desalination and wastewater recovery. Water Res 88:225–234. https://doi.org/10.1016/j.watres.2015.10.017

Madadi R, Bester K (2021) Fungi and biochar applications in bioremediation of organic micropollutants from aquatic media. Mar Pollut Bull 166:112247. https://doi.org/10.1016/j.marpolbul.2021.112247

Manai I, Miladi B, El Mselmi A, Hamdi M, Bouallagui H (2017) Improvement of activated sludge resistance to shock loading by fungal enzyme addition during textile wastewater treatment. Environ Technol 38:880–890. https://doi.org/10.1080/09593330.2016.1214623

Martínez-Alcalá I, Guillén-Navarro JM, Fernández-López C (2017) Pharmaceutical biological degradation, sorption and mass balance determination in a conventional activated-sludge wastewater treatment plant from Murcia, Spain. Chem Eng J 316:332–340. https://doi.org/10.1016/j.cej.2017.01.048

Martins DS, Estevam BR, Perez ID, Américo-Pinheiro JHP, Isique WD, Boina RF (2022) Sludge from a water treatment plant as an adsorbent of endocrine disruptors. J Environ Chem Eng 108090. https://doi.org/10.1016/j.jece.2022.108090

Mojiri A, Zhou J, Vakili M, Van Le H (2020) Removal performance and optimisation of pharmaceutical micropollutants from synthetic domestic wastewater by hybrid treatment. J Contam Hydrol 235:103736. https://doi.org/10.1016/j.jconhyd.2020.103736

Montaño-Medina CU, Lopéz-Martínez LM, Ochoa-Terán A, López-Maldonado EA, Salazar-Gastelum MI, Trujillo-Navarrete B, Pérez-Sicairos S, Cornejo-Bravo JM (2023) New pyridyl and aniline-functionalized carbamoylcarboxylic acids for removal of metal ions from water by coagulation-flocculation process. Chem Eng J 451:138396. https://doi.org/10.1016/j.cej.2022.138396

Mora-Cabrera K, Peña-Guzmán C, Trapote A, Prats D (2021) Use of combined UASB + eMBR treatment for removal of emerging micropollutants and reduction of fouling. J Water Supply Res Technol 70:984–1001. https://doi.org/10.2166/aqua.2021.058

Morsi R, Bilal M, Iqbal HMN, Ashraf SS (2020) Laccases and peroxidases: The smart, greener and futuristic biocatalytic tools to mitigate recalcitrant emerging pollutants. Sci Total Environ 714:136572. https://doi.org/10.1016/j.scitotenv.2020.136572

Nguyen LN, Hai FI, Kang J, Price WE, Nghiem LD (2013) Removal of emerging trace organic contaminants by MBR-based hybrid treatment processes. Int Biodeterior Biodegradation 85:474–482. https://doi.org/10.1016/j.ibiod.2013.03.014

Paździor K, Bilińska L, Ledakowicz S (2019) A review of the existing and emerging technologies in the combination of AOPs and biological processes in industrial textile wastewater treatment. Chem Eng J 376:120597. https://doi.org/10.1016/j.cej.2018.12.057

Pisharody L, Gopinath A, Malhotra M, Nidheesh PV, Kumar MS (2022) Occurrence of organic micropollutants in municipal landfill leachate and its effective treatment by advanced oxidation processes. Chemosphere 287:132216. https://doi.org/10.1016/j.chemosphere.2021.132216

Prado M, Borea L, Cesaro A, Liu H, Naddeo V, Belgiorno V, Ballesteros F (2017) Removal of emerging contaminant and fouling control in membrane bioreactors by combined ozonation and sonolysis. Int Biodeterior Biodegradation 119:577–586. https://doi.org/10.1016/j.ibiod.2016.10.044

Pushkar B, Sevak P, Parab S, Nilkanth N (2021) Chromium pollution and its bioremediation mechanisms in bacteria: A review. J Environ Manage 287:112279. https://doi.org/10.1016/j.jenvman.2021.112279

Ramos BP, Perez ID, Aliprandini P, Boina RF (2022) Cu^{2+}, Cr^{3+}, and Ni^{2+} in mono- and multi-component aqueous solution adsorbed in passion fruit peels in natura and physicochemically modified: A comparative approach. Environ Sci Pollut Res. https://doi.org/10.1007/s11356-021-18132-8

Ramos BP, Perez ID, Paiano MS, Vieira MGA, Boina RF (2021a) Activated carbons from passion fruit shells in adsorption of multimetal wastewater. Environ Sci Pollut Res. https://doi.org/10.1007/s11356-021-15449-2

Ramos BP, Perez ID, Wessling M, Boina RF (2021b) Metal recovery from multi-elementary electroplating wastewater using passion fruit powder. J Sustain Metall. https://doi.org/10.1007/s40831-021-00398-4

Rani J, Pandey KP, Kushwaha J, Priyadarsini M, Dhoble AS (2022) Antibiotics in anaerobic digestion: Investigative studies on digester performance and microbial diversity. Bioresour Technol 361:127662. https://doi.org/10.1016/j.biortech.2022.127662

Rempel A, Gutkoski JP, Nazari MT, Biolchi GN, Cavanhi VAF, Treichel H, Colla LM (2021) Current advances in microalgae-based bioremediation and other technologies for emerging contaminants treatment. Sci Total Environ 772:144918. https://doi.org/10.1016/j.scitotenv.2020.144918

Roccuzzo S, Beckerman AP, Trögl J (2021) New perspectives on the bioremediation of endocrine disrupting compounds from wastewater using algae-, bacteria- and fungi-based technologies. Int J Environ Sci Technol 18:89–106. https://doi.org/10.1007/s13762-020-02691-3

Ryu B, Wong KT, Choong CE, Kim J-R, Kim H, Kim S-H, Jeon B-H, Yoon Y, Snyder SA, Jang M (2021) Degradation synergism between sonolysis and photocatalysis for organic pollutants with different hydrophobicity: A perspective of mechanism and application for high mineralization efficiency. J Hazard Mater 416:125787. https://doi.org/10.1016/j.jhazmat.2021.125787

Saidulu D, Gupta B, Gupta AK, Ghosal PS (2021) A review on occurrences, eco-toxic effects, and remediation of emerging contaminants from wastewater: Special emphasis on biological treatment based hybrid systems. J Environ Chem Eng 9:105282. https://doi.org/10.1016/j.jece.2021.105282

Saleh IA, Zouari N, Al-Ghouti MA (2020) Removal of pesticides from water and wastewater: Chemical, physical and biological treatment approaches. Environ Technol Innov 19:101026. https://doi.org/10.1016/j.eti.2020.101026

Samaei SM, Gato-Trinidad S, Altaee A (2018) The application of pressure-driven ceramic membrane technology for the treatment of industrial wastewaters – A review. Sep Purif Technol 200:198–220. https://doi.org/10.1016/j.seppur.2018.02.041

Shah MP (2020) Microbial Bioremediation & Biodegradation. Springer.

Shah MP (2021a) Removal of Emerging Contaminants Through Microbial Processes. Springer.

Shah MP (2021b) Removal of Refractory Pollutants from Wastewater Treatment Plants. CRC Press.

Singa PK, Isa MH, Lim J-W, Ho Y-C, Krishnan S (2021) Photo-fenton process for removal of polycyclic aromatic hydrocarbons from hazardous waste landfill leachate. Int J Environ Sci Technol 18:3515–3526. https://doi.org/10.1007/s13762-020-03010-6

Song X, Huang D, Zhang L, Wang H, Wang L, Bian Z (2020) Electrochemical degradation of the antibiotic chloramphenicol via the combined reduction-oxidation process with Cu-Ni/graphene cathode. Electrochim Acta 330:135187. https://doi.org/10.1016/j.electacta.2019.135187

Song K, Ren X, Zhang Q, Xu L, Liu D (2022) Electrochemical treatment for leachate membrane retentate: Performance comparison of electrochemical oxidation and electro-coagulation technology. Chemosphere 303:134986. https://doi.org/10.1016/j.chemosphere.2022.134986

Stazi V, Tomei MC (2018) Enhancing anaerobic treatment of domestic wastewater: State of the art, innovative technologies and future perspectives. Sci Total Environ 635:78–91. https://doi.org/10.1016/j.scitotenv.2018.04.071

Tiquia-Arashiro SM (2018) Lead absorption mechanisms in bacteria as strategies for lead bioremediation. Appl Microbiol Biotechnol 102:5437–5444. https://doi.org/10.1007/s00253-018-8969-6

Tran T-K, Chiu K-F, Lin C-Y, Leu H-J (2017) Electrochemical treatment of wastewater: Selectivity of the heavy metals removal process. Int J Hydrogen Energy 42:27741–27748. https://doi.org/10.1016/j.ijhydene.2017.05.156

Tu J-W, Li T, Gao Z-H, Xiong J, Miao W (2022) Construction of CdS-tetrahymena thermophila hybrid system by efficient cadmium adsorption for dye removal under light irradiation. J Hazard Mater 439:129683. https://doi.org/10.1016/j.jhazmat.2022.129683

Vassalle L, García-Galán MJ, Aquino SF, Afonso RJCF, Ferrer I, Passos F, R Mota C (2020) Can high rate algal ponds be used as post-treatment of UASB reactors to remove micropollutants? Chemosphere 248:125969. https://doi.org/10.1016/j.chemosphere.2020.125969

Vieira WT, de Farias MB, Spaolonzi MP, da Silva MGC, Vieira MGA (2020) Removal of endocrine disruptors in waters by adsorption, membrane filtration and biodegradation. A review. Environ Chem Lett 18:1113–1143. https://doi.org/10.1007/s10311-020-01000-1

Vieira WT, de Farias MB, Spaolonzi MP, da Silva MGC, Vieira MGA (2021) Latest advanced oxidative processes applied for the removal of endocrine disruptors from aqueous media – A critical report. J Environ Chem Eng 9:105748. https://doi.org/10.1016/j.jece.2021.105748

Zhang H, Lin H, Li Q, Cheng C, Shen H, Zhang Z, Zhang Z, Wang H (2021) Removal of refractory organics in wastewater by coagulation/flocculation with green chlorine-free coagulants. Sci Total Environ 787:147654. https://doi.org/10.1016/j.scitotenv.2021.147654

Zonato RO, Estevam BR, Perez ID, Aparecida dos Santos Ribeiro V, Boina RF (2022) Eggshell as an adsorbent for removing dyes and metallic ions in aqueous solutions. Clean Chem Eng 100023. https://doi.org/10.1016/J.CLCE.2022.100023

7 Application of Activated Carbon and Its Modifications in Wastewater Treatment

Ketiyala Ao and Latonglila Jamir

7.1 INTRODUCTION

The quality of water is deteriorating exponentially due to various contaminants. Both point and non-point sources are polluting the water resources as a result of tremendous population growth, modern industrialization, civilization, domestic and agricultural activities, and other geological, environmental, and global changes. Hence, water treatment and recycling of wastewater are the best approaches to get safe water for routine activities. One of the biggest issues in recent years has been the contamination of water resources with various unwanted substances and the subsequent remediation. Among several hazardous substances detected in various water resources, persistent endocrine disrupting chemicals (EDCs) have raised significant public and environmental concerns due to their toxic, mutagenic, and carcinogenic properties (Supong et al., 2019). The occurrence of new/emerging microcontaminants in polluted water/wastewater has rendered existing conventional treatment plants ineffective in meeting environmental standards. The release of these substances into the aquatic environment has had an impact on every living thing. The idea of using activated carbon for water purification has been long known. Activated carbons are highly microporous, with both high internal surface area and porosity, and are commercially the most common adsorbents used for the removal of organic compounds from air and water streams. Due to its simplicity of design, convenience of operation, and insensitivity to harmful compounds, it has been determined that its efficacy in removing pollutants through the adsorption concept is superior to many other ways (Salihi et al., 2018). Adsorption has a wide range of applications and is simple to use, making it the best wastewater treatment method. It is also regarded as a universal method for water treatment and restoration since it can be used to remove biological, inorganic, and organic pollutants with an efficiency of up to 99%.

Traditional materials and treatment methods, such as oxidation, activated sludge, nanofiltration (NF), and reverse osmosis (RO) membranes, are ineffective for treating complex and intricately contaminated waters that contain pharmaceuticals, personal care products, surfactants, different industrial additives, and chemicals. The removal of a broad range of harmful compounds and pathogenic microorganisms from raw water cannot be satisfactorily addressed by standard water treatment procedures.

DOI: 10.1201/9781003381327-7

It has been reported that bacteria preferably adhere to the solid support made up of carbon material, or they may also grow on the activated carbon during the purification process. Numerous studies have been done using various types of silver-coated materials to evaluate their antibacterial effectiveness, as silver is well known for its antimicrobial qualities. Various chlorine-based disinfection techniques currently in use effectively eliminate microbial pathogens but could also produce carcinogenic disinfection byproducts (DBPs) in the presence of organic matter from the environment, anthropogenic contaminants, bromide, and iodide, which remain available in the water, making them unsuitable. Henceforth, recently, it was shown that the biological activated carbon (BAC) treatment, which simultaneously involves biological breakdown and adsorption, has an advantage over the conventional activated carbon treatment for the removal of organic pollutants in achieving this goal (Suzuki et al., 1995; Takeuchi et al., 1995; Mochidzuki, 1995; El-Aassar et al., 2013). As a result, this analysis offers relevant data on the potential difficulties with such adsorption methods and the feasibility of various techniques acting as significant bactericides.

7.2 APPLICATIONS OF ACTIVATED CARBON

7.2.1 Synthesis of Activated Carbon and Its Removal Efficiency

The characteristics of the precursor, particle size (granules or powder), and production method affect the adsorption capacity of activated carbon. The physical adsorption and superficial area are used to assess its quality. A large portion of activated carbon's efficacy in these fields can be attributed to the variety of chemical functions that can be altered both before and after its production process and its large surface area. The process of making activated carbon from biomass often begins with pre-treatment steps, including crushing, drying (at a temperature of about 100°C), and sieving, to obtain small particles of a particular size. Following this, biomass is carbonized in a dry, inert atmosphere at a temperature between 300 and 500°C, facilitating the removal of tars and volatile substances and resulting in the formation of biochar (Wong et al., 2018). Several researchers and experts have been studying wastewater treatment by utilizing numerous biowastes. These waste materials are biodegradable, with a good efficiency of producing activated carbon. Bandara et al. (2020) studied biochars produced in a commercial-scale mobile pyrolizer from feedstocks: poultry litter, lucerne shoot, vetch shoot, canola shoot, wheat straws, and sugar-gum wood. They were tested in a liquid-based system to examine the immobilization of Cd(II) and Cu(II) from aqueous solutions and contaminated mine water and were characterized by Fourier transform electron infrared spectroscopy (FTIR), X-ray photoelectron spectroscopy, and X-ray diffraction (XRD) before and after the mine water treatment, which provided an efficient removal of heavy metals from aqueous solutions. Likewise, biowastes such as Brazil nutshell (Lima et al., 2019) utilized for the removal of acetaminophen and synthetic hospital effluents showed good removal efficiency; cherry kernels were studied by Pap et al. (2016) for the removal of Pb^{2+}, Cd^{2+}, and Ni^{2+} from aqueous systems; human hair to remove hexavalent chromium from aqueous solution was studied by Mondal and Basu (2019); and granular activated carbon (GAC) from animal horns was compared with the available commercial GAC for the removal of Zn (II) ions by Aluyor and Oboh (2013). Simply put, the use of activated carbon as a removal agent has had a great impact on social and economic fields. But over the course of time, several experts with much research have found that activated carbon, when applied in water filtration applications, shows an issue with bacterial contamination.

7.2.2 Chlorine as Purifying Agent

In addition to the efficient use of activated carbon, people have also been using chlorine as one of the major disinfectants. It has significantly improved public health by reducing the spread of infectious diseases. Chlorination is regarded as one of the most ubiquitous and economical types of disinfection used globally, which helps to control water-borne diseases. Chlorine reduces the quantity of the organic and other oxidizable substances contained in the water, or even up to the extent of removing them. The most widely used disinfectants for treating water are chlorine and synthetic derivatives, which are popular as they have a higher oxidizing potential, offer a minimal level of chlorine residual throughout the distribution system, and prevent microbiological recontamination (Sadiq and Rodriguez, 2004; Gopal et al., 2007). But chlorine, despite its high impact as a disinfectant, with continuous usage and due course of time, resulted in deterioration of the water quality, especially when used for drinking purposes. According to Gopal et al. (2007), chlorine persists in the water for as long as it is not consumed; disinfection via chlorination is the most crucial stage in water treatment for public supply. But chlorine also interacts with the natural organic matter (NOM) in the water, creating a multitude of byproducts that have detrimental long-term impacts. Chlorination disinfection byproducts such as trihalomethanes and haloacetic acids (Mazhar et al., 2020) can be found when chlorine comes in contact with the NOM in raw water, resulting in a substratum for bacterial growth.

7.2.3 Contaminant Removal with Biowaste-Derived Activated Carbon

The major characteristics of activated carbon having increased removal effectiveness of pollutants from wastewater include a large volume, micropores, and mesopores, as well as a high surface area. Hence, a wide number of studies have been made on activated carbon as an adsorbent for various pollutants such as pharmaceutical wastes, heavy metals, dyes, and organic pollutants.

7.2.4 Pharmaceutical Wastes

Pharmaceuticals in wastewater are an increasing problem since these compounds are not biodegradable. They are recognized as emerging contaminants as there is insufficient evidence that links these compounds to animal and human health after long-term exposure. About 30–90% of antibiotic doses remain undigested in the human or animal body, eventually excreting as an active molecule (Darweesh and Ahmed, 2017b). Therefore, the practice of using and disposing of pharmaceutical waste jeopardizes environmental safety while also having potentially serious effects on human health (Sasu et al., 2011). Many drugs, such as paracetamol, tramadol hydrochloride, diazepam, and acetaminophen, have used activated charcoal as an adsorbent, according to Hassen et al. (2018). Activated carbon from *Phoenix dactylifera* L. stones was analyzed by Darweesh and Ahmed (2017b) for the attraction of pollutants like ciprofloxacin and norfloxacin, which revealed an adsorption capacity of 2.094 and 1.992 mg/g, respectively. Similarly, Thue et al. (2017) successfully studied activated carbon from Sapelli sawdust for the adsorption of bisphenol A (adsorption capacity of 334.28 mg/g), paracetamol (226.71), caffeine (256.29), 2-naphtol (329.00), 2-nitrophenol (294.10), 4-nitrophenol (272.28), resorcinol (150.87), and hydroquinone (254.36). The GAC made from date (*P. dactylifera* L.) stones through microwave KOH activation was used for the adsorption of levofloxacin, which resulted in an adsorption capacity of 100.3 mg/g when compared to synthetic zeolites, biochars, and clays (Darweesh and Ahmed, 2017a). Close monitoring of potential contaminants such as naproxen, diclofenac, and ibuprofen that are detrimental to human health can uplift the status, considering the presence of various contaminants in complex mixtures with varied physicochemical characteristics, coupled with their varying affinities to affect a variety of molecular and cellular pathways.

7.2.5 Heavy Metals

The occurrence of heavy metals in the environment can be emitted from both natural and anthropogenic activities, where the major source is anthropogenic, like mining sites. The accumulation of heavy metal ions in the environment that exceeds the tolerable limit, leading to environmental imbalance, is achieved by their inability to break down into specific non-hazardous metabolites. The disposal of electronic waste in soil and water is also one of the major reasons for heavy metal pollution. Aluminum, copper, zinc, lead, nickel, chromium, and iron are the major heavy metals found in industrial wastewaters and e-wastes disposal areas (Debnath et al., 2016). Different methods have been applied for the removal of heavy metals, which include membrane filtration, chemical precipitation, electrochemical reduction, adsorption, ion exchange, coagulation/flocculation, and flotation (Carolin et al., 2017). But after much study, different experts have said that the most effective method for removing heavy metals from wastewaters is adsorption (Acharya et al., 2009; Cronje et al., 2011; Kumar and Jena, 2017). As such, Van Thuan et al. (2016) reported that activated carbon derived from banana peel can be used for the adsorption of Cu^{2+}, Ni^{2+}, and Pb^{2+}, where the maximum adsorption capacity resulted in Cu^{2+} (14.3 mg/g), Ni^{2+} (27.4 mg/g), and Pb^{2+} (34.5 mg/g), respectively. Chemical activation with $ZnCl_2$, the activated carbon made from fox nutshell, was used in the adsorption of Cr(VI) by Kumar and Jena (2017), showing a high removal efficiency of 99.08% of 10 mg/L concentration with an adsorption capacity of 43.45 mg/g. Likewise, camel bone-based charcoal was used as an adsorbent in removing Hg (II) from wastewater with a removal efficiency of 95.8–98.5%. It was also compared with various adsorbents previously used to remove Hg(II) from wastewater effluents and demonstrate the carbon's exceptional effectiveness over several other adsorbents (Hassan et al., 2008). Despite its effectiveness in removing pollutants from wastewater, commercial activated carbon is inadequate due to its limited affinity for heavy metals (Sigdel et al.,

2017; Shah, 2021b). Hence, the development of adsorbents from biomass is vital to reduce the introduction of heavy metals into waste bodies.

7.2.6 Dyes

With the ever-increasing population, the demand for the clothing and textile industries has hit its peak. Hence, coloring agents and dyes have been in continuous use, and their effects on the environment have risen. Water is used extensively in the textile industry throughout the supply chain, from fibers to finished products. It has been slammed for polluting water with its wastes, in addition to absorbing a considerable part of the available freshwater (Kabir et al., 2019). Several techniques, such as aerobic and anaerobic microbial degradation, membrane separation, chemical oxidation, dilution, electrochemical treatment, adsorption, and other procedures, are used for the removal of color from water. However, the majority of these technologies have high operational and maintenance costs, low removal efficiency, a lack of selectivity, and other drawbacks (Akpen et al., 2011). Adsorption by activated carbon has been the best option for dye removal due to its high thermal stability, effective internal porous structure, low acid/base reactivity, chemical structure, and surface area. Studies are being carried out on the removal of dyes, particularly methylene blue (MB), which creates difficulties in the eyes, skin, and brain functions (Ardekani et al., 2017; Shah, 2021a). Anionic acid dyes (acid red 1 (AR1), acid blue 45 (AB45), and acid yellow 127 (AY127)) were successfully removed from the fish scales (FS)-derived activated carbon by Kabir et al. (2019), where after different physiochemical analyses, the adsorption capacity was found to be 1.8, 2.7, and 3.4 mg/g, and the removal efficiencies as AR1 (63.5%), AB45 (89.3%), and AY127 (93%), respectively. $ZnCl_2$ activation of acorn shell was used for the adsorption of MB from aqueous solution and found the adsorption capacity to be 330.0 mg/g for AC and 357.1 mg/g in the case of Fe-AC adsorbents (Altıntıg et al., 2017).

7.2.7 Organic Pollutants

High concentrations of organic and inorganic salts, organic compounds, and heavy metals are commonly found in landfill leachate (Azmi et al., 2014). Increasingly affluent lifestyles, as well as continued industrial and commercial growth, have been accompanied by a substantial increase in both municipal and industrial solid waste disposal in the last decade (Renou et al., 2008). Annually, tonnes of nitro-aromatic explosives have been in use for more than 50 years for military and civilian purposes, with the most common being 2,4,6-trinitrotoluene (TNT) (Xu et al., 2017; Shah, 2020). Persistent organic pollutants are mainly classified into three groups, viz. pesticides, especially organochlorine pesticides (OCPs) such as dichlorodiphenyltrichloroethane (DDT) and its metabolites; secondly is the industrial and technical compounds such as perfluorooctanesulfonate (PFOS), polychlorinated biphenyls (PCBs), and polybrominated diphenyl ethers; and third form of POPs are polyaromatic hydrocarbons (PAHs), polychlorinated dibenzofurans (PCDFs), and polychlorinated dibenzo-p-dioxins (PCDDs), which are the byproducts of industrial processes (Guo et al., 2019; Aziz et al., 2021). Varied methods and techniques have come up for the adsorption of such pollutants, in which action with activated carbon is the most efficient.

Recent studies in support of the said statement have provided evidence in different organic compounds. COD, color, and NH_3-N from landfill leachate were successfully removed by sugarcane bagasse-based activated carbon. And after many studies and analyses, the adsorption capacity for COD was 126.58 mg/L, color was 555.56 Pt-Co/g, and NH_3-N was 14.62 mg/g, with a good removal efficiency of 83.61% for COD, 94.74% for color, and 46.65% for NH_3-N, respectively (Azmi et al., 2014). DDT, despite being one of the most popular pesticides for crops, has eventually affected the soil quality as well as human health, targeting the immune system, endocrine disorder, nervous system dysfunction, reproductive system, etc. As a result, several studies have proved that activated carbon is extremely versatile in removing such compounds. The microwave irradiation technique with CO_2 gasification successfully synthesized coconut-shell-based AC (CSAC) for the adsorption of DDT. With the help of different analyses and isotherm studies, the pollutant adsorption capacity was found to be 14.51 mg/g with an 84.83% removal efficiency (Aziz et al., 2021).

TABLE 7.1
Adsorption of Pollutants on Adsorbents Derived From Different Biowastes

Adsorbent	Adsorbate	Adsorption Capacity, Qe (mg/g)	Reference
Date (*Phoenix dactylifera* L.) stones (500°C/ K_2CO_3/microwave)	Ciprofloxacin and norfloxacin	2.094 and 1.992	Darweesh and Ahmed (2017b)
Sawdust ($ZnCl_2$/ microwave)	Bisphenol A, paracetamol, caffeine, 2-naphtol, 2-nitrophenol, 4-nitrophenol, resorcinol, and hydroquinone	334.28, 226.71, 256.29, 329.00, 294.10, 272.28, 150.87, and 254.36	Thue et al. (2017)
Date (*P. dactylifera* L.) stones (microwave KOH activation)	Levofloxacin	100.3	Darweesh and Ahmed (2017a)
Waste tea residue (H_3PO_4/450°C) (stream, air, N_2)	Oxytetracycline	273.7, 242.6, 175.0	Kan et al. (2017)
Chicken manure (600°C)	Phenol (Ph) and 2,4-dinitrophenol	106.2 and 148.1	Thang et al. (2019)
Tithonia diversifolia (700°C/KOH)	Bisphenol A	15.69	Supong et al. (2019)
Banana peel (500°C/ KOH)	Cu^{2+}, Ni^{2+}, and Pb^{2+}	14.3, 27.4, and 34.5	Van Thuan et al. (2016)
Fox nutshell (600°C/ $ZnCl_2$)	Cr(VI)	43.45	Kumar and Jena (2017)
Fish scales (*Oreochromis mossambicus*)	Lead	344.8	Liu et al. (2017)

(Continued)

TABLE 7.1 ***(Continued)***
Adsorption of Pollutants on Adsorbents Derived From Different Biowastes

Adsorbent	Adsorbate	Adsorption Capacity, Qe (mg/g)	Reference
Camel bone	Hg(II)	28.24	Hassan et al. (2008)
Tamarind wood (100°C/ $ZnCl_2$)	Lead(II)	43.85	Acharya et al. (2009)
Fish scales	Zinc	13	Ribeiro et al. (2018)
Fish scales (*Rutilus kutum*) (80°C)	Cu^{2+}	FS (61.73), MFS (103.1)	Ahmadifar and Dadvand Koohi (2018)
Fish scales (600°C/ NaOH)	Methylene blue (MB)	184.40	Marrakchi et al. (2017)
Fish scales (600°C/N_2)	Acid red 1 (AR1), acid blue 45 (AB45), and acid yellow 127 (AY127)	1.8, 2.7, and 3.4	Kabir et al. (2019)
Acorn shell (700°C/ $ZnCl_2$)	Methylene blue (MB)	330.0 (AC) and 357.1 (Fe-AC)	Altıntıg et al. (2017)
Fish (*Labeo rohita*) scales (343 K)	Methylene blue (MB) and crystal violet (CV)	58.67 (MB) and 74.39 (CV)	Chakraborty et al. (2012)
Sugarcane bagasse (700°C/KOH/600°C)	COD, color, and NH_3-N from landfill	126.58, 555.56 (Pt-Co/g), and 14.62	Azmi et al. (2014)
Coconut shell (microwave, CO_2)	DDT	14.51	Aziz et al. (2021)
Reed canary grass, flax, hemp fiber	Oily wastewater	3300	Pasila (2004)

7.2.8 Equilibrium Isotherm and Adsorption Kinetics

Measurement of adsorption isotherms is generally done with a given quantity of adsorbate and varied dosages of the adsorbent. The mixture is then stirred at a fixed temperature until it reaches equilibrium, after which the adsorbate concentration in the aqueous phase is monitored and the adsorption capacity at equilibrium is estimated. To describe this relationship, several isotherm studies like the Langmuir isotherm, the Freundlich isotherm, Temkin, Dubinin–Radushkevich, Sips, Toth, Radke–Prausnitz, and Vieth–Sladek were initiated. Likewise, kinetic studies in adsorption are crucial for understanding the operation and its mechanism. Such analysis can provide the solute uptake rate that defines the retention time needed for the completion of adsorption. The adsorbent's physical or chemical properties determine the nature of the adsorption process (Kongsri et al., 2013). Different kinetic models have been studied for wastewater treatment such as pseudo first order and second order, Elovich, and intraparticle diffusion models.

The adsorption rate of selenite was modeled using pseudo-second-order kinetic equations when fish (*Tilapia nilotica*) scale waste was activated using the alkaline heat treatment method. Freundlich isotherm ($R^2 > 0.91$) was also recorded as the best-fit isotherm study (Kongsri et al., 2013). In a batch system, the viability of using FN as a new biosorbent for removing textile dyes, specifically MB and crystal violet (CV), was examined in which Langmuir isotherm model showed the best-fit adsorption isotherm while pseudo second order gave the better result when compared with pseudo first order (Chakraborty et al., 2012). *Oreochromis niloticus* FS were used as biosorbents for the removal of zinc, where the Langmuir–Freundlich isotherm best fitted to characterize the biosorption equilibrium data, and the parameter ($1/n > 1$) verified the heterogeneity hypothesis in contrast to cooperative sorption. The Elovich model resulted in the best fit in the kinetic study when compared with pseudo first order, second order, and Elovich models (Ribeiro et al., 2018). Adsorption of heavy metals (Zn, Fe) by activated carbon made from Mozambique tilapia (*M. tilapia*) FS was studied by Othman et al., (2016). The good biosorbent efficiency of the carbon material is supported by the Langmuir isotherm of $R^2 = 0.997$ (Zn) and R^2 0.990 (Fe). Pseudo second order with $R^2 = 0.999$ (Zn) and $R^2 = 0.998$ (Fe) also exhibited the best-fit kinetic study, hence showing it as a promising adsorbent for removing heavy metals from wastewater. Removal of Bisphenol A was done by *Tithonia diversifolia* biomass, in which different characterization and ANOVA analyses were studied. Besides, isotherm studies such as Langmuir, Freundlich, Temkin, and Dubinin–Radushkevich were employed, in which the Langmuir isotherm (R^2 0.9967) resulted in the best fit. Also, four kinetic models such as pseudo first order, second order, Elovich, and intraparticle diffusion were used to study the kinetics and found that the adsorption best fit pseudo second order (Supong et al., 2019). Some of the best-fit isotherm and kinetic models are presented in Table 7.2.

TABLE 7.2
Isotherm and Kinetic Modeling

Adsorbate	Best Fit Isotherm Model	Best Fit Kinetic Model	Reference
Selenite	Freundlich isotherm ($R^2 > 0.91$)	Pseudo second order	Kongsri et al. (2013)
Methylene blue (MB) and crystal violet (CV)	Langmuir isotherm	Pseudo second order	Chakraborty et al. (2012)
Methylene blue (MB)	Langmuir isotherm	Pseudo second order and intraparticle diffusion models	Ardekani et al. (2017)
Crystal violet	Langmuir isotherm	Pseudo second order	Ahmad (2009)
Methylene blue	Langmuir and Freundlich isotherm	Pseudo first order	Sharma and Uma (2009)
Basic blue 3 and Reactive orange 16	Langmuir and Freundlich (BB3 and RO16)	Pseudo second order	Ong et al. (2007)

(Continued)

TABLE 7.2 *(Continued)*
Isotherm and Kinetic Modeling

Adsorbate	Best Fit Isotherm Model	Best Fit Kinetic Model	Reference
Cu(II)	Langmuir isotherm	Pseudo second order	Jaman et al. (2009) and Ahmadifar and Dadvand Koohi (2018)
Lead(II)	Langmuir and Freundlich isotherm	Pseudo second order	Acharya et al. (2009)
Zinc	Langmuir-Freundlich isotherm	Elovich model	Ribeiro et al. (2018)
Zn, Fe	Langmuir isotherm	Pseudo second order	Othman et al. (2016)
Cd(II)	Langmuir isotherm	Pseudo second order	Kumar and Bandyopadhyay (2006)
Bisphenol A	Langmuir isotherm	Pseudo second order	Supong et al. (2019)
DDT	Langmuir isotherm	Pseudo second order	Aziz et al. (2021)
COD, color, and NH_3-N from landfill	Langmuir isotherm	Pseudo second order	Azmi et al. (2014)
COD	Freundlich isotherm	Pseudo second order	Xu et al. (2017)
Ibuprofen, ketoprofen, naproxen, and diclofenac	Langmuir isotherm	Pseudo second order	Baccar et al. (2012)
2,4-Dichlorophenol	Freundlich and Langmuir Dubinin–Radushkevich	Pseudo first order	Akhtar et al. (2006)

7.2.9 Modifications of Activated Carbon

Activated carbon, despite being regarded as one of the most vital techniques for the removal of numerous pollutants, has been reported to have good biocompatibility for bacterial growth. Henceforth, new approaches have been studied in which activated carbon substrates derived from various biowastes were subjected to certain chemical modifications for which further in-depth studies were enunciated. In compliance with this objective, the recently found BAC treatment (Suzuki et al., 1995; Takeuchi et al., 1995; Mochidzuki, 1995; El-Aassaret al., 2013) has paved the way for microbial reduction. As it prolongs the life of carbon, it has garnered a lot of attention as one of the most effective water treatment techniques (Suzuki et al., 1996; El-Aassar et al., 2013). Bioadsorption can lead to the development of bacteria colonies on the AC due to (i) the adsorptive properties of carbon, which enhance the concentration of nutrients and oxygen while also removing disinfectant compounds; (ii) the porous structure of the carbon particles, providing a protective environment; and (iii) the presence of a wide range of functional groups present on the surface, which improves adhesion (Rivera-Utrilla et al., 2003; Camper et al., 1986; Stewart et al., 1990; Yin et al., 2007). The authors classified the procedures based on thorough literature surveys into three major divisions, such as change of chemical, physical, and biological features, which are further divided into their relevant treatment techniques (Yin et al., 2007).

7.2.10 Approaches for Manufacturing Biosorbents

For safe drinking water, several techniques/methods have been initiated. As such, a sand-supported column (Pongener et al., 2017), inhibition zone, impregnation, and column techniques (El-Aassar et al., 2013), plate counting method (Tuan et al., 2011; Zhao et al., 2013; Bartram, 2013; Altintig et al., 2016; Martí et al., 2018; Joshi et al., 2022), hydrothermal technique and green synthesis technique (Pandey et al., 2019), agar cup double-diffusion methods (Al-Gaashani et al., 2021), homemade water purifying apparatus (Vukčević et al., 2008), disk diffusion method (Altintig et al., 2022), etc. are some of the most efficient methods applied recently.

7.2.11 Morphological Studies

Various analytical studies, such as XRD, scanning electron microscope (SEM), and energy dispersive X-ray, have been undertaken to understand the morphological characteristics of biowastes used for the microbial study. For instance, the crystal characteristics of activated carbon derived from corncob were studied by Altintig et al. (2016) using XRD, FTIR, and surface morphology using a SEM. The specific surface area was characterized using Brunauer–Emmett–Teller theory (BET) and was determined to be 970.38 m^2/g at 700°C (Altintig et al., 2016). Likewise, a mild hump ranging from 20 to 25 indicating the amorphous structure of AC by Pandey et al. (2019) was shown through XRD, while FE-SEM images and EDX (Fig. 7.1) analysis confirm the formation of composite particles (Pandey et al., 2019). Al-Gaashani et al. (2021) reported that both amorphous and crystalline phases can be seen on AC when doped with 10% Al_2O_3 and 5 wt% Ag on XRD analysis (Fig. 7.2), SEM, energy dispersive X-ray spectroscopy (EDS), and thermal gravimetric analysis (TGA), which were also used to characterize the required material (Al-Gaashani et al., 2021). The composition and crystallinity of silver deposits were examined by XRD of a silver-doped carbon monolith, while the specific surface area was examined by N_2 adsorption using the BET method (Vukčević et al., 2008). SEM micrographs of activated carbon derived from *Mucuna pruriens* and

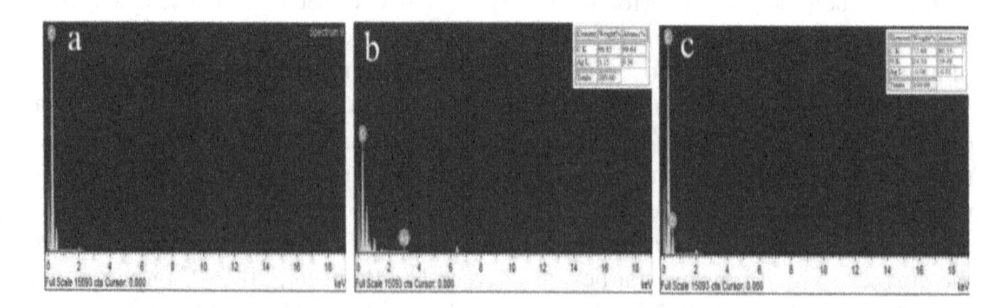

FIGURE 7.1 EDX mapping of pristine AC (A), Ag/AC composite prepared via hydrothermal process (B), and Ag/AC composite prepared via green synthesis (C) (Pandey et al., 2019).

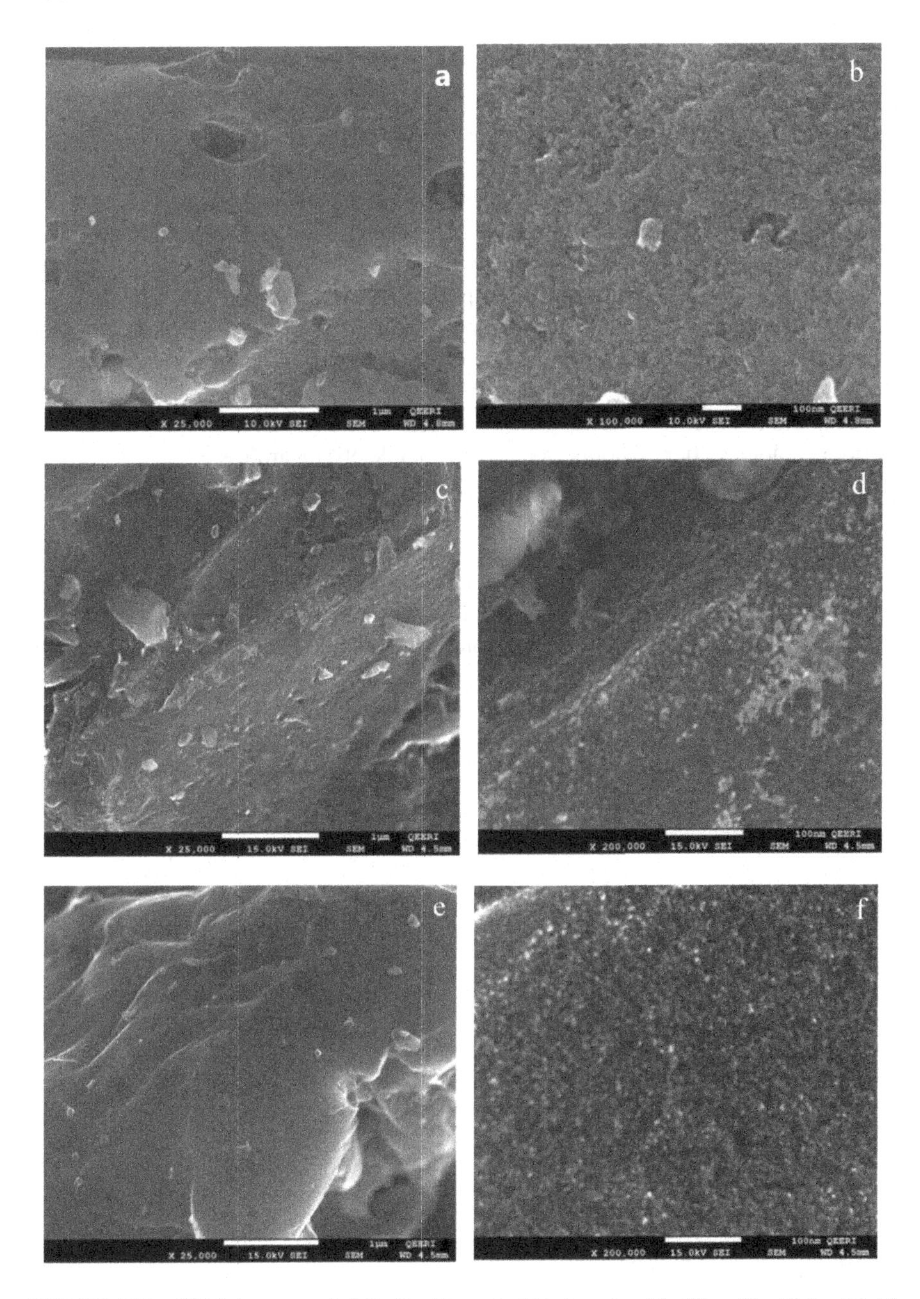

FIGURE 7.2 SEM images of AC (A, B) and AC doped with 10 wt% Al_2O_3 (C–F) (Al-Gaashani et al., 2021).

Manihot esculenta plants provide a vivid picture of an adsorbent's porosity as well as complex disordered surface structures with varying open pore sizes, shapes, and diameters. The SEM micrographs were obtained at various magnifications between 1200 and 2000 resolutions (Pongener et al., 2017).

7.2.12 Antimicrobial Mechanism and Its Removal Efficiency

Activated carbons have been shown to be effective at removing pollutants from aqueous solutions such as heavy metals, dyes, pharmaceutical and personal care products (PPCPs), and organic pollutants. As the recent studies suggest, activated carbon derived from different biowastes doped with certain chemical agents proved to be very efficient in the removal of microbes. Altintig et al. (2016) studied that activated carbon derived from corncobs coated with Ag resulted in an effective adsorbent toward *Escherichia coli.* In 40 mL of a phosphate buffer solution for bacterial culture, 0.05 g and 0.1 g of activated carbon-supported silver (AC/Ag) powders were blended. It was then incubated anaerobically at 37°C for 24 hours. Then, 0.5 mL of the aforementioned slurry was grown on an agar plate and then incubated at 37°C for 24 hours to observe bacterial growth. As shown in Fig. 7.3, gram-negative *E. coli* bacterial colonies after 24 hours were killed completely, regardless of the 0.1 g of AC/Ag powder used (Altintig et al., 2016). Commercial activated carbon-coated silver nanoparticles were studied by Pandey et al. (2019) for the adsorption of *E. coli* and *Staphylococcus aureus.* To achieve the desired bacterial density of 10^7 CFU/mL, the standardized suspension was diluted, and 0.4 g/L of various composites was then added to the diluted mixture while continuously being shaken for 120 minutes at room temperature. 1 mL of the suspension was taken at certain intervals and serially diluted in distilled water. Agar plates made with nutritional agar were used for bacterial counting, which was done after 24 hours of incubation at 37°C. By undergoing a one-pot hydrothermal process and green synthesis, both bacteria were efficiently removed. Similarly, activated carbon derived from *M. pruriens* and *M. esculenta* plants was used for the removal of coliform bacteria and *E. coli* Pongener et al. (2017)

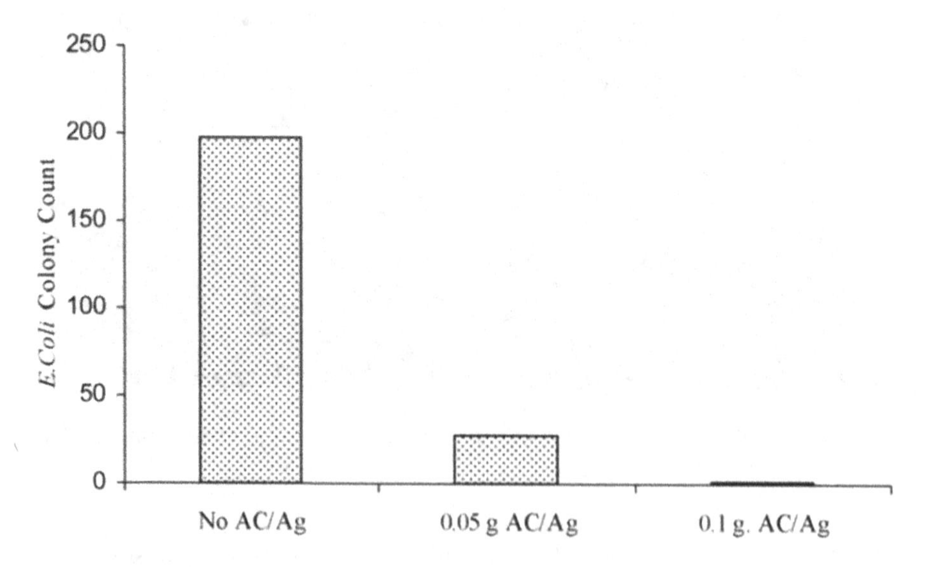

FIGURE 7.3 Antibacterial performance of non-activated corncob carbon (AC/Ag), with 0.05 g AC/Ag and 0.1 g AC/Ag (Altintig et al., 2016).

FIGURE 7.4 Setup of a sand-supported column (Pongener et al., 2017).

using a sand-supported column (Fig. 7.4). In a laminar flow setting, 1 mL of the sample obtained was placed on a petri dish, premade agar solution was gently poured over it, and was then incubated for 24 hours at 37°C, after which it was counted by a digital colony counter (Pongener et al., 2017). Al-Gaashani et al. (2021) investigated the antibacterial activity of the prepared samples against *E. coli* and *Bacillus subtilis* bacteria. The microorganisms were cultivated in nutrient agar overnight at 37°C. Bacterial inoculants had cell concentrations of 120–230 colony-forming units (CFU)/mL. The carbon-based components were administered to the petri dishes, which were then incubated for 24 hours at 37°C. The zones of inhibition were examined after incubation by measuring the diameter of the zone created around the materials, which resulted in a notable antibacterial adsorbent. Silver nanoparticles doped with active carbon spheres (ACS(Ag)) reported by Joshi et al. (2022) were studied for the removal of *E. coli*, *B. subtilis*, and *S. aureus*. 100 μL of an overnight culture (OD-0.6 nm) was inoculated for 12 hours to prepare 50 mL of nutrient broth. On the nutrient agar plate, 100 μL of freshly prepared bacterial culture (*E. coli* and *B. subtilis*) was disseminated. Following solidification, using a sterile cork borer, a 6 mm diameter was punched. And the well was loaded with 10 mg of ACS and varied levels of silver-doped ACS (2, 4, 6, and 8 mg), after which the plates were incubated for 12 hours at 37°C, and the diameter of the zone of inhibition was measured, resulting in a growth inhibition of 94% for *E. coli* and 93% for *B. subtilis*. While *S. aureus* was tested according to ASTM: E2149-01 method, biocidal activity of up to 73% was observed (Joshi et al., 2022). Hazelnut shells were coated with silver ions for the adsorption of two types of bacteria: gram-negative *E. coli* and gram-positive *S. aureus*. The antibacterial impact on the AC/Ag composite changes as the bacteriostatic time increases. Although the bacteriostatic period of the AC/Ag composite was 30 minutes, the antibacterial rate was over 90%, and the microbial effect showed 10^5 CFU/mL of *E. coli*, which was fully suppressed in 30 minutes (Altintig et al., 2022).

7.3 CONCLUSION

This paper outlines a few modification approaches that have been investigated in order to enhance the adsorption potential of activated carbon for various types of contaminants in water and wastewater. In addition to the good absorbance of pristine activated carbon, the study revealed that such carbon materials, when doped with certain modifications, can be more efficient at removing pollutants such as microbes. According to the literature review, the impacts of various modification strategies on AC adsorption of selected contaminant species from aqueous solutions were typically profound, while certain effects may be deleterious to chemical species uptake. Furthermore, there is very little information available regarding the use of these surface-modified activated carbons at column, pilot, or full size, and the majority of this research is only conducted at the lab scale (batch tests). For a larger range of applications, it is crucial to evaluate the efficiency of modified ACs in column, pilot, or full-scale conditions. Hence, in-depth studies and improving the current methodologies, keeping in mind not to overlook the final disposal of the spent carbon samples, can be achieved to a great extent.

REFERENCES

Acharya, J., J. N. Sahu, C. R. Mohanty, and B. C. Meikap. (2009). Removal of lead(II) from wastewater by activated carbon developed from Tamarind wood by zinc chloride activation. *Chemical Engineering Journal*, *149*(1–3), 249–262. https://doi.org/10.1016/j.cej.2008.10.029

Ahmad, R. (2009). Studies on adsorption of crystal violet dye from aqueous solution onto coniferous pinus bark powder (CPBP). *Journal of Hazardous Materials*, *171*(1–3), 767–773. https://doi.org/10.1016/j.jhazmat.2009.06.060

Ahmadifar, Z., and A. Dadvand Koohi. (2018). Characterization, preparation, and uses of nanomagnetic Fe_3O_4 impregnated onto fish scale as more efficient adsorbent for Cu^{2+} ion adsorption. *Environmental Science and Pollution Research*, *25*(20), 19687–19700. https://doi.org/10.1007/s11356-018-2058-3

Akhtar, M., M. I. Bhanger, S. Iqbal, and S. M. Hasany. (2006). Sorption potential of rice husk for the removal of 2,4-dichlorophenol from aqueous solutions: Kinetic and thermodynamic investigations. *Journal of Hazardous Materials*, *128*(1), 44–52. https://doi.org/10.1016/j.jhazmat.2005.07.025

Akpen, G. D., I. L. Nwaogazie, and T. G. Leton. (2011). Optimum conditions for the removal of colour from waste water by mango seed shell based activated carbon. *Indian Journal of Science and Technology*, *4*(8), 890–894. https://doi.org/10.17485/ijst/2011/v4i8.16

Al-Gaashani, R., D. Almasri, B. Shomar, and V. Kochkodan. (2021). Preparation and properties of novel activated carbon doped with aluminum oxide and silver for water treatment. *Journal of Alloys and Compounds*, *858*, 158372. https://doi.org/10.1016/j.jallcom.2020.158372

Altıntıg, E., H. Altundag, M. Tuzen, and A. Sarı. (2017). Effective removal of methylene blue from aqueous solutions using magnetic loaded activated carbon as novel adsorbent. *Chemical Engineering Research and Design*, *122*, 151–163. https://doi.org/10.1016/j.cherd.2017.03.035

Altintig, E., G. Arabaci, and H. Altundag. (2016). Preparation and characterization of the antibacterial efficiency of silver loaded activated carbon from corncobs. *Surface and Coatings Technology*, *304*, 63–67. https://doi.org/10.1016/j.surfcoat.2016.06.077

Altintig, E., B. Sarıcı, and S. Karataş. (2022). Prepared activated carbon from hazelnut shell where coated nanocomposite with Ag^+ used for antibacterial and adsorption properties. *Environmental Science and Pollution Research*, *30*(5), 13671–13687. https://doi.org/10.1007/s11356-022-23004-w

Aluyor, E. O., and I. O. Oboh. (2013). A comparative study of zinc (II) ions removal by a locally produced granular activated carbon. *Covenant Journal of Physical and Life Sciences 1* (1): 14–18.

Ardekani, P. S., H. Karimi, M. Ghaedi, A. Asfaram, and M. K. Purkait. (2017). Ultrasonic assisted removal of methylene blue on ultrasonically synthesized zinc hydroxide nanoparticles on activated carbon prepared from wood of cherry tree: Experimental design methodology and artificial neural network. *Journal of Molecular Liquids*, *229*, 114–124. https://doi.org/10.1016/j.molliq.2016.12.028

Aziz, A., M. N. Nasehir Khan, M. F. Mohamad Yusop, E. Mohd Johan Jaya, M. A. Tamar Jaya, and M. A. Ahmad (2021). Single-stage microwave-assisted coconut-shell-based activated carbon for removal of dichlorodiphenyltrichloroethane (DDT) from aqueous solution: Optimization and batch studies. *International Journal of Chemical Engineering*, *2021*, 1–15. https://doi.org/10.1155/2021/9331386

Azmi, N. Bt., M. J. K. Bashir, S. Sethupathi, and C. A. Ng. (2014). Anaerobic stabilized landfill leachate treatment using chemically activated sugarcane bagasse activated carbon: Kinetic and equilibrium study. *Desalination and Water Treatment*, *57*(9), 3916–3927. https://doi.org/10.1080/19443994.2014.988660

Baccar, R., M. Sarrà, J. Bouzid, M. Feki, and P. Blánquez. (2012). Removal of pharmaceutical compounds by activated carbon prepared from agricultural by-product. *Chemical Engineering Journal*, *211–212*, 310–317. https://doi.org/10.1016/j.cej.2012.09.099

Bandara, T., J. Xu, I. D. Potter, A. Franks, J. B. A. J. Chathurika, and C. Tang. (2020). Mechanisms for the removal of Cd(II) and Cu(II) from aqueous solution and mine water by biochars derived from agricultural wastes. *Chemosphere*, *254*, 126745. https://doi.org/10.1016/j.chemosphere.2020.126745

Bartram, J. (2013). Heterotrophic plate counts and drinking-water safety: The significance of HPCs for water quality and human health. *Water Intelligence Online*, *12*. https://doi.org/10.2166/9781780405940

Camper, A. K., M. W. Lechevallier, S. C. Broadaway, and G. A. McFeters. (1986). Bacteria associated with granular activated carbon particles in drinking water. *Applied and Environmental Microbiology*, *52*(3), 434–438. https://doi.org/10.1128/aem.52.3.434-438.1986

Carolin, C. F., P. S. Kumar, A. Saravanan, G. J. Joshiba, and Mu Naushad. (2017). Efficient techniques for the removal of toxic heavy metals from aquatic environment: A review. *Journal of Environmental Chemical Engineering*, *5*(3), 2782–2799. https://doi.org/10.1016/j.jece.2017.05.029

Chakraborty, S., S. Chowdhury, and P. D. Saha. (2012). Fish (*Labeo rohita*) scales as a new biosorbent for removal of textile dyes from aqueous solutions. *Journal of Water Reuse and Desalination*, 2(3), 175–184. https://doi.org/10.2166/wrd.2012.074

Cronje, K. J., K. Chetty, M. Carsky, J. N. Sahu, and B. C. Meikap. (2011). Optimization of chromium(VI) sorption potential using developed activated carbon from sugarcane bagasse with chemical activation by zinc chloride. *Desalination*, *275*(1-3), 276–284. https://doi.org/10.1016/j.desal.2011.03.019

Darweesh, T. M., and M. J. Ahmed. (2017a). Batch and fixed bed adsorption of levofloxacin on granular activated carbon from date (*Phoenix dactylifera* L.) stones by KOH chemical activation. *Environmental Toxicology and Pharmacology*, *50*, 159–166. https://doi.org/10.1016/j.etap.2017.02.005

Darweesh, T. M., and M. J. Ahmed. (2017b). Adsorption of ciprofloxacin and norfloxacin from aqueous solution onto granular activated carbon in fixed bed column. *Ecotoxicology and Environmental Safety*, *138*, 139–145. https://doi.org/10.1016/j.ecoenv.2016.12.032

Debnath, B., P. Roychowdhury, and R. Kundu. (2016). Electronic components (EC) reuse and recycling – A new approach towards WEEE management. *Procedia Environmental Sciences*, *35*, 656–668. https://doi.org/10.1016/j.proenv.2016.07.060

El-Aassar, A. H. M., M. M. Said, A. M. Abdel-Gawad, and H. A. Shawky. (2013). Using silver nanoparticles coated on activated carbon granules in columns for microbiological pollutants water disinfection in Abu Rawash area, Great Cairo, Egypt. *Australian Journal of Basic and Applied Sciences 7* (1): 422–432.

Gopal, K., S. S. Tripathy, J. L. Bersillon, and S. P. Dubey. (2007). Chlorination byproducts, their toxicodynamics and removal from drinking water. *Journal of Hazardous Materials*, *140*(1), 1–6. https://doi.org/10.1016/j.jhazmat.2006.10.063

Guo, W., B. Pan, S. Sakkiah, G. Yavas, W. Ge, W. Zou, W. Tong, and H. Hong. (2019). Persistent organic pollutants in food: Contamination sources, health effects and detection methods. *International Journal of Environmental Research and Public Health*, *16*(22). https://doi.org/10.3390/ijerph16224361

Hassan, S. S. M., N. S. Awwad, and A. H. A. Aboterika. (2008). Removal of mercury(II) from wastewater using camel bone charcoal. *Journal of Hazardous Materials*, *154*(1-3), 992–997. https://doi.org/10.1016/j.jhazmat.2007.11.003

Hassen, J. H., Y. M. Farhan, and A. H. Afyan. (2018). Comparative study of levocetrizine elimination by pristine and potassium permanganate modified activated charcoal. *International Journal of Pharamaceutical Sciences and Research*, *9*(12), 5155–5160. https://doi.org/10.13040/IJPSR.0975-8232.9(12).5155-60

Joshi, H. C., D. Dutta, N. Gaur, G. S. Singh, R. Dubey, and S. K. Dwivedi. (2022). Silver-doped active carbon spheres and their application for microbial decontamination of water. *Heliyon 8* (4): e09209.

Kabir, S. M. F., R. Cueto, S. Balamurugan, L. D. Romeo, J. T. Kuttruff, B. D. Marx, and I. I. Negulescu. (2019). Removal of acid dyes from textile wastewaters using fish scales by absorption process. *Clean Technologies*, *1*(1), 311–324. https://doi.org/10.3390/cleantechnol1010021

Kan, Y., Q. Yue, D. Li, Y. Wu, and B. Gao. (2017). Preparation and characterization of activated carbons from waste tea by H_3PO_4 activation in different atmospheres for oxytetracycline removal. *Journal of the Taiwan Institute of Chemical Engineers*, *71*, 494–500. https://doi.org/10.1016/j.jtice.2016.12.012

Kongsri, S., K. Janpradit, K. Buapa, S. Techawongstien, and S. Chanthai. (2013). Nanocrystalline hydroxyapatite from fish scale waste: Preparation, characterization and application for selenium adsorption in aqueous solution. *Chemical Engineering Journal*, *215-216*, 522–532. https://doi.org/10.1016/j.cej.2012.11.054

Kumar, U., and M. Bandyopadhyay. (2006). Sorption of cadmium from aqueous solution using pretreated rice husk. *Bioresource Technology*, *97*(1), 104–109. https://doi.org/10.1016/j.biortech.2005.02.027

Kumar, A., and H. M. Jena. (2017). Adsorption of Cr(VI) from aqueous phase by high surface area activated carbon prepared by chemical activation with $ZnCl_2$. *Process Safety and Environmental Protection*, *109*, 63–71. https://doi.org/10.1016/j.psep.2017.03.032

Lima, D. R., A. Hosseini-Bandegharaei, P. S. Thue, E. C. Lima, Y. R. T. de Albuquerque, G. S. dos Reis, C. S. Umpierres, S. L. P. Dias, and H. N. Tran. (2019). Efficient Acetaminophen Removal From Water and Hospital Effluents Treatment by Activated Carbons Derived From Brazil Nutshells. *Colloids and Surfaces A*, *583*(123966). https://doi.org/10.1016/j.colsurfa.2019.123966

Liu, W. -K., B. -S. Liaw, H. -K. Chang, Y. -F. Wang, and P. -Y Chen. (2017). From waste to health: Synthesis of hydroxyapatite scaffolds from fish scales for lead ion removal. *JOM*, *69*(4), 713–718. https://doi.org/10.1007/s11837-017-2270-5

Marrakchi, F., M. Auta, W. A. Khanday, and B. H. Hameed. (2017). High-surface-area and nitrogen-rich mesoporous carbon material from fishery waste for effective adsorption of methylene blue. *Powder Technology*, *321*, 428–434. https://doi.org/10.1016/j.powtec.2017.08.023

Martí, M., B. Frígols, and A. Serrano-Aroca. (2018). Antimicrobial characterization of advanced materials for bioengineering applications. *Journal of Visualized Experiments: JoVE*, *138*. https://doi.org/10.3791/57710

Mazhar, M. A., N. A. Khan, S. Ahmed, A. H. Khan, A. Hussain, Rahisuddin, F. Changani, M. Yousefi, S. Ahmadi, and V. Vambol. (2020). Chlorination disinfection by-products in municipal drinking water – A review. *Journal of Cleaner Production*, *273*, 123159. https://doi.org/10.1016/j.jclepro.2020.123159

Mochidzuki, K. (1995). *Studies on the biological activated carbon treatment of wastewater in which organic substance and heavy metal ions co-exist* [Master Thesis], Meiji University.

Mondal, N. K., and S. Basu. (2019). Potentiality of waste human hair towards removal of chromium(VI) from solution: Kinetic and equilibrium studies. *Applied Water Science*, *9*(3). https://doi.org/10.1007/s13201-019-0929-5

Ong, S. T., C. K. Lee, and Z. Zainal. (2007). Removal of basic and reactive dyes using ethylenediamine modified rice hull. *Bioresource Technology*, *98*(15), 2792–2799. https://doi.org/10.1016/j.biortech.2006.05.011

Othman, N., A. Abd-Kadir, and N. Zayadi. (2016). Waste fish scale as cost effective adsorbent in removing zinc and ferum ion in wastewater. *ARPN Journal of Engineering and Applied Sciences*, *11*(3). https://www.researchgate.net/publication/298711155

Pandey, P., B. Karki, B. Lekhak, A. R. Koirala, R. K. Sharma, and H. R. Pant. (2019). Comparative antibacterial study of silver nanoparticles doped activated carbon prepared by different methods. *Journal of the Institute of Engineering*, *15*(1), 187–194. https://doi.org/10.3126/jie.v15i1.27729

Pap, S., J. Radonić, S. Trifunović, D. Adamović, I. Mihajlović, M. V. Miloradov, and M. T. Sekulić. (2016). Evaluation of the adsorption potential of eco-friendly activated carbon prepared from cherry kernels for the removal of Pb^{2+}, Cd^{2+} and Ni^{2+} from aqueous wastes. *Journal of Environmental Management*, *184*(2), 297–306. https://doi.org/10.1016/j.jenvman.2016.09.089

Pasila, A. (2004). A biological oil adsorption filter. *Marine Pollution Bulletin*, *49*(11-12), 1006–1012. https://doi.org/10.1016/j.marpolbul.2004.07.004

Pongener, C., P. Bhomick, S. Upasana Bora, R. L. Goswamee, A. Supong, and D. Sinha. (2017). Sand-supported bio-adsorbent column of activated carbon for removal of coliform bacteria and *Escherichia coli* from water. *International Journal of Environmental Science and Technology*, *14*(9), 1897–1904. https://doi.org/10.1007/s13762-017-1274-6

Renou, S., J. G. Givaudan, S. Poulain, F. Dirassouyan, and P. Moulin. (2008). Landfill leachate treatment: Review and opportunity. *Journal of Hazardous Materials, 150*(3), 468–493. https://doi.org/10.1016/j.jhazmat.2007.09.077

Ribeiro, C., F. B. Scheufele, H. J. Alves, A. D. Kroumov, F. R. Espinoza-Quiñones, A. N. Módenes, and C. E. Borba. (2018). Evaluation of hybrid neutralization/biosorption process for zinc ions removal from automotive battery effluent by dolomite and fish scales. *Environmental Technology, 40*(18), 2373–2388. https://doi.org/10.1080/09593330.2018.1441332

Rivera-Utrilla, J., I. Bautista-Toledo, M. A. Ferro-García, and C. Moreno-Castilla. (2003). Bioadsorption of Pb(II), Cd(II), and Cr(VI) on Activated Carbon From Aqueous Solutions. *Carbon, 41*(2), 323–330. https://doi.org/10.1016/s0008-6223(02)00293-2

Sadiq, R., and M. Rodriguez. (2004). Disinfection by-products (DBPs) in drinking water and predictive models for their occurrence: A review. *Science of the Total Environment, 321*(1-3), 21–46. https://doi.org/10.1016/j.scitotenv.2003.05.001

Salihi, I. U., S. R. M. Kutty, and H. H. M. Ismail. (2018). Copper metal removal using sludge activated carbon derived from wastewater treatment sludge. *MATEC Web of Conferences, 203*, 03009. https://doi.org/10.1051/matecconf/201820303009

Sasu, S., K. Kümmerer, and M. Kranert. (2011). Assessment of pharmaceutical waste management at selected hospitals and homes in Ghana. *Waste Management & Research, 30*(6), 625–630. https://doi.org/10.1177/0734242x11423286

Sharma, Y, and C. Uma. (2009). Optimization of parameters for adsorption of methylene blue on a low-cost activated carbon. *Journal of Chemical & Engineering Data, 55*(1), 435–439. https://doi.org/10.1021/je900408s

Sigdel, A., W. Jung, B. Min, M. Lee, U. Choi, T. Timmes, S.-J. Kim, C.-U. Kang, R. Kumar, and B.-H Jeon. (2017). Concurrent removal of cadmium and benzene from aqueous solution by powdered activated carbon impregnated alginate beads. *CATENA, 148*, 101–107. https://doi.org/10.1016/j.catena.2016.06.029

Shah, M. P. (2020). *Microbial bioremediation & biodegradation*. Springer.

Shah, M. P. (2021a). Removal of emerging contaminants through microbial processes. Springer.

Shah M. P. (2021b). Removal of refractory pollutants from wastewater treatment plants. CRC Press.

Stewart, M. H., R. L. Wolfe, and E. G. Means. (1990). Assessment of the bacteriological activity associated with granular activated carbon treatment of drinking water. *Applied and Environmental Microbiology, 56*(32), 3822–3829. https://doi.org/10.1128/aem.56.12.3822-3829.1990

Supong, A., P. C. Bhomick, M. Baruah, C. Pongener, U. B. Sinha, and D. Sinha. (2019). Adsorptive removal of bisphenol a by biomass activated carbon and insights into the adsorption mechanism through density functional theory calculations. *Sustainable Chemistry and Pharmacy, 13*, 100159. https://doi.org/10.1016/j.scp.2019.100159

Suzuki, Y., K. Mochidzuki, Y. Takeuchi, Y. Yagishita, T. Fukuda, H. Amakusa, and H. Abe. (1996). Biological activated carbon treatment of effluent water from wastewater treatment processes of plating industries. *Separations Technology, 6*(2), 147–153. https://doi.org/10.1016/0956-9618(96)00150-6

Suzuki, Y., Y. Takeuchi, and H. Amakusa. (1995). Treatment technology of industrial wastewater containing sodium salts and sulfides of high concentrations. *Kemikaru Enginiaringu (Chemical Engineering in Japanese) 40* (6): 316–321.

Takeuchi, Y., Y. Suzuki, K. Mochidzuki, Y. Yagishita, T. Fukuda, H. Amakusa, and H. Abe. (1995). Biological activated carbon treatment of waste water containing organics and heavy metal ion. In: M. D. LeVan (Ed.), *Fundamentals of Adsorption*, 937–944, Springer.

Thang, P. Q., K. Jitae, B. L. Giang, N. M. Viet, and P. T. Huong. (2019). Potential application of chicken manure biochar towards toxic phenol and 2,4-dinitrophenol in wastewaters. *Journal of Environmental Management*, *251*, 109556. https://doi.org/10.1016/j.jenvman.2019.109556

Thuan, T. V., B. T. P. Quynh, T. D. Nguyen, V. T. T. Ho, and L. G. Bach. (2016). Response surface methodology approach for optimization of Cu^{2+}, Ni^{2+} and Pb^{2+} adsorption using KOH-activated carbon from banana peel. *Surfaces and Interfaces*, *6*, 209–217. http://dx.doi.org/10.1016/j.surfin.2016.10.007

Thue, P. S., E. C. Lima, J. M. Sieliechi, C. Saucier, S. L. P. Dias, J. C. P. Vaghetti, F. S. Rodembusch, and F. A. Pavan. (2017). Effects of first-row transition metals and impregnation ratios on the physicochemical properties of microwave-assisted activated carbons from wood biomass. *Journal of Colloid and Interface Science*, *486*, 163–175. http://dx.doi.org/10.1016/j.jcis.2016.09.070

Tuan, T. Q., N. V. Son, H. T. K. Dung, N. H. Luong, B. T. Thuy, N. T. V. Anh, N. D. Hoa, and N. H. Hai. (2011). Preparation and properties of silver nanoparticles loaded in activated carbon for biological and environmental applications. *Journal of Hazardous Materials*, *192*(3), 1321–1329. https://doi.org/10.1016/j.jhazmat.2011.06.044

Vukčević, M., A. Kalijadis, S. Dimitrijević-Branković, Z. Laušević, and M. Laušević. (2008). Surface characteristics and antibacterial activity of a silver-doped carbon monolith. *Science and Technology of Advanced Materials*, *9*(1), 015006. https://doi.org/10.1088/1468-6996/9/1/015006

Wong, S., N. Ngadi, I. M. Inuwa, and O Hassan. (2018). Recent advances in applications of activated carbon from biowaste for wastewater treatment: A short review. *Journal of Cleaner Production*, *175*, 361–375. https://doi.org/10.1016/j.jclepro.2017.12.059

Xu, W., Q. Zhao, R. Wang, Z. Jiang, Z. Zhang, X. Gao, and Z. Ye. (2017). Optimization of organic pollutants removal from soil eluent by activated carbon derived from peanut shells using response surface methodology. *Vacuum*, *141*, 307–315. https://doi.org/10.1016/j.vacuum.2017.04.031

Yin, C. Y., M. K. Aroua, and W. M. A. W. Daud. (2007). Review of modifications of activated carbon for enhancing contaminant uptakes from aqueous solutions. *Separation and Purification Technology*, *52*(3), 403–415. https://doi.org/10.1016/j.seppur.2006.06.009

Zhao, Y., Q. Z. Wang, X. Zhao, W. Li, and X. S. Liu. (2013). Antibacterial action of silver-doped activated carbon prepared by vacuum impregnation. *Applied Surface Science* *266*: 67–72.

8 Advanced Biomass Strategies for Aerobic and Anaerobic Treatment Systems

Insights from Practical Applications

Tsenbeni N. Lotha and Latonglila Jamir

8.1 INTRODUCTION

Industrial growth, which is an important factor in the development of the world economy, has been sparked by the tremendous rise in human population. The issue of environmental contamination has recently gained more attention due to the swift economic growth of both developing and industrialized nations. Industrial wastewater production, mainly from the food, distillery, and paper sectors, among others, is related to the manufacture of chemicals and is expanding quickly in middle- and low-income nations. The majority of chemicals that are discharged into the aquatic ecosystem pose a serious risk to the environment and harm to human health. Rising wastewater pollution, harmful compounds, and water scarcity could slow economic growth and increase the likelihood of diseases, famine, and poverty. Reducing the impact of water constraints and contamination is one of the finest strategies and is found to be more significant in light of urban development (Brack et al. 2022). In line with this, the use of biological methods for wastewater treatment has proven to be straightforward, inexpensive, quite effective, and environmentally friendly. However, the production of sludge during biological treatment might cause disposal problems and increase the cost of the process. The use of microbial fuel cells, in which microorganisms interact with electrodes that utilize electrons, and bioaugmentation technology are further technologies that enhance pollution breakdown. Utilizing microorganisms with the capacity to remove contaminants, either native to a contaminated area or isolated from other places, is known as bioaugmentation. The type of pollutant, the features of the microorganism, and environmental conditions frequently have an impact on bioaugmentation. In addition to being able to breakdown particular pollutants, the microorganisms chosen to create an inoculum must also be competitive and continue to exist after inoculation. When microorganisms work together as a consortium, it will be easier for them to establish themselves and operate properly in the

 DOI: 10.1201/9781003381327-8

environment, so they can be used in the bioremediation process (Custodio et al. 2022). The most popular biological method for treating wastewater today is the "activated sludge process," but it produces a sizeable volume of debris that needs to be disposed of. Despite being widely used, it still undergoes modifications in several areas of its design and usage circumstances. The conventional activated sludge (CAS) process necessitates a sizable space, intricate design solutions, and aeration, which has high operational expenses and makes up around 45–75% of the energy expenditures of a wastewater treatment plant (WWTP) (Brockmann et al. 2021). This sludge produces enormous quantities of residual solids and incurs considerable disposal costs since it has high volatile solids (VS) fractions and retains a significant quantity of water until potentially drying (>95% by weight). As a result, the traditional approach converts the problem of water pollution into that of solid waste disposal.

Due to the rate-limiting cell lysis stage, the aerobic or anaerobic digestion of waste activated sludge (WAS) is frequently delayed. To enhance the aerobic and anaerobic biodegradation of solid wastes, a number of methods integrating biological and physico-chemical treatment have been investigated. Ultrasonic treatment is acknowledged as a promising technology to lower sludge formation, and its potential applications have grown in both number and diversity of devices (Ødegaard 2004). The CAS process performance varies depending on the type of wastewater used; for instance, the process biochemical oxygen demand (BOD) removal efficiency can range from 39 to 86% (Joel, Ezekiel, and Mwamburi 2021), while the removal efficiencies of chemical oxygen demand (COD), total suspended solids (TSS), nitrate, and phosphate are respectively 50, 20, 75, and 80%, respectively (Muhammed et al. 2021). According to earlier studies, using sonication before aerobic degradation (i.e., on the recycling loop) or anaerobic digestion also significantly reduced the generation of sludge (Yoon, Kim, and Lee 2004). Release and solubilization of organic substances such as COD, proteins, nucleic acids, and extracellular polysaccharides (EPS) (Zhang, Zhang, and Wang 2007), shrinkage of the floc size (Nah et al. 2000; Gonze et al. 2003), and increased biodegradability (Bougrier, Carrère, and Delgenès 2005) are the primary physico-chemical impacts of sonication on sludge properties. A pollutant is removed from sediment, soil, and water using microorganisms, enzymes, or plants, as these treatments minimize toxicity and the harmful effects on the environment. In order to treat wastewater from various sources, microbial treatments are frequently used. Recently, the use of mixed microbial cultures in wastewater treatment has grown significantly (Al-Dhabi and Arasu 2022). Pollution can be reduced by microbes using pollutants as carbon sources, where the diversity of microbial species in the wastewater system causes different microbial sources to target the effluent molecules, and the produced product may act as a substrate for further microbial strains. There are still a number of restrictions, despite numerous improvements in the effectiveness of the current biological therapeutic procedures. In general, combining multiple treatment approaches is advised to get past technical challenges (Dhandayuthapani et al. 2022). Due to the fact that methane gas is produced during anaerobic digestion and may be used as renewable energy, biological treatment of wastewater seems to be a promising technique for generating income from Certified Emission Reduction (CER) credits, also known as carbon credits from the CDM. Almost all wastewaters with biodegradable components and a BOD/COD ratio of 0.5 or above can be easily treated by biological means with the right analysis and environmental control. It also has the advantages of less expensive treatment and no secondary pollution compared to other wastewater treatment systems. In the former, microorganisms (aerobes) use free or dissolved oxygen to break down organic wastes into biomass and CO_2, whereas in the latter, complex organic wastes are broken down into methane, CO_2, and H_2O through three basic steps (hydrolysis, acidogenesis, including acetogenesis, and methanogenesis) in the absence of oxygen. While anaerobic biotechnology has made significant progress in waste treatment based on the concept of resource recovery and utilization while still achieving the objective of pollution control, aerobic biological processes are frequently used in the treatment of organic wastewaters to achieve high levels of treatment efficiency. Different techniques are used for the wastewater treatment process; however, they all have various drawbacks. There are three different types of treatment: primary treatment, secondary treatment with nutrient elimination (chemical and biological), and advanced tertiary treatment (Figure 8.1).

The characteristics of wastewater are influenced by the type of industry, population, and preservation practises. In general, the majority of the parameters are arranged and connected (e.g., dissolved gases and microbes' actions in wastewater). It has to do with people, the environment, local enterprises, and preservation efforts. Turbidity, temperature, and colour are the three most important physical characteristics of wastewater (solid content, sediments, floating content, and soluble chemicals). The chemical characteristics include both organic and inorganic matter, with the organic constituents being essential for the development of wastewater and lowering the standard of quality. Examples of organic constituents include hydrogen, carbon, oxygen, nitrogen, phosphorous, iron, and sulphur (wastewater). The natural fading process increases the quantity of inorganic chemicals present. The first treatment procedure mostly involves turning wastewater through physical and mechanical means into both liquid and solid trash (Haixia et al. 2013). Microorganisms mostly carry the liquid sewage into the secondary treatment procedure. It is used in two main methods: aerobic conditions and anaerobic conditions. The bacteria in liquid waste breakdown the biological material in aerobic circumstances. The tertiary treatment process, which is the most sophisticated, is required to disinfect the wastewater.

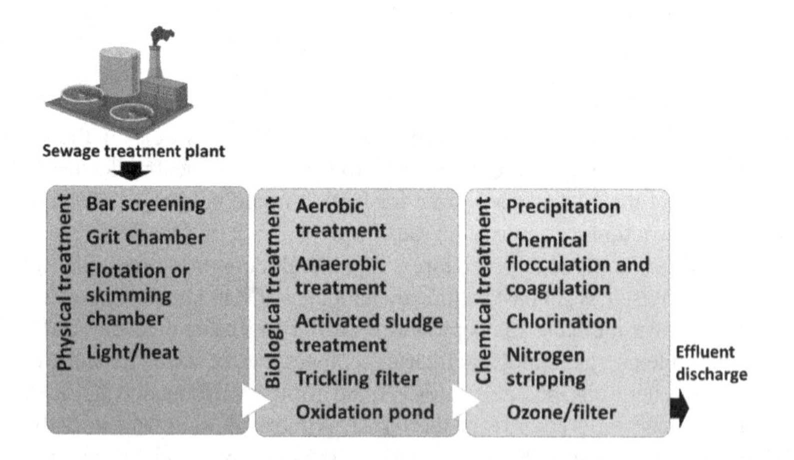

FIGURE 8.1 Different stages of sewage treatment.

TABLE 8.1
Comparison of Aerobic and Anaerobic Treatments (Leslie Grady, Daigger, and Lim 1999; Yeoh 1995; Chan et al. 2009)

Feature	Aerobic Treatment	Anaerobic Treatment
Start-up time	2–4 weeks	2–4 months
Temperature sensitivity	Low	High
Organic removal efficiency	High	High
Quality of the effluent	Excellent	Moderate to poor
Organic loading rate	Moderate	High
Sludge production	High	Low
Nutrient requirement	High	Low
Alkalinity requirement	Low	High for certain industrial waste
Energy requirement	High	Low to moderate
Odour	Less opportunity for odours	Potential odour problems
Mode of treatment	Total (depending on feedstock characteristics)	Essentially pre-treatment
Bioenergy and nutrient recovery	No	Yes

Aerobic bacteria are crucial to the wastewater treatment process because they may break down the harmful pollutants in wastewater by using the free oxygen present in the water. The oxygen must be mechanically provided; otherwise, the bacteria will grow and multiply on their food source. Most commonly, anaerobic bacteria are utilized in the secondary treatment process to reduce the amount of sludge, produce methane gas, and provide other benefits such as phosphate removal from wastewater. Depending on the environment, facultative bacteria play an intermediate function between aerobic and anaerobic processes (Zou, Lu, and Abualhail 2014). The advantages of both treatments in various ways are shown in Table 8.1, and very high organic removal efficiencies can be attained by both techniques. Anaerobic systems are best suited for treating high-strength wastewaters (biodegradable COD concentrations exceeding 4000 mg/L), whereas aerobic systems are best suited for treating low-strength wastewaters (biodegradable COD concentrations less than 1000 mg/L).

8.2 INDUSTRIAL WASTEWATER: EFFECTS ON ENVIRONMENT AND HUMAN HEALTH

Industrial wastewater discharge into rivers, lakes, and coastal areas leads to major water pollution issues, has an adverse effect on the ecology, and endangers both humans and other living things. Different forms of contaminants, including organic materials, suspended solids (SS), inorganic dissolved salts, petroleum hydrocarbons, heavy metals, surfactants, and detergents, are carried by industrial discharge. These pollutants may contaminate receiving bodies of water, making them unfit for drinking and irrigation. They also may have a negative impact on people, animals, plants, and aquatic life. Numerous synthetic materials, many of which are resistant in nature and are non-biodegradable or disintegrate very slowly, are produced by industries. As a result of their protracted persistence in the environment, such chemicals may gradually increase in concentration. The resistant compounds can build up in an organism's tissues and are poisonous, mutagenic, or carcinogenic. These contaminants impact people and other living things as they move up the food chain through biomagnification. It has become the main disaster in several nations throughout the world and is causing millions of people to suffer. In residential areas, toxic chemicals and detergents flushed down drains and toilets embarrass families. Other common causes of water pollution include the use of unsafe fertilizers by farmers, improper sewage disposal, radioactive contaminants, oil dumping, and the disposal of plastics in the ocean. There is a lot of evidence showing how contaminated water affects human strength. In many nations that do not use the systems, the tainted water leads to a water shortage. Current information on water contamination shows that it affects many individuals daily. Numerous individuals often drink contaminated water on a daily basis, which might result in pathogenic diseases. Drinking contaminated water, in particular, has been linked to a number of illnesses, including cholera, typhoid, dysentery, hepatitis (A and B), botulism, parasite infections, and traveller's diarrhoea (Garg et al. 2022).

8.3 ROLE OF MICROBES IN WASTEWATER TREATMENT

To break down the various types of wastewater, a variety of microbiological sources are utilized. The sources consist of strong bacteria and fungi that are very important in the breakdown of inorganic components (Velusamy et al. 2022). The most powerful strains of streptomyces have been shown to degrade the odour and inorganic chemical components of the effluent (BOD, COD, phosphorous, and nitrogen). Microbes have an impact on the outcome of the wastewater treatment process in both positive and negative ways. The removal of harmful pollutants, including the chemical elements nitrite, ammonia, phosphate, and hydrogen sulphide, as well as feed degradation and aquatic animal nutrition, are among the benefits (fish, shrimp, and chicken). Numerous health advantages and the sustainability of aquatic ecology are two of microorganisms' essential functions. The fact that bacteria and protozoa have played a larger part in the removal of phosphate and nitrogen indicates that the treatment process is biological. Bacteria have a significant role in the conversion of organic matter and carry out the majority of industrial and municipal wastewater treatment vegetation activities. Due to the production of bioenergy (biogas), this treatment process is more significant from an economic and environmental standpoint. Plants and microorganisms work together to clean wastewater. Roots contain the desired microbial groups, which play the best part in the breakdown of hazardous pollutants in wastewater. The microorganisms have increased plant growth, decreased metal toxicity, and increased the bioavailability of harmful chemicals (Shahid et al. 2020). Any wastewater treatment facility that processes wastewater from cities and industry, consumes soluble organic contaminants, or uses a mixture of several types of wastewater sources must include biological treatment as a crucial and integral component (Figure 8.2). The biological treatment completes other management processes (chemical and thermal oxidation), resulting in a combined wastewater treatment facility with a discernible economic benefit in relation to capital assets and operating costs. Over the past century, extensive research on the prospective aerobic activated sludge biological treatment method has been published in exercise. The cumulative stress is to experience increasingly stringent discharge

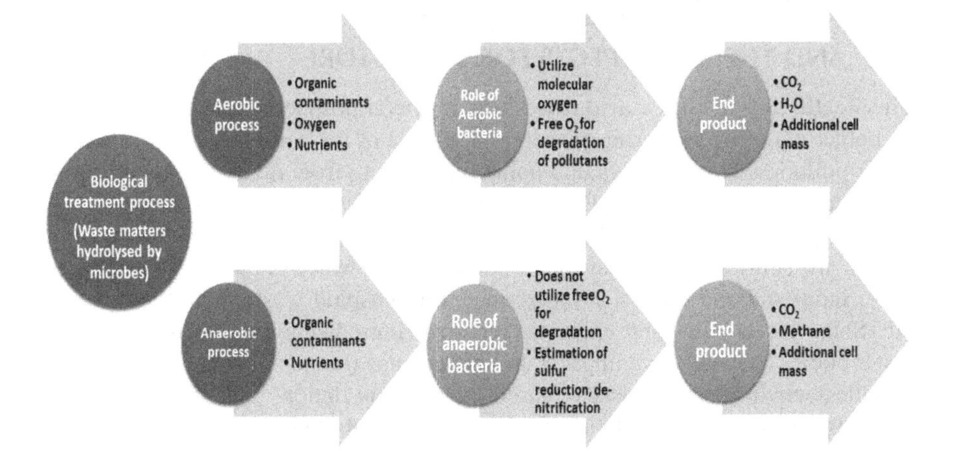

FIGURE 8.2 Role of microbes in aerobic and anaerobic treatment of wastewater.

values or be prohibited from releasing treated effluent samples. There were two ways that the biological wastewater treatment process worked (aerobic and anaerobic). Both methods have a close relationship with a certain bacterial or microbial species.

The elimination of organic contaminants from wastewater and handling of the fermenter are primarily handled by the bacteria. Because bacteria use free air to integrate pollutants, the aerobic treatment process can only begin in the presence of oxygen. Under aerobic treatment conditions, organic pollutants are transformed into CO_2, H_2O, and biomass. On the other hand, anaerobic treatment is performed without the use of oxygen (microbes do not utilize free air to integrate pollutants). The organic contaminants are converted to CO_2, methane, and biomass during the anaerobic treatment process, as shown in Figure 8.3.

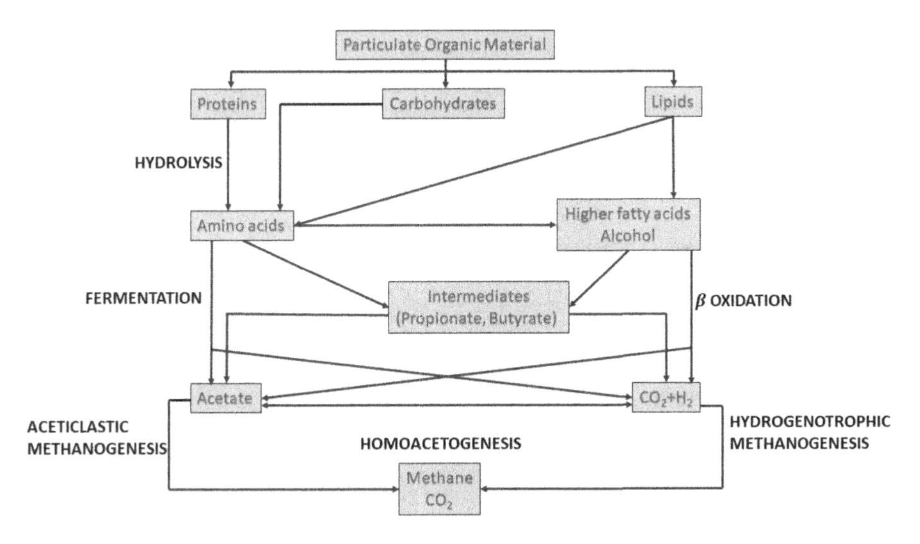

FIGURE 8.3 Reactive scheme for the anaerobic degradation of particulate organic matters (Gujer and Zehnder 2018).

8.4 VARIOUS WASTEWATER GENERATING SOURCES AND THEIR PROSPECTS FOR THE FUTURE

The weathering of rocks, acid deposition in the atmosphere, volcanic eruptions, and other naturally occurring processes are some of the natural sources of wastewater pollution. However, organic pollutants coming from domestic, agricultural, and industrial processing are the primary origins of water contamination (Akhtar et al. 2021). Marine birds, fish, and other marine species that are used as food for people are being killed by the contaminated water, which is due to the release of some industrial and domestic waste, including hospital waste, household waste, etc. Some insecticides and chemical fertilizer ingredients that are blended with water bodies have negative impacts on living things (humans). There are currently countless environmental issues caused by water pollution, and some infectious diseases are mostly brought on by hazardous contaminants (cholera, typhoid, immune suppression, respiratory infection, infertility, skin diseases, typhoid, cholera, tuberculosis, etc.) (Owa 2013). The risk of polluted water feeding and its sanitation issues are daily concerns in most developing countries due to rapid urbanization and industrial development. The growing issue of water scarcity has a significant fabricated impact on global economic development, quality of life for people, and environmental excellence. Therefore, protecting water from contamination or developing practical methods for its protection has become crucial in today's environment (Abdel-Raouf et al. 2019). Wastewater is treated using biological sources, such as bacteria, microalgae, and enzymes, and includes wastewater from the pesticide industry, olive oil mills, domestic wastewater, sewage wastewater, aquaculture wastewater, petrochemical wastewater, paper industry wastewater, and wastewater from gold mines (Table 8.2).

TABLE 8.2
Treatment of Different Wastewater by Biological Methods Using Biomass

Source	Class	Example	Microbial Biomass	References
Textile dye wastewater	Dyes	Malachite green, crystal violet, Methyl orange	Bacterial Isolates (*Bacillus megaterium*)	Moyo, Makhanya, and Zwane (2022)
Petrochemical wastewater treatment	Organic compound	Benzene, toluene, ethylbenzene	*Firmicutes* and *Proteobacteria*	Biswas et al. (2022)
Pharmaceutical wastewater	Analgesic and antipyretic drug	Acetaminophen (paracetamol)	*Pseudomonas*	Poddar et al. (2022)
Marine aquaculture wastewater	Inorganic compound	Ammonia	*Isochrysis galbana* and *Chlorella* sp.	Qian et al. (2022)

(Continued)

TABLE 8.2 *(Continued)*
Treatment of Different Wastewater by Biological Methods Using Biomass

Source	Class	Example	Microbial Biomass	References
Pesticide Industry wastewater	Pesticide	Glyphosate	*Providencia vermicola*	Gupta, Pandey, and Verma (2022)
Paper industries wastewater	Organic compound	Carbohydrates	*Cloacibacterium* and *Aerococcus*	Tyagi et al. (2022)
Industrial wastewater	Organic compounds	Hexadecane and derivatives, Dodecane and derivatives	Proteobacteria, Bacteroidota, Nitrospirota	Yang et al. (2022)
Olive oil mill wastewater	Organic matter	Hydroxytyrosol, tyrosol	Bacteria	Sar and Yesilcimen Akbas (2022)
Domestic and non-domestic wastewaters	Surfactant	Sodium dodecyl sulphate (SDS)	*Pseudomonas mendocina* and *Bacillus*	Najim, Ismail, and Hummadi (2022)
Sewage wastewater	Organic compound	Benzene, chloroform, naphthalene, phenols, toluene	Enzyme	Siddiqui and Dahiya (2022)
Artisanal gold mine wastewater	Organic compound	Cyanide	*Bacillus subtilis*	Rosario et al. (2022)

8.5 APPLICATION OF BIOMASS IN TREATMENT OF INDUSTRIAL WASTEWATER

The type of wastewater treatment required is primarily determined by the wastewater's effluent requirements. Pollutants released from industrial wastewaters have a harmful impact on all areas of the ecosystem, including water, air, and land. Industrial effluent treatment is often difficult due to the presence of numerous contaminants, large organic matter levels, and poorly biodegradable components. Large solids/particles are eliminated in primary treatment, whereas bioremediation of organic pollutants occurs in secondary treatment by the action of microorganisms. Treatment of agro-industrial and industrial wastewaters is generally difficult because wastewater compositions vary and can include high organic matter contents and poorly biodegradable components. Only municipal wastewater treatment typically necessitates tertiary treatment since disinfection is required before the water is released into bodies of water. A bar screen, a grit and scum removal chamber, and a settling tank or clarifier for coagulation flocculation are all employed. Bacteria feed on the organic substances in wastewater and emit greenhouse gases (GHGs) such as carbon dioxide, hydrogen sulphide, and methane as a result. Microalgae, biofilms, anaerobic systems, and membrane filtration are currently used as wastewater treatment techniques for agro-industrial and industrial wastewater.

8.6 AEROBIC GRANULAR SLUDGE IN WASTEWATER TREATMENT

One of the most highly developed biological techniques for wastewater treatment is aerobic granular sludge (AGS). Microorganisms self-immobilize throughout the bio granulation process. Granular sludge has superior settling qualities to traditional activated sludge, and it is more resilient to changes in environmental factors including temperature, organic load, and pH. The microbial makeup of the biomass has a significant impact on granulation. Agglomerates, which serve as the foundation for granulation, are produced by filamentous bacteria. Depending on the type of carbon source available, these bacteria may disappear or remain in the biomass throughout the later stages of granulation, and their presence impacts granule compactness (Liu and Liu 2006). Additionally, using aerobic granules reduces energy usage. According to Pronk and co-workers (Pronk et al. 2015), in comparison to a typical traditional activated sludge plant, the energy utilisation in a full-scale wastewater treatment facility using aerobic granules to treat domestic wastewater was 58–68% lower. The biomass's microbial structure affects the efficacy and stability of wastewater treatment. Based on the results of denaturant gradient gel electrophoresis, with mature aerobic granules, Flavobacterium sp., Aquabacterium sp., and Thauera sp. occurred in granules, and Flavobacterium sp. supported granule formation by EPS production, as reported on aerobic granulation in wastewater treatment have primarily focused on technological research (Li et al. 2014). Using fluorescent in situ hybridization, Pronk and co-workers investigated the microbial structure of aerobic granules from a full-scale experiment and found that phosphate-accumulating organisms (PAOs) made up a sizable portion of the population, while glycogen-accumulating microorganisms were only moderately abundant (Pronk et al. 2015).

Additionally, hazardous metal ions from various industrial and municipal wastes can be eliminated via aerobic granulation. Due to their large surface area, high porosity, and good settleability, aerobic granules have a very promising future in the sorption of harmful substances. Aerobic granular biomass has the potential to be used as an efficient biosorbent material for extracting uranium from diluted nuclear waste (Nancharaiah et al. 2006). Prior research on granulation technology primarily concentrated on general aspects of granule formation, effects of substrate type, loading rate, filling patterns, shear stress, dissolved oxygen concentration, COD:N:P ratio, and settling velocity on granulation, as well as characteristics of granules, including granule size, morphology, Sludge Volume Index (SVI), density, hydrophobicity, and conversion processes under various process conditions (JH et al. 2003).

8.7 ANAEROBIC GRANULAR SLUDGE BIOREACTOR TECHNOLOGY

Among the most widely utilized techniques for waste treatment are anaerobic treatment systems, which utilize unidentified microbial populations to lessen the pollution potential of waste and wastewater. High levels of active biomass must be retained within the system to ensure the technology is successfully applied. This can be done either by immobilizing the microbes on inert carriers, as in the anaerobic filter, or by allowing them to self-immobilize as sludge aggregates or granules, as in the upflow anaerobic sludge blanket. The use of compact and affordable wastewater treatment facilities is made possible by the retention of a high biomass content within the system, which permits the use of very high organic loading rates. The creation and management of well-settling sludge granules inside the reactor is the basis for the great majority of full-scale anaerobic waste treatment plants currently in use. In the absence of a support material, anaerobic bacteria self-immobilize to produce particle biofilms known as anaerobic granules. The key to the efficient operation of the modern, high-rate anaerobic digester is found in these dense particles, which are made up of an entangled combination of the symbiotic anaerobic microorganisms that collaborate in methane production (McCarty 2018). The several microorganisms required for the methanogenic breakdown of organic materials are all included in one functional unit called a granule. Per gramme of biomass, a typical granule may contain millions of organisms. However, no one species in these micro-ecosystems is able to fully degrade influent wastes (Liu and Tay 2002). Due to the necessity of competitive and cooperative connections between the granule's component microorganisms, the consortium creates a distinctive microbial ecology within a few millimetres of an aggregate (McHugh et al. 2003). To address syntrophic and competitive interactions in methanogenic granules and demonstrate connections between microbial structure and function, a variety of approaches, including immunological (Schmidt and Ahring 1999), microscopic (Sekiguchi et al. 1998), rRNA-based molecular techniques, histochemical analysis, conventional enumeration techniques, and bacterial activity (Colleran and Pistilli 1994) studies, are necessary. As the distance between microorganisms, such as acetogens and methanogens, must be sufficiently short to provide and maintain a low hydrogen partial pressure, a precise spatial orientation of microbial consortia within the anaerobic granule is crucial (Harmsen et al. 1996).

Microbiological, other biotic, and abiotic factors all interact in this complicated process. Our knowledge of the microbiological properties of upflow anaerobic sludge bed granules and the interaction between various species in the granules have advanced dramatically over the past two decades. Bacterial adhesion to inert substances, inorganic precipitates, and to one another through physico-chemical interactions and syntrophic relationships may all be factors in the beginning of granulation. For future bacterial development, these chemicals serve as the fundamental building blocks (carriers or nuclei). Other filamentous bacteria, such as *methanosaeta*, may also aid in the creation of matrices that further embed other cells. Then, these loosely attached bacterial clumps are strengthened using extracellular polymers produced by bacteria (Schmidt and Ahring 1996).

8.8 AEROBIC AND ANAEROBIC – SYSTEMS USING HIGH-RATE BIOREACTORS

Regarding COD and pollutant removal with the possibility for energy savings or even energy-positive scenarios, high-rate aerobic and anaerobic processes seem to be a viable technique for wastewater treatment. The secret to successfully managing industrial and municipal wastewater is to use the proper mix and order of treatment techniques, i.e., diverse anaerobic and aerobic bioreactor combinations that have been used to treat a variety of industrial wastes. The treatment is carried out in two different bioreactors that are connected in series by the anaerobic-aerobic systems, which use high-rate bioreactors. The anaerobic–aerobic systems using high-rate bioreactors reviewed in a study reported by Chan et al. (2009) attain significant COD elimination (above 70%) with a brief hydraulic retention time (HRT) (ranging from a few hours to a few days). Therefore, treating industrial and municipal wastewater using anaerobic-aerobic processes is effective.

One of the granular sludge bioreactors utilized for the quick biotransformation of organic matter into methane with the aid of granulated microbial aggregates is the upflow anaerobic sludge-fixed film (UASFF) reactor. A fixed film part that is upflowing over an anaerobic sludge blanket part makes up this hybrid reactor. The prolonged start-up period is the main issue with UASB reactors (2–4 months). In a study by Najafpour et al. (2006), in order to reduce the start-up time at low hydraulic retention time (HRT) for palm oil mill effluent (POME) treatment, a UASFF bioreactor with tubular flow behaviour was designed. The reactor was run at 38°C for 1.5 and 3 days of HRT. Chemical oxygen demand (COD)/l day for organic loading was steadily increased from 2.63 to 23.15 g. Within 20 days, granular sludge had grown quickly. Granules grew in size from their original pinpoint size to a maximum of 2 mm. At HRT of 1.5 and 3 days, high COD removals of 89 and 97%, respectively, were attained. The highest organic input rate resulted in a methane output of 0.346 L CH_4/g COD removed. The flocculated biomass precipitated over the sludge blanket due to the usage of an internal upflow anaerobic fixed film section. To create a spatial arrangement of microbial species in the process of mature bio-granulation, the precipitated biomass functioned as an appropriate and natural hydrophobic core. It shows that the UASFF bioreactor, a unique hybrid bioreactor with a high organic load and SS concentration, was successful in treating POME. It was a smart move to use the UASFF reactor to speed up anaerobic granulation and obtain high COD removal efficiency quickly.

8.9 METHODS ENHANCING ANAEROBIC AND AEROBIC DIGESTION CONTROLLED CONDITIONS

WAS can be broken up using ultrasound prior to feeding it to an anaerobic digester in order to improve performance. A relatively recent development in the treatment of sewage sludge is the use of high-power ultrasound for cavitation-induced disintegration (Tiehm et al. 2009). As a result of mechanical cavitation's strong effects, which cause microbial cell walls to be broken when the cavitation bubbles implode, the bacterial cells in the sludge disintegrate and are eventually destroyed. The surviving living microorganisms have a greater bioavailability of substrate as a result of the cell's contents being released into the medium. This mechanical disintegration of the sludge effectively replaces and catalyses enzymatic-biological hydrolysis, the first and rate-limiting step in the anaerobic food chain (Shimizu, Kudo, and Nasu 1993).

Sonicating activated sludge before reintroducing it to the activated sludge tank (referred to as return activated sludge) has developed as a novel and creative method to boost denitrification in order to improve the aerobic digestion process. The use of ultrasonically treated sludge to enhance anaerobic digestion has generated useful leads in the field of research, despite the fact that the mechanisms driving the reactions that ultimately lead to this improvement are not fully understood or known. The sludge's bacteria experience stress during sonication, which causes the colonies to disintegrate. The detrimental effects of filamentous colonies are significantly reduced or eliminated by the strong shear pressures produced by the implosion of cavitation bubbles. The release of intercellular material from the damaged cell is also thought to accelerate the biological processes that take place in the sludge. Particularly, it is believed that released enzymes play a vital part in promoting many biological processes. The cellular debris that has been destroyed itself acts as a non-foreign carbon source for the sludge's remaining viable microbes (Neis, Nickel, and Lunden 2008).

8.10 SEQUENTIAL BIODEGRADATION OF WASTEWATER

8.10.1 Biodegradation of Azo Dyes in a Sequential Anaerobic–Aerobic System

Numerous substances that are challenging to break down aerobically appear to be easily broken down anaerobically. The by-products of anaerobic biotransformation, in turn, are ideal substrates for aerobic biodegradation while resisting further anaerobic mineralization. A sequential anaerobic-aerobic treatment technique is therefore thought to be the most effective for the entire mineralization of many stubborn contaminants (Zitomer and Speece 1993). The microbial breakdown of sulphonated aromatics is frequently carried out via mixed culture because resistant molecules with xenobiotic properties require special catabolic activities. An anaerobic-aerobic technique can be used to sequentially breakdown pollutants like azo dyes (Figure 8.4). Using mixed bacterial cultures, glucose as a co-substrate, and an H-donor, researchers found that azo dyes entirely lost their colour while amine was mineralized in the aerobic portion of a two-stage system. The ability of organisms to use aniline as a source of both nitrogen and carbon for microorganisms and as their sole supply of

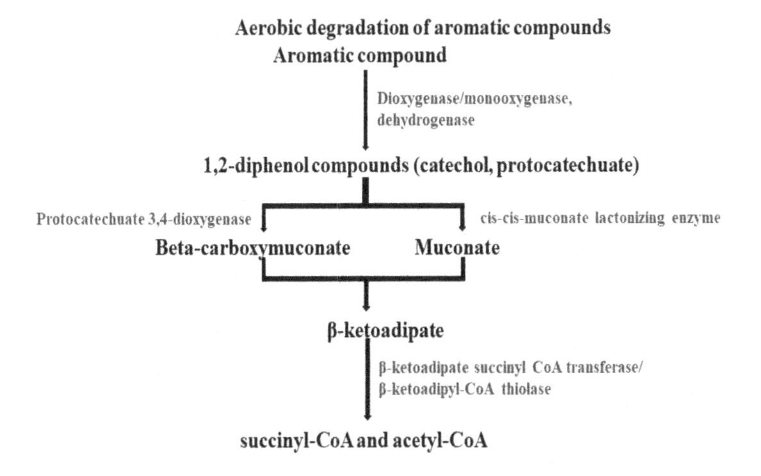

FIGURE 8.4 Aerobic degradation of aromatic compounds.

nitrogen, the observed growth and subsequent stabilization of the microbial biomass in the fixed-bed column and the agitated aerobic tank over the course of the experimental period of more than 10 months, in which no additional nitrogen source other than the test dye was added to the feed solution of the anaerobic column, show that the organisms used in the study by Rajaguru et al. are capable of utilizing azo dyes. It proved that the highly electrophilic sulphonated azo dyes could be reduced in the anaerobic pre-treatment and that the reduction products were suitable for aerobic mineralization. Due to a lack of facilities, it was impossible to determine the metabolic pathway for the dyes used in mixed cultures or whether complete mineralization actually took place. However, the development and maintenance of the biomass, the uninterrupted operation of the two-stage treatment system for more than a year, and the disappearance of amine in the aerobic tank indirectly suggest either that all the dyes are fully mineralized or that the nature of the end products is not inhibitory or toxic (Rajaguru et al. 2000).

8.11 TREATMENT OF INDUSTRIAL WASTEWATER USING BACTERIAL BIOMASS OR CONSORTIA

8.11.1 Diary Industry Wastewater

In an experiment by Custodio et al. (2022) at laboratory size, a reactor was used to treat diary wastewater aerobically with bacterial consortia. The order *Burkholderiales* performed exceptionally well at removing COD in aerobic reactors. The orders *Lactobacillales*, *Thermoanaerobacterales*, *Clostridiales*, *Selenomonadales*, and *Veillonellales* performed better in anaerobic reactors at removing TSS and BOD. The makeup of the bacterial communities in the consortiums may change based on the circumstances unique to the reactor type configuration, according to generalized linear model (GLM) research. This demonstrates that the examined bacterial consortiums can be an environmentally sound and sustainable wastewater treatment method that effectively removes organic contaminants. However, more study is required to determine how bacterial consortiums behave in non-sterile dairy wastewater and in uncontrolled environments.

8.12 PIGGERY WASTEWATER

In areas with the highest concentrations of animal farms, the growing livestock husbandry for the production of human food poses a significant environmental issue. One of the most damaging agro-industrial wastewaters is piggery wastewater (PWW) (Godos et al. 2009). Additionally, improper wastewater management can pollute the local environment with unpleasant odours. The treatment and commercialization of PWW have been proposed and researched using a variety of technologies. Piggery effluent has typically been treated primarily using aerobic, anaerobic, or even a combination of these procedures. The use of single-stage or multi-stage anaerobic lagoons to store and treat PWW is one of the strategies for PWW management that is most frequently used (Montefiore et al. 2022). High amounts of

GHGs such as CO_2, CH_4, NH_3, and H_2S are released into the open atmosphere, which has a number of negative consequences on the ecosystem. Despite the fact that the decanted manure from these anaerobic lagoons may include antibiotics and heavy metals that collect and build up in the food chain, it is often employed as fertilizer due to its high concentration of nitrogen and phosphorus. Additionally, these effluents have a high concentration of germs, which are released when manure is distributed, contaminating groundwater and surface water (Campagnolo et al. 2002). Anaerobic digestion is a traditional method of PWW therapy. In the absence of oxygen or oxidized forms of nitrogen, this activity is carried out by the symbiotic interaction of bacteria and archaea. These microorganisms break down the organic matter in PWW to produce mainly methane (CH_4) at concentrations between 40 and 75%, carbon dioxide (CO_2) at concentrations between 15 and 60%, and other gases at lower concentrations like hydrogen (H_2), ammonium (NH_3), oxygen (O_2), and hydrogen sulphide (H_2S) at higher concentrations (Struk et al. 2023). PWW can be transformed using this kind of technology into biogas, a renewable gas energy product. However, anaerobic digestion can only break down the carbon, leaving behind liquid effluents that are rich in dissolved nitrogen and phosphorus that the anaerobic microbial population is unable to absorb. Additionally, this procedure produces solid sludge that must be dumped in landfills before composting, primarily made up of the non-biodegradable portion of PWW and the anaerobic biomass created throughout the process. Despite having a high energy need, the aerobic technique of activated sludge treatment is frequently employed for PWW treatment (Montes et al. 2015). This method begins with the removal of SS, then moves on to an aerobic bioreactor using activated sludge, and ends with settling to gather the biomass produced (Suzuki et al. 2010). The aerobic tank, where ammonium is converted to nitrate or nitrate and carbon to CO_2 and H_2O, may be followed by an anoxic tank for heterotrophic denitrification (NO_3 to N_2). Eukaryotes (protozoa and fungi), archaea, and viruses are frequently present in this kind of process; however, bacteria predominate in activated sludge systems. The treatment of domestic wastewater using this method is the most researched and widely used in the world. Unfortunately, this technique necessitates mechanical aeration to augment the air to ensure enough oxygenation using blowers or turbines, which has high operational costs due to the related high energy consumption. Another drawback of activated sludge methods comes from the fact that carbon and nitrogen are primarily removed through volatilization into the atmosphere. In fact, nitrogen in PWW is turned into N_2 through nitrification and denitrification processes or volatilizes as NH_3, while organic matter is transformed into CO_2, which aids in the formation of GHGs. Finally, the biomass produced, which is dumped in landfills, retains the heavy metals, such as copper and zinc, which are found in PWW.

8.13 WASTEWATER TREATMENT BY MICROALGAE

Because of their metabolic flexibility, microalgae have the ability to perform photoautotrophic, mixotrophic, or heterotrophic metabolism, making them a promising biological system for treating a variety of sources of wastewater. Microalgae biomass produced from wastewater streams offers a great potential for sustainable bio-based products, such as proteins, fatty acids, pigments, biofertilizers/biochar, and animal feed (dependent on national legislation on reusing microalgae biomass/bioproducts) (Subashchandrabose et al. 2013). This is especially true in the context of a circular and bio-based economy and the advancement of biorefinery concepts. Microalgae are used in WWTPs due to the direct uptake or transformation of water contaminants and the enhancement of bacterial purification performance by supplying additional oxygen from photosynthesis, thereby lowering the overall energy costs of direct or indirect oxygen supply. Green, red, and brown algae, as well as cyanobacteria, are only a few examples of this diverse group of photosynthetic microorganisms. They are promising biological tools that will create bio-based energy carriers and goods due to their capacity to fix carbon dioxide (CO_2) utilizing light as the only energy source (Cuellar-Bermudez et al. 2017). For the treatment of industrial wastewater with a high organic load, certain microalgal species are also capable of chemoheterotrophic or mixotrophic metabolism in addition to photoautotrophy. *Chlorella vulgaris*, one of the most researched conventional microalgal strains, has recently been tested for its capacity to produce biomass from a variety of industrial effluents. Using synthetic wastewater, Zhai et al. (2017) revealed the high N (81.51%) and P (80.52%) removal effectiveness of the commonly used cyanobacterium *Arthrospira platensis*.

Most wastewater treatment methods involve stirred ponds with activated bacterial sludge, particularly for urban and industrial wastewater. However, the ability of activated sludge to sufficiently remove N and P or eliminate heavy metals without the use of chemical precipitation is restricted. Microalgal use in wastewater treatment is linked to more complex technological needs for photobioreactor (PBR) systems. This is mostly due to photoautotrophic activities, which necessitate a significant amount of light energy and CO_2 (Christenson and Sims 2011). The municipal, industrial, and agricultural wastewater streams can all be treated by Algal Enterprises' (Australia) algae WWTP. In a closed PBR system, the primary energy source for local algae species is photosynthetically active radiation. Produced algal biomass is combined with anaerobic digestion to create a methane-rich biogas that is then employed for generating energy (Montingelli, Tedesco, and Olabi 2015).

8.14 MICROALGAE CULTIVATION SYSTEM

Different types of growth systems are now used for both small- and large-scale microalgal farming for a variety of objectives. However, the selection of microalgal species, nutrient availability, and microalgal biomass for end use typically influence the microalgal production technique chosen. Microalgae cultivation systems can be divided into open and closed cultivation systems based on their varying relevance. In order to produce microalgal biomass at the best cost, several microalgal growth systems have led to the development of effective methods. The rate of biomass production varies across all cultivation approaches, and it continues to alter as a result of various operational circumstances. However, in order to highlight the microalgae cultivation technology that is better suited for the generation of biofuel and wastewater treatment, critical comparative studies are advised. The main pros and cons of microalgal cultivation systems are shown in Table 8.3.

TABLE 8.3
Main Advantages and Disadvantages of Different Microalgae Cultivation Systems (Muñoz and Guieysse 2006; Mehariya et al. 2021)

Cultivation System	Advantage	Disadvantage
Open system	Construction and operation are simple. Lower costs as a result of fewer energy use and large open spaces are frequently used to raise microalgae for biofuel production from wastewater.	Insufficient use of light. Because of the lower cell density, biomass harvesting is difficult and expensive.
Closed system	Optimal growth parameters can be controlled, and there is an increase in biomass contamination with less contamination.	Capital outlays (CAPEX) and operating expenses (OPEX) are both substantial. Biofouling and overheating upscaling are extremely costly.
Algal turf scrubber (ATS)	Fuel production costs are quite low. Over an open cultivation method, biomass productivity is higher and biomass collection is easier. Low upkeep and monitoring of environmental carbon sequestration.	Space is required, but it is less than in an open cultivation system. Lower wastewater handling capacity necessitates extensive infrastructure.
Hybrid cultivation system (HCS)	Increased biomass productivity. Lower chance of contamination. Low upkeep and monitoring of environmental carbon sequestration	Higher process of CAPEX and OPEX requires big infrastructure, high maintenance, and frequent monitoring.

It can be stated that open cultivation systems are commonly used for wastewater treatment due to their capacity for larger volume treatment, which can reduce overall wastewater treatment costs. The close cultivation approach is primarily used to increase biomass productivity in order to produce high-value-added chemicals from algal biomass (Mehariya et al. 2021). However, the development of an advanced hybrid culture technique can boost biomass yield, allowing the overall cost to be offset. Microalgal biomass is a rich source of many compounds and chemicals that can be converted into various types of biofuels. This can be generated by the microbial conversion of biomass derived from various sources and can help to meet the growing need for fossil fuels. Because of its higher lipid and sugar content, algal biomass is a popular choice for biofuel production among the various types of biomass. It has the potential to produce a wide range of biofuels via various routes. Thus, it can be converted into biodiesel by lipid transesterification, bioethanol, and other processes (Mohsenpour et al. 2021). Microalgae-based biofuel production has several advantages, including no need for arable land for mass cultivation of microalgae, higher photosynthesis activity than terrestrial plants, resulting in higher biomass productivity as well as higher CO_2 mitigation, no need for agriculture land, which is not permitted to be planted with food crops, no need for fertilizer or the addition of extra nutrients, and the ability to remove pollutants from wastewater (Kant et al. 2022).

8.15 MICROALGAE AND BACTERIA CONSORTIA FOR BIOREMEDIATION OF INDUSTRIAL WASTEWATER

Due to its additional advantages, biological wastewater treatment employing microalgae has been gaining popularity. The use of microalgae has been regarded as both environmentally and commercially advantageous. In addition to providing a renewable supply of biomass, the bio-fixation of CO_2 by microalgae in wastewater treatment is both practical and affordable. Mixotrophic microalgae are used to clean wastewater because they can utilize the organic and inorganic carbon, as well as the inorganic N and P, that are present in wastewater for their growth, which lowers the concentration of these elements in the water. Algae can aid in the sequestration of carbon and can produce valuable products like biofuels like ethanol, biodiesel, methane, and hydrogen by utilizing nutrients from wastewater outflow (Pacheco et al. 2020). Phycoremediation is the technique of utilizing microalgae to remove contaminants, such as xenobiotics, and nutrients from wastewater or effluents before they are released or reused. In prior investigations, the following types of microalgae for wastewater treatment were looked at: *Scenedesmus abundans* and *Chlorella* (Park, Craggs, and Shilton 2011).

The non-aerated microalgal-bacterial granular sludge (MBGS) process has received a lot of interest. Instead of using external aeration, sustainable solar energy may drive the creation of self-oxygen through photosynthesis, and microbial absorption may serve as the primary method for removing contaminants from MBGS. As a result, wastewater's organics and nutrients can be improved in the compact solid phase from the loose liquid phase, favouring the possible recovery of energy and resources such as biofertilizer, biodiesel, and natural pigments (Wang et al. 2021). Non-aerated MBGS can produce effluent of outstanding quality while emitting hardly any GHGs (Ji et al. 2020).

Through the process of photosynthesis, microalgae create oxygen molecules. These oxygen molecules are essential for aerobic bacteria's development and respiration in order for them to break down and remove pollutants from wastewater. In exchange, it would be possible to use the carbon dioxide that bacteria make during respiration to grow microalgae. Microalgae species could adsorb and use nitrogen, phosphorus, and other trace elements in nutrient-rich organic effluent to grow and synthesize desired biomolecules (such as polysaccharides, carboxylic acids, and polyphosphate) through photosynthesis. Both bacteria and microalgae secrete enzymes into the cultivation system throughout their growth, including phosphatase, sulphatase, glucosidase, and galactosidase. These enzymatic agents can hydrolyse large macromolecules like proteins and polysaccharides into smaller molecules that bacteria and microalgae can consume and use. In the natural world, bacteria can provide growth factors to aid in the growth of microalgae while they interact with other microorganisms to offer a habitat for bacteria. Because of this connection, the microalgae-bacteria consortium's energy system can be recovered, which lowers the cost of additional energy sources needed for production (Chan et al. 2022). In water and wastewater treatment systems, many algal strains have proven beneficial. For instance, microalgal cells absorb nutrients like N and P to build biomass (Fallahi et al. 2020).

8.16 CHALLENGES RELATED WITH BIOLOGICAL TREATMENT OF WASTEWATER

Industrial effluent poses a serious barrier to environmentally sustainable growth, and its remediation is a difficult challenge. Although there has been a lot of study done in the past few years, there are still many obstacles to be addressed (Muñoz and Guieysse 2006). For dye removal to be effectively applied to industries, it is highly desirable to integrate various treatment technologies.

8.17 CHALLENGES RELATED WITH TREATMENT OF DYES USING AEROBIC AND ANAEROBIC TREATMENT

The incomplete degradation of aromatic amines produced during the process prevents the complete removal of azo dyes, which is one of the main drawbacks of anaerobic-aerobic sequential dye degradation methods. Anthraquinone, phthalocyanine, and triphenylmethane dyes depend on the removal of aromatic amines during aerobic treatment (Harrelkas et al. 2008). Although many other types of dyes are partially biodegraded because of their complicated structures, azo dyes are biodegraded satisfactorily by successive processes. Because of this, the sequential anaerobic-aerobic process requires additional combination treatments, such as advanced oxidation processes (AOPs), and takes more time and money to complete. Additionally, sulphonated azo dye compounds like o-aminohydroxynaphthalenes can engage in autoxidation reactions when exposed to oxygen. Because sulphonated aminohydroxynaphthalene can produce stable products such as 1,2,7-triamino-8-hydroxynaphthalene-3,6-disulphonate (TAHNDS) and 1-amino2-hydroxynaphthalene-6-sulphonate (AHNS) under aerobic conditions, its behaviour is crucial for the sequential anaerobic-aerobic processes. When all of the contaminants in the wastewater are biodegradable, biological wastewater treatment is typically the most cost-effective. Due to the presence of biorefractory compounds, this isn't always the case. The sluggish rate of biodegradation of specific pollutants, the limited biodegradability of principally xenobiotic compounds, and the toxicity or inhibitory effects of pollutants on the microbial population are the main causes of the limitations of biological processes. Industrial wastewaters frequently contain xenobiotic contaminants. In these situations, additional treatment methods must be used, which typically entails a sharp rise in investment and operating expenses. Despite the fact that some substances are generally biodegradable, the needed residence times are too long, which results in high investment costs. Wastewater from industry and agriculture frequently contains harmful substances. If they decompose, careful feeding might enable a stable operation (Ji et al. 2010).

8.18 CHALLENGES RELATED WITH MICROALGAE

The use of microalgae to treat agro-industrial and industrial wastewaters has both advantages and disadvantages. Carbon is not a limiting element for microalgae growth because CO_2 can be collected from air and flue gases, and organic carbon is abundant in nutrient-rich wastewaters. Microalgae absorb nutrients from wastewater, decreasing the nutritional content of the wastewater and, as a result, the BOD and COD. However, in order to achieve these reductions in the long run, the microalgae must be removed from the wastewater after the treatment process is complete, as the microalgae biomass contributes to the organic content. The extraction of biomass from the growing media and biomolecules is a difficult issue in microalgal biomass production in the wastewater stream. The type of microalgae used, the wastewater parameters, and the intended applications all influence the choice of growing system for bulk production of biomass. An ideal culture system should be simple to maintain, inexpensive to build, provide adequate lighting, be capable of effective liquid and gas exchanges, and pose little risk of contamination. The downstream processing of the microalgal suspension and the rearing of microalgae are two difficulties associated with using microalgae for the treatment of industrial and agro-industrial effluent. The difficulties include internal shading, turbidity of the wastewater, harvesting of the microalgae, suspended particulates in the wastewater, and others (Chan et al. 2022).

8.18.1 Turbidity and Suspended Solids in Wastewater

Large levels of suspended particles in agro-industrial and industrial wastewaters may hinder the growth of microalgae. Light will enter wastewater less readily if there are solids present. The water column's ability to transmit light is further constrained by high turbidity. This decrease in light will slow photosynthesis because microalgae deeper in the wastewater will not receive enough sunlight because not all microalgae can float on the surface to gather sunlight. To remove the most SS from wastewater, flocculants can be used as a pre-treatment. Adding agitation to the wastewater is another approach to overcoming this difficulty. Turbulence increases the productivity of the microalgal treatment of wastewater by exposing the microalgae to light for at least a brief amount of time. The operation expenses will be higher if wastewater is treated in a closed system with additional light. Raceway ponds are one example of a solution to this issue. Rotating discs add turbulence to raceway ponds, ensuring that the nutrients are dispersed evenly throughout the pond and that all microalgae have access to CO_2 and sunlight.

8.18.2 Microalgae Recovery and Harvesting

A critical step in separating microalgae from wastewater is harvesting. Engineers frequently run into difficulties when trying to harvest microalgae; thus, different techniques are utilized depending on the type of microalgae. Centrifugation, flocculation, flotation, filtering, gravity sedimentation, and ultrasonication are some of the harvesting techniques. The size, density, and yield of the desired products are factors that influence how well microalgae can be separated from water. For instance, because of its small size (often less than 20 Fluid Thioglycollate Medium (Ftm)) and negative charge, *Chlorella* sp. is exceedingly challenging to recover. Filtration, sedimentation, and microstraining are among the common separation methods that are hampered by this combination. The effectiveness of harvesting methods is inversely correlated with their cost; inefficient methods create biomass that is a mixture of microalgae and chemical flocculants and take a long time to separate (gravity sedimentation) (Mehariya et al. 2021).

8.18.3 Internal Shading

One of the causes of the low photosynthetic activity of microalgae is internal shadowing. Algal biomass can double in 24 hours, and during the log phase, it can double in as little as 3.5 hours. Due to the thick culture in the upper portion, which lowers light's capacity to penetrate into the water, the quick increase in cell concentration will reduce the amount of light received by areas of the effluent. A raceway pond or physically based rendering (PBR) can be used to solve this issue. A PBR ensures that the microalgae receive enough light by lighting the outer portion of the PBR. Air being sparged into the system will allow light to enter the PBR system. The PBR's internal air flow will rotate the microalgae, ensuring that they are mixed toward the side closest to the light source. In a raceway pond, the microalgae are circulated around the entire pond by the rotation of the paddle shift, and those from the bottom are circulated toward the upper part, allowing them an opportunity to absorb light energy while near the surface.

8.19 CHALLENGES RELATED WITH MICROALGAE AND BACTERIA CONSORTIA

The utilization of microalgae-bacteria consortia for co-metabolism in organic wastewater treatment may lead to adverse effects on both microalgae and bacteria. The elevated concentration of induced matrix and refractory organic materials influenced microbial activity. Various antibiotic dosages act as either a hindrance, inhibiting the growth of bacteria and microalgae, or a stimulant, promoting microalgae growth by enhancing protease synthesis and gene expression within the cell. Antibiotics have the potential to be absorbed, bioaccumulated, and biodegraded by bacteria and microalgae at low concentrations. However, when present in high concentrations within the microalgae-bacteria consortia, antibiotics may exert a growth-inhibitory effect on the system. Elevated antibiotic levels in the wastewater treatment system could alter the composition of microalgae cells, leading to reduced cell membrane integrity, solubility, and permeability, as well as decreased chlorophyll, protein content, and other cellular components, thereby suppressing gene expression and growth. This disruption in the replication and transcription processes of essential proteins and enzymes could significantly impede microalgae growth, particularly affecting crucial functions such as electron transfer required for photosynthesis, nitrogen and phosphorus uptake, as well as oxygen and carbon dioxide absorption and transportTop of Form (S. S. Chan et al. 2022).

Microalgae have demonstrated their ability to compete in the biotreatment of wastewater, nutrient recovery, and biomass production. However, costly biomass harvesting and subpar nutrient removal efficacy are significant barriers to the widespread use of microalgae technology (Kant et al. 2022). Microalgae-bacterial consortia were developed based on the direct or indirect symbiotic interaction between microalgae and bacteria, including substrate exchange, intercellular communication, and horizontal gene transfer, to improve wastewater treatment efficiency and biomass production (Chen et al. 2022). Microalgae-bacteria consortia, aerobic heterotrophic biosystems, and heterotrophic nitrifying bacteria can all achieve high N removal efficiency rate, however they cannot support sustainable N management due to their 8% to 90% N loss. Although microalgae and combined photosynthetic bacteria (PSB) technologies have the capacity to completely recover nitrogen from wastewater, their poor N removal rates and lack of environmental adaptation make them insufficient for the massive flow required for effective wastewater treatment (Capson-Tojo et al. 2021). Additionally, in open systems, competition for nutrients and space by ambient microbes and microalgae/PSB may cause a decrease in the amount of microalgae and PSB (Mohsenpour et al. 2021). For PSB systems and microalgae-bacteria consortia, light conditions are a difficulty. The daily, weekly, and even yearly variations in natural light lead to more unpredictable performance. Although they increase energy costs, artificial light sources can guarantee the effectiveness of healing and recuperation. Last but not least, the effectiveness of nutrient recovery, wastewater treatment, and production cost can all be significantly impacted by liquid-solid separation efficiency. There are a few microalgae species that respond differently depending on the source of the wastewater, even though the majority of microalgae are accustomed to or acclimatised in wastewater. According to studies, wastewater with high levels of COD, total phosphorus (TP), and total nitrogen (TN) cannot support the growth of microalgae. As a result, algal biomass production and the extent to which COD, TP, and TN are removed from the effluent vary (Huo et al. 2020).

8.20 CONCLUSION AND FUTURE PROSPECTIVE

The presence of diverse inorganic and organic hazardous substances in wastewater discharged from various sectors pose significant threat to both the environment and human health. However, achieving effective pollutant removal poses a considerable challenge, necessitating thorough wastewater characterization and treatment. Moreover, the utilization of treated wastewater in agricultural or other domestic applications can lead to soil salinity, metal toxicity, and nitrogen imbalance. Therefore, prior to its use in irrigation, it is imperative to assess and analyse its toxicity levels, as it has the potential to bioaccumulate and directly harm human internal organs. The primary treatment procedure typically involves screening and the sedimentation of large elements from industrial, domestic, and municipal waste streams.

Enhancing the biodegradability of organic matter remains a significant challenge in both aerobic and anaerobic treatments, necessitating a deeper understanding and tighter control, especially concerning pre-treatment management practices. While aerobic treatment has shown promise in reducing direct greenhouse gas emissions, anaerobic digestion stands out for its superior effectiveness and adaptability. Nevertheless, ongoing research is essential to optimize management practices and enhance productivity. Additionally, the ability to effectively remove emerging pollutants, such as antibiotics and estrogenic compounds, is a crucial factor in selecting treatment systems, highlighting the need for research-backed data in this area. Microalgae have shown promise in removing various pollutants through bioaccumulation, biodegradation, and bio-adsorption, yet transitioning from laboratory-based investigations to field applications remains a challenge. Thus, further studies are warranted to leverage the biodegradation capabilities of microalgal species for effective bioremediation. Scaling up microbial wastewater treatment processes, conducting cost analyses, utilizing treated effluent to reduce freshwater demand, and developing cost-effective strategies for industrial sector adoption are key areas requiring additional research. Moreover, toxicity studies on degraded products are imperative to ensure environmental safety, emphasizing the importance of assessing the safety of degradation products for recipient organisms. To enhance overall efficiency, comprehensive research efforts are needed to identify suitable microorganisms, optimize experimental parameters, determine optimal bioremediation locations, elucidate organic pollutant degradation pathways, and reevaluate previous bioremediation studies. Ultimately, integrating techniques and scaling up laboratory successes to treat real industrial effluents should be a priority for future endeavours in environmental remediation.

REFERENCES

Abdel-Raouf, Manar Elsayed, Nermine E Maysour, Reem Kamal Farag, and Abdul-Raheim Mahmoud Abdul-Raheim. 2019. "Wastewater Treatment Methodologies, Review Article." *International Journal of Environment & Agricultural Science* 3 (1): 1–26. https://www.researchgate.net/publication/332183222.

Akhtar, Naseem, Muhammad Izzuddin Syakir Ishak, Showkat Ahmad Bhawani, and Khalid Umar. 2021. "Various Natural and Anthropogenic Factors Responsible for Water Quality Degradation: A Review." *Water* 13 (19): 1–35. https://doi.org/10.3390/w13192660.

Al-Dhabi, Naif Abdullah, and Mariadhas Valan Arasu. 2022. "Biosorption of Hazardous Waste From the Municipal Wastewater by Marine Algal Biomass." *Environmental Research* 204 (3): 112115. https://doi.org/10.1016/j.envres.2021.112115.

Biswas, Tethi, Srimoyee Banerjee, Amrita Saha, Abhishek Bhattacharya, Chaitali Chanda, Lalit Mohan Gantayet, Punyasloke Bhadury, and Shaon Ray Chaudhuri. 2022. "Bacterial Consortium Based Petrochemical Wastewater Treatment: From Strain Isolation to Industrial Effluent Treatment." *Environmental Advances* 7 (4): 100132. https://doi.org/10.1016/j.envadv.2021.100132.

Bougrier, C., H. Carrère, and J. P. Delgenès. 2005. "Solubilisation of Waste-Activated Sludge by Ultrasonic Treatment." *Chemical Engineering Journal* 106 (2): 163–169. https://doi.org/10.1016/j.cej.2004.11.013.

Brack, Werner, Damia Barcelo Culleres, Alistair B.A. Boxall, Hélène Budzinski, Sara Castiglioni, Adrian Covaci, and Valeria Dulio, et al. 2022. "One Planet: One Health. A Call to Support the Initiative on a Global Science–Policy Body on Chemicals and Waste." *Environmental Sciences Europe* 34 (1): 1–10. https://doi.org/10.1186/s12302-022-00602-6.

Brockmann, Doris, Yves Gérand, Chul Park, Kim Milferstedt, Arnaud Hélias, and Jérôme Hamelin. 2021. "Wastewater Treatment Using Oxygenic Photogranule-Based Process Has Lower Environmental Impact than Conventional Activated Sludge Process." *Bioresource Technology* 319 (11): 1–9. https://doi.org/10.1016/j.biortech.2020.124204.

Campagnolo, Enzo R., Kammy R. Johnson, Adam Karpati, Carol S. Rubin, Dana W. Kolpin, Michael T. Meyer, and J. Emilio Esteban, et al. 2002. "Antimicrobial Residues in Animal Waste and Water Resources Proximal to Large-Scale Swine and Poultry Feeding Operations." *Science of the Total Environment* 299 (1–3): 89–95. https://doi.org/10.1016/S0048-9697(02)00233-4.

Capson-Tojo, Gabriel, Shengli Lin, Damien J. Batstone, and Tim Hulsen. 2021. "Purple Phototrophic Bacteria Are Outcompeted by Aerobic Heterotrophs in the Presence of Oxygen." *Water Research* 194 (4): 116941. https://doi.org/10.1016/j.watres.2021.116941.

Chan, Yi Jing, Mei Fong Chong, Chung Lim Law, and D. G. Hassell. 2009. "A Review on Anaerobic-Aerobic Treatment of Industrial and Municipal Wastewater." *Chemical Engineering Journal* 155 (1–2): 1–18. https://doi.org/10.1016/j.cej.2009.06.041.

Chan, Sook Sin, Kuan Shiong Khoo, Kit Wayne Chew, Tau Chuan Ling, and Pau Loke Show. 2022. "Recent Advances Biodegradation and Biosorption of Organic Compounds from Wastewater : Microalgae-Bacteria Consortium - A Review." *Bioresource Technology* 344 (1): 126159. https://doi.org/10.1016/j.biortech.2021.126159.

Chen, Zhipeng, Yue Xie, Shuang Qiu, Mengting Li, Wenqi Yuan, and Shijian Ge. 2022. "Granular Indigenous Microalgal-Bacterial Consortium for Wastewater Treatment: Establishment Strategy, Functional Microorganism, Nutrient Removal, and Influencing Factor." *Bioresource Technology* 353 (6): 127130. https://doi.org/10.1016/j.biortech.2022.127130.

Christenson, Logan, and Ronald Sims. 2011. "Production and Harvesting of Microalgae for Wastewater Treatment, Biofuels, and Bioproducts." *Biotechnology Advances* 29 (6): 686–702. https://doi.org/10.1016/j.biotechadv.2011.05.015.

Colleran, E., and A. Pistilli. 1994. "Activity Test System for Determining the Toxicity of Xenobiotic Chemicals to the Methanogenic Process." *Health and Environmental Research Online* 44 (1): 1–20.

Cuellar-Bermudez, Sara P., Gibran S. Aleman-Nava, Rashmi Chandra, J. Saul Garcia-Perez, Jose R. Contreras-Angulo, Giorgos Markou, Koenraad Muylaert, Bruce E. Rittmann, and Roberto Parra-Saldivar. 2017. "Nutrients Utilization and Contaminants Removal. A Review of Two Approaches of Algae and Cyanobacteria in Wastewater." *Algal Research* 24 (6): 438–449. https://doi.org/10.1016/j.algal.2016.08.018.

Custodio, María, Richard Peñaloza, Ciro Espinoza, Wilson Espinoza, and Juana Mezarina. 2022. "Treatment of Dairy Industry Wastewater Using Bacterial Biomass Isolated from Eutrophic Lake Sediments for the Production of Agricultural Water." *Bioresource Technology Reports* 17 (2): 1–9. https://doi.org/10.1016/j.biteb.2021.100891.

Dhandayuthapani, K., P. Senthil Kumar, Wen Yi Chia, Kit Wayne Chew, V. Karthik, H. Selvarangaraj, P. Selvakumar, P. Sivashanmugam, and Pau Loke Show. 2022. "Bioethanol from Hydrolysate of Ultrasonic Processed Robust Microalgal Biomass Cultivated in Diary Waste Wastewater Under Optimal Strategy." *Energy* 244 (1): 122604. https://doi.org/10.1016/j.energy.2021.122604.

Fallahi, Alireza, Nima Hajinajaf, Omid Tavakoli, and Mohammad Hossein Sarrafzadeh. 2020. "Cultivation of Mixed Microalgae Using Municipal Wastewater: Biomass Productivity, Nutrient Removal, and Biochemical Content." *Iranian Journal of Biotechnology* 18 (4): 88–97. https://doi.org/10.30498/IJB.2020.2586.

Garg, Shivani, Zaira Zaman Chowdhury, Abu Nasser Mohammad Faisal, Nelson Pynadathu Rumjit, and Paul Thomas. 2022. Impact of Industrial Wastewater on Environment and Human Health. In S. Roy, A. Garg, S. Garg, T. A. Tran (Eds.), *Advanced Industrial Wastewater Treatment and Reclamation of Water* (pp. 197–209). Springer Cham. https://doi.org/10.1007/978-3-030-83811-9.

Godos, Ignacio De, Saúl Blanco, Pedro A. García-encina, Eloy Becares, and Raúl Muñoz. 2009. "Bioresource Technology Long-Term Operation of High Rate Algal Ponds for the Bioremediation of Piggery Wastewaters at High Loading Rates." *Bioresource Technology* 100 (19): 4332–4339. https://doi.org/10.1016/j.biortech.2009.04.016.

Gonze, E., S. Pillot, E. Valette, Y. Gonthier, and A. Bernis. 2003. "Ultrasonic Treatment of an Aerobic Activated Sludge in a Batch Reactor." *Chemical Engineering and Processing* 42 (12): 965–975. https://doi.org/10.1016/S0255-2701(03)00003-5.

Gujer, W., and A. J. B. Zehnder. 2018. "Conversion Processes in Anaerobic Digestion." *Water Science and Technology* 15 (8–9): 127–167. https://doi.org/10.2166/wst.1983.0164.

Gupta, Priyanka, Komal Pandey, and Nishith Verma. 2022. "Augmented Complete Mineralization of Glyphosate in Wastewater via Microbial Degradation Post CWAO Over Supported Fe-CNF." *Chemical Engineering Journal* 428 (8): 1–11. https://doi.org/10.1016/j.cej.2021.132008.

Haixia, Zhao, Duan Xuejun, Becky Stewart, You Bensheng, and Jiang Xiawei. 2013. "Spatial Correlations between Urbanization and River Water Pollution in the Heavily Polluted Area of Taihu Lake Basin, China." *Journal of Geographical Sciences* 23 (4): 735–752. https://doi.org/10.1007/s11442-013-1041-7.

Harmsen, Hermie J. M., Antoon D. L. Akkermans, Alfons J. M. Stams, and Willem M. de Vos. 1996. "Population Dynamics of Propionate-Oxidizing Bacteria Under Methanogenic and Sulfidogenic Conditions in Anaerobic Granular Sludge." *Applied and Environmental Microbiology* 62 (6): 2163–2168.

Harrelkas, F., A. Paulo, M. M. Alves, L. El Khadir, O. Zahraa, M. N. Pons, and F. P. Van der Zee. 2008. "Photocatalytic and Combined Anaerobic–Photocatalytic Treatment of Textile Dyes." *Chemosphere* 72 (11): 1816–1822. https://doi.org/10.1016/j.chemosphere.2008.05.026.

Huo, Shuhao, Miao Kong, Feifei Zhu, Jingya Qian, Daming Huang, Paul Chen, and Roger Ruan. 2020. "Co-Culture of Chlorella and Wastewater-Borne Bacteria in Vinegar Production Wastewater: Enhancement of Nutrients Removal and Influence of Algal Biomass Generation." *Algal Research* 45 (1): 101744. https://doi.org/10.1016/j.algal.2019.101744.

Ji, Yulan, Yanhong Wang, Jinsheng Sun, Tingyan Yan, Jing Li, Tingting Zhao, Xiaohong Yin, and Changjiang Sun. 2010. "Enhancement of Biological Treatment of Wastewater by Magnetic Field." *Bioresource Technology* 101 (22): 8535–8540. https://doi.org/10.1016/j.biortech.2010.05.094.

Ji, Bin, Meng Zhang, Jun Gu, Yingqun Ma, and Yu Liu. 2020. "A Self-Sustaining Synergetic Microalgal-Bacterial Granular Sludge Process Towards Energy-Efficient and Environmentally Sustainable Municipal Wastewater Treatment." *Water Research* 179 (7): 115884. https://doi.org/10.1016/j.watres.2020.115884.

Joel, Chebor, K. Ezekiel, and Lizzy A. Mwamburi. 2021. "Effects of Seasonal Variation on Performance of Conventional Wastewater Treatment System." *Journal of Applied & Environmental Microbiology* 5 (1): 1–7. https://doi.org/10.7176/jees/11-7-06.

Kant, Shashi, Vishal Ahuja, Neha Chandel, Sanjeet Mehariya, Pradeep Kumar, and Yung-hun Yang. 2022. "An Overview on Microalgal-Bacterial Granular Consortia for Resource Recovery and Wastewater Treatment." *Bioresource Technology* 351 (3): 127028. https://doi.org/10.1016/j.biortech.2022.127028.

Leslie Grady, Jr, C. P., G. T. Daigger, and H.C. Lim. 1999. "Biological Wastewater Treatment." In *Revised and Expanded* (2nd ed.). CRC Press.

Li, Jun, Li-bin Ding, Ang Cai, Guo-Xian Huang, and Harald Horn. 2014. "Aerobic Sludge Granulation in a Full-Scale Sequencing Batch Reactor." *Hindawi* 2014 (4): 1–13. https://doi.org/10.1155/2014/268789.

Liu, Yu, and Qi-Shan Liu. 2006. "Causes and Control of Filamentous Growth in Aerobic Granular Sludge Sequencing Batch Reactors." *Biotechnology Advances* 24 (1): 115–127. https://doi.org/10.1016/j.biotechadv.2005.08.001.

Liu, Yu, and Joo-Hwa Tay. 2002. "The Essential Role of Hydrodynamic Shear Force in the Formation of Biofilm and Granular Sludge." *Water Research* 36 (4): 1653–1665. https://doi.org/10.1016/S0043-1354(01)00379-7.

McCarty, P. L. 2018. "The Development of Anaerobic Treatment and Its Future." *Water Science and Technology* 44 (8): 149–156. https://doi.org/10.2166/wst.2001.0487.

McHugh, Sharon, Micheal Carton, Therese Mahony, and Vincent O'Flaherty. 2003. "Methanogenic Population Structure in a Variety of Anaerobic Bioreactors." *FEMS Microbiology Letters* 219 (2): 297–304. https://doi.org/10.1016/S0378-1097(03)00055-7.

Mehariya, Sanjeet, Rahul Kumar Goswami, Pradeep Verma, Roberto Lavecchia, and Antonio Zuorro. 2021. "Integrated Approach for Wastewater Treatment and Biofuel Production in Microalgae Biorefineries." *Energies* 14 (8): 2282. https://doi.org/10.3390/en14082282.

Mohsenpour, Seyedeh Fetemeh, Sebastian Hennige, Nicholas Willoughby, Adebayo Adeloye, and Tony Gutierrez. 2021. "Integrating Micro-Algae into Wastewater Treatment: A Review." *Science of the Total Environment* 752 (1): 142168. https://doi.org/10.1016/j.scitotenv.2020.142168.

Montefiore, Lise R., Natalie G. Nelson, Amanda Dean, and Mahmoud Sharara. 2022. "Reconstructing the Historical Expansion of Industrial Swine Production from Landsat Imagery." *Scientific Reports* 12 (1): 1–12. https://doi.org/10.1038/s41598-022-05789-5.

Montes, Nuria, Marta Otero, Ricardo N. Coimbra, Rosa Mendez, and Javier Martin-Villacorta. 2015. "Removal of Tetracyclines from Swine Manure at Full-Scale Activated Sludge Treatment Plants." *Environmental Technology* 36 (15): 1966–1973. https://doi.org/10.1080/09593330.2015.1018338.

Montingelli, M. E., S. Tedesco, and A. G. Olabi. 2015. "Biogas Production from Algal Biomass : A Review." *Renewable and Sustainable Energy Reviews* 43 (3): 961–972.

Moyo, Senelisile, Bukisile P. Makhanya, and Pinkie E. Zwane. 2022. "Use of Bacterial Isolates in the Treatment of Textile Dye Wastewater: A Review." *Heliyon* 8 (6): 1–13. https://doi.org/10.1016/j.heliyon.2022.e09632.

Muhammed, Adil, Avatar N. Poduval, Piyush Oonnikrishnan, Pranav K. Narayanan, and K. Yaduraj. 2021. "The Oxygenic Photogranule for Wastewater Treatment Process." *IOP Conference Series: Materials Science and Engineering* 1114 (1): 1–8. https://doi.org/10.1088/1757-899x/1114/1/012090.

Muñoz, Raul, and Benoit Guieysse. 2006. "Algal-Bacterial Processes for the Treatment of Hazardous Contaminants: A Review." *Water Research* 40 (15): 2799–2815. https://doi.org/10.1016/j.watres.2006.06.011.

Nah, In Wook, Yun Whan Kang, Kyung Yub Hwang, and Woong Ki Song. 2000. "Mechanical Pretreatment of Waste Activated Sludge for Anaerobic Digestion Process." *Water Research* 34 (8): 2362–2368. https://doi.org/10.1016/S0043-1354(99)00361-9.

Najafpour, G. D., A. A. L Zinatizadeh, A. R. Mohamed, and M. Hasnain Isa. 2006. "High-Rate Anaerobic Digestion of Palm Oil Mill Effluent in an Upflow Anaerobic Sludge-Fixed Film Bioreactor." *Process Biochemistry* 41 (2): 370–379. https://doi.org/10.1016/j.procbio.2005.06.031.

Najim, Aya A., Zainab Z. Ismail, and Khalid K. Hummadi. 2022. "Biodegradation Potential of Sodium Dodecyl Sulphate (SDS) by Mixed Cells in Domestic and Non-Domestic Actual Wastewaters: Experimental and Kinetic Studies." *Biochemical Engineering Journal* 180 (3): 108374. https://doi.org/10.1016/j.bej.2022.108374.

Nancharaiah, Y. V., H. M. Joshi, T. V. K. Mohan, V. P. Venugopalan, and S. V. Narasimhan. 2006. "Aerobic Granular Biomass : A Novel Biomaterial for Efficient Uranium Removal." *Current Science* 91 (8): 503–509.

Neis, Uwe, Klaus Nickel, and Anna Lunden. 2008. "Improving Anaerobic and Aerobic Degradation by Ultrasonic Disintegration of Biomass." *Journal of Environmental Science and Health Part A* 43 (10): 1541–1545. https://doi.org/10.1080/10934520802293701.

Ødegaard, Hallvard. 2004. "Sludge Minimization Technologies - An Overview." *Water Science and Technology* 49 (10): 31–40. https://doi.org/10.2166/wst.2004.0602.

Owa, F. D. 2013. "Water Pollution: Sources, Effects, Control and Management." *Mediterranean Journal of Social Sciences* 4 (8): 65–68. https://doi.org/10.5901/mjss.2013.v4n8p65.

Pacheco, Diana, Ana Cristina Rocha, Leonel Pereira, and Tiago Verdelhos. 2020. "Microalgae Water Bioremediation: Trends and Hot Topics." *Applied Sciences* 10 (5): 1886.

Park, J. B. K., R. J. Craggs, and A. N. Shilton. 2011. "Bioresource Technology Wastewater Treatment High Rate Algal Ponds for Biofuel Production." *Bioresource Technology* 102 (1): 35–42. https://doi.org/10.1016/j.biortech.2010.06.158.

Poddar, Kasturi, Debapriya Sarkar, Debatri Chakraborty, Pritam Bajirao Patil, Sourav Maity, and Angana Sarkar. 2022. "Paracetamol Biodegradation by Pseudomonas Strain PrS10 Isolated from Pharmaceutical Effluents." *International Biodeterioration & Biodegradation* 175 (11): 105490. https://doi.org/10.1016/j.ibiod.2022.105490.

Pronk, M., M. K. De Kreuk, B. De Bruin, P. Kamminga, R. V. Kleerebezem, and M. C. Van Loosdrecht. 2015. "Full Scale Performance of the Aerobic Granular Sludge Process for Sewage Treatment." TU Delft University. https://doi.org/10.1016/j.watres.2015.07.011.

Qian, Zhang, Li Na, Wang Bao-Long, Zhang Tao, Ma Peng-Fei, Zhang Wei-Xiao, Nusrat Zahan Sraboni, Ma Zheng, Zhang Ying-Qi, and Ying LI. 2022. "Capabilities and Mechanisms of Microalgae on Nutrients and Florfenicol Removing from Marine Aquaculture Wastewater." *Journal of Environmental Management* 320 (10): 115673. https://doi.org/10.1016/j.jenvman.2022.115673.

Rajaguru, P., K. Kalaiselvi, M. Palanivel, and V. Subburam. 2000. "Biodegradation of Azo Dyes in a Sequential Anaerobic-Aerobic System." *Applied Microbiology and Biotechnology* 54 (8): 268–273. https://doi.org/10.1007/s002530000322.

Rosario, Carlos Gonzalo Alvarez, Amzy Tania Vallenas-Arevalo, Santiago Justo Arevalo, Denise Crocce Romano Espinosa, and Jorge Alberto Soares Tenorio. 2022. "Biodegradation of Cyanide Using a Bacillus Subtilis Strain Isolated from Artisanal Gold Mining Tailings." *Brazilian Journal of Chemical Engineering* 40 (3): 129–136. https://doi.org/10.1007/s43153-022-00228-4.

Sar, Taner, and Meltem Yesilcimen Akbas. 2022. "Potential Use of Olive Oil Mill Wastewater for Bacterial Cellulose Production." *Bioengineered* 13 (3): 7659–7669. https://doi.org/10.1080/21655979.2022.2050492.

Schmidt, Jens E., and Birgitte K. Ahring. 1996. "Granular Sludge Formation in Upflow Anaerobic Sludge Blanket (UASB) Reactors." *Biotechnology and Bioengineering* 49 (3): 229–246. https://doi.org/10.1002/(SICI)1097-0290.

Schmidt, Jens Ejbye, and Birgitte Kjaer Ahring. 1999. "Immobilization Patterns and Dynamics of Acetate-Utilizing Methanogens Immobilized in Sterile Granular Sludge in Upflow Anaerobic Sludge Blanket Reactors." *Applied and Environmental Microbiology* 65 (3): 1050–1054. https://doi.org/10.1128/AEM.65.3.1050-1054.1999.

Sekiguchi, Yuji, Yoichi Kamagata, Kazuaki Syutsubo, Akiyoshi Ohashi, Hideki Harada, and Kazunori Nakamura. 1998. "Phylogenetic Diversity of Mesophilic and Thermophilic Granular Sludges Determined by 16s RRNA Gene Analysis." *Microbiology* 144 (9): 2655–2665. https://doi.org/10.1099/00221287-144-9-2655.

Shahid, Munazzam Jawad, Ameena A. Al-surhanee, Fayza Kouadri, Shafaqat Ali, Neeha Nawaz, Muhammad Afzal, Muhammad Rizwan, Basharat Ali, and Mona H. Soliman. 2020. "Role of Microorganisms in the Remediation of Wastewater in Floating Treatment Wetlands : A Review." *Sustainability* 12 (7): 1–29. https://doi.org/10.3390/su12145559.

Shimizu, Tatsuo, Kenzo Kudo, and Yoshikazu Nasu. 1993. "Anaerobic Waste-Activated Sludge Digestion-A Bioconversion Mechanism and Kinetic Model." *Biotechnology and Bioengineering* 41 (11): 1082–1091. https://doi.org/10.1002/bit.260411111.

Siddiqui, Nahid Masood, and Praveen Dahiya. 2022. *Enzyme-Based Biodegradation of Toxic Environmental Pollutants. Development in Wastewater Treatment Research and Processes: Microbial Degradation of Xenobiotics through Bacterial and Fungal Approach.* https://doi.org/10.1016/B978-0-323-85839-7.00003-7.

Struk, Martin, Cristian A. Sepulveda-Munoz, Ivan Kushkevych, and Raul Munoz. 2023. "Photoautotrophic Removal of Hydrogen Sulfide from Biogas Using Purple and Green Sulfur Bacteria." *Journal of Hazardous Materials* 443 (6): 130337. https://doi.org/10.1016/j.jhazmat.2022.130337.

Subashchandrabose, Suresh R., Balasubramanian Ramakrishnan, Mallavarapu Megharaj, Kadiyala Venkateswarlu, and Ravi Naidu. 2013. "Mixotrophic Cyanobacteria and Microalgae as Distinctive Biological Agents for Organic Pollutant Degradation." *Environment International* 51 (1): 59–72. https://doi.org/10.1016/j.envint.2012.10.007.

Suzuki, Kazuvoshi, Mivoko Waki, Tomoko Yasuda, Yasuvuki Fukumoto, Kazutaka Kuroda, Takahiro Sakai, Naoto Suzuki, Ryoji Suzuki, and Kenji Matsuba. 2010. "Bioresource Technology Distribution of Phosphorus, Copper and Zinc in Activated Sludge Treatment Process of Swine Wastewater." *Bioresource Technology* 101 (23): 9399–9404. https://doi.org/10.1016/j.biortech.2010.07.014.

Tay, J. H., S. T. L. Tay, V. Ivanov, S. Pan, H. L. Jiang, and Q. S. Liu. 2003. "Biomass and Porosity Profiles in Microbial Granules Used for Aerobic Wastewater Treatment." *Letters in Applied Microbiology* 36 (5): 297–301. https://doi.org/10.1046/j.1472-765X.2003.01312.x.

Tiehm, A., K. Nickel, M. Zellhorn, and U. Neis. 2009. "Ultrasonic Waste Activated Sludge Disintegration for Improving Anaerobic Stabilization." *Water Research* 35 (8): 2003–2009. https://doi.org/10.1016/S0043-1354(00)00468-1.

Tyagi, I., K. Tyagi, K. Chandra, and Vikas Kumar. 2022. "Characterization of Bacterial Diversity in Wastewater of Indian Paper Industries with Special Reference to Water Quality." *International Journal of Environmental Science and Technology* 19 (5): 3669–3684. https://doi.org/10.1007/s13762-021-03249-7.

Velusamy, Karthik, Selvakumar Periyasamy, Ponnusamy Senthil Kumar, Femina Carolinc Car, Thanikachalam Jayaraj, M. Gokulakrishnan, and P. Keerthana. 2022. "Transformation of Aqueous Methyl Orange to Green Metabolites Using Bacterial Strains Isolated from Textile Industry Effluent." *Environmental Technology & Innovation* 25 (2): 102126. https://doi.org/10.1016/j.eti.2021.102126.

Wang, Shulian, Bin Ji, Baihui Cui, Yingqun Ma, Dabin Guo, and Yu Liu. 2021. "Cadmium-Effect on Performance and Symbiotic Relationship of Microalgal-Bacterial Granules." *Journal of Cleaner Production* 282 (2): 125383. https://doi.org/10.1016/j.jclepro.2020.125383.

Yang, Lei, Xijun Xu, Hui Wang, Jin Yan, Xu Zhou, Nangi Ren, Duu-Jong Lee, and Chuan Chen. 2022. "Biological Treatment of Refractory Pollutants in Industrial Wastewaters Under Aerobic or Anaerobic Condition: Batch Tests and Associated Microbial Community Analysis." *Bioresource Technology Reports* 17 (2): 100927. https://doi.org/10.1016/j.biteb.2021.100927.

Yeoh, B. G. 1995. "Anaerobic Treatment of Industrial Wastewaters in Malaysia." In *Post Conference Seminar on Industrial Wastewater Management in Malaysia, Kuala Lumpur, Malaysia*.

Yoon, Seong Hoon, Hyung Soo Kim, and Sangho Lee. 2004. "Incorporation of Ultrasonic Cell Disintegration into a Membrane Bioreactor for Zero Sludge Production." *Process Biochemistry* 39 (12): 1923–1929. https://doi.org/10.1016/j.procbio.2003.09.023.

Zhai, Jun, Xiaoting Li, Wei Li, Md Hasibur Rahaman, Yuting Zhao, Bubo Wei, and Haoxuan Wei. 2017. "Optimization of Biomass Production and Nutrients Removal by Spirulina Platensis from Municipal Wastewater." *Ecological Engineering* 108 (11): 83–92. https://doi.org/10.1016/j.ecoleng.2017.07.023.

Zhang, Panyue, Guangming Zhang, and Wei Wang. 2007. "Ultrasonic Treatment of Biological Sludge: Floc Disintegration, Cell Lysis and Inactivation." *Bioresource Technology* 98 (1): 207–210. https://doi.org/10.1016/j.biortech.2005.12.002.

Zitomer, Daniel H., and Richard E. Speece. 1993. "Sequential Environments for Enhanced Biotransformation of Aqueous Contaminants." *Environmental Science and Technology* 27 (2): 226–244. https://doi.org/10.1021/es00039a001.

Zou, Haiming, Xiwu Lu, and Saad Abualhail. 2014. "Characterization of Denitrifying Phosphorus Removal Microorganisms in a Novel Two-Sludge Process by Combining Chemical with Microbial Analysis." *Journal of Chemistry* 2014 (1): 1–8. https://doi.org/10.1155/2014/360503.

9 Microbial Treatment of Industrial Wastewater

Concept, Methodology, Economic, and Sustainability Inference

Hera Fatma, Nishu Pundir, Pratyaksha Srivastava, and Kalpana Katiyar

9.1 INTRODUCTION

The differentiation basis for various types of wastewater treatment plants mainly shows their dependency on the type and pattern of waste that needs to be treated. Depending on the kind and degree of contamination, a variety of techniques can be employed to treat wastewater (Cyprowski et al., 2018). Physical, chemical, and biological methodologies are included in the wastewater treatment steps. Microorganisms (also known as aerobes) that use molecular/free oxygen to assimilate organic pollutants or convert them into carbon dioxide, water, and biomass are used in aerobic treatment processes, as shown in Figure 9.1, which take place in the presence of air. On the other hand, anaerobic treatment procedures are carried out by microbes (also known as anaerobes) that do not require air to digest organic pollutants (David, 2017).

Domestic sewage, industrial sewage, and storm sewage are the three different categories of wastewater or sewage. Sanitary sewage is another name for domestic sewage, which carries wastewater from homes and apartments. Wastewater from manufacturing and chemical operations is referred to as industrial sewage. Runoff from precipitation that is gathered in a network of pipes or open channels is known

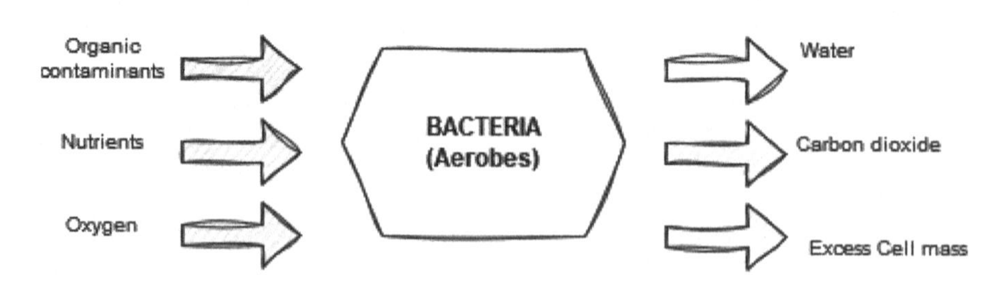

FIGURE 9.1 Overview of microbial treatment.

DOI: 10.1201/9781003381327-9

as storm sewage or storm water. Some of the major important types of wastewater treatment processes are as follows:

1. Effluent treatment plants (ETPs)
2. Sewage treatment plants (STPs)
3. Common and combined effluent treatment plants (CETPs) (Singh Asiwal et al., 2016)

Prominent pharmaceutical and chemical industries cleanse water and remove any hazardous and non-hazardous contaminants or chemicals from it using efficient ETPs. All businesses and industrial companies employ these plants to safeguard the environment. An ETP is an effective unit where industrial effluents and wastewaters are treated (Rajasulochana & Preethy, 2016). The anticipated volume of sewage produced by homes, businesses, and industries linked to sewer systems, as well as the projected inflows and infiltration, determine the size and capacity of wastewater treatment facilities.

9.2 HAZARDOUS WASTEWATER TREATMENT TECHNIQUES

9.2.1 Physical Treatment

Physical barriers and naturally existing forces like gravity, electrostatic attraction, and Van der Waal forces are used as physical methodologies of wastewater treatment to remove pollutants.

Typically, the physical treatment procedures do not alter the chemical structure of the target compounds. In certain circumstances, physical state is altered, as in the context of vapourisation, and frequently dispersed substances are made to clump together collectively called as agglomerate, such as in the case of filtration.

In the case of physical treatment, wastewater is subjected to pass through screens or filter media, or by air flotation or gravity settling in order to eliminate solids particles from it. Air-entrapped particles float to the top and can be taken out manually from the surface. They still serve as the foundation for the majority of wastewater treatment method flow systems (Sharma et al., 2021). Tanks and other structures created to confine and regulate the flow of wastewater to encourage the removal of impurities are employed in physical techniques of treatment (Roesler & Jenny, 2021). The first phase in the treatment of sanitary and industrial effluents is physical wastewater treatment, which not only improves the effectiveness of subsequent steps but also guards against equipment damage during chemical and biological treatment. Depending on the kind of effluents and the desired level of wastewater quality, different machinery and operations are opted for physical wastewater treatment. The tactics of mass transfer method serve as the foundation for the physical methodology.

The fundamental and crucial benefit of adopting physical treatments is that they are flexible in terms of technology, need little equipment and machinery load, and they can be easily adapted to a variety of treatment formats. Furthermore, when compared to other approaches, the generation of solid waste is substantially lower (Ahmed et al., 2021). It is essential to provide a solution for the removal of very fine sand during the last stages of the physical treatment of wastewater. Floating with tools like a classifier is among the better options in these circumstances and termed to be as one of the best solutions. It is clear and hence concluded that applying these cutting-edge techniques will significantly improve the wastewater treatment operations. However, due to their own limitations, these techniques can only be applied in specific cases (Sharma et al., 2021).

Equipment like the following is needed for the physical management of commercial and municipal effluents, which is based on the numerous procedures and steps indicated above:

- Different trash types (including grating, mechanical, and manual)
- Sludge bridges and settling pools for gathering sludge that has been dumped
- Classifier and various granular to get rid of small grains like sands (Piaskowski et al., 2018)

9.2.2 Chemical Treatment

Chemicals are employed in a variety of wastewater treatment techniques to speed up disinfection or sterilisation. These chemical processes, also known as chemical unit processes, cause chemical reactions and are utilised in association with physical and biological cleansing procedures to meet the necessary water standards. At wastewater treatment plants, specialised chemicals like chlorine, hydrogen peroxide, sodium chlorite, and sodium hypochlorite (bleach) serve as agents to disinfect, sterilise, and aid in the purification of wastewater. Chemical coagulation, chemical precipitation, chemical oxidation, advanced oxidation, ion exchange, and chemical neutralisation and stabilisation are only a few of the numerous chemical unit processes that can be used to purify wastewater. Chemical techniques used in wastewater treatment involve chemical disinfection (often using chlorine, ozone, or carbon dioxide, depending on the application), chemical precipitation, and ion exchange (Piaskowski et al., 2018). Chemical treatment is reportedly seen as a tertiary step that is more extensively described as "treatment of wastewater using a chemical treatment method."

Chemical-based technologies are used for both drinking water processing and wastewater treatment (Table 9.1). Chemical precipitation, neutralisation, adsorption, disinfection (using chlorine, ozone, and ultraviolet radiation), and ion exchange are the chemical treatment methods that succeeded in finding their way and are termed to be most commonly utilised in current time (Ochando-Pulido et al., 2017). A proportion of the hazardous compounds and heavy metals will react during chemical treatment or tertiary treatment, while a different proportion of the polluting substances will remain unaffected. This treatment method is also inadequate due to the high pricing of chemical additives and the environmental issue of disposing of significant amounts of chemical sludge. Both the treatment of wastewater and the processing of drinking water involve chemical-based technologies.

In fact, conventional treatments like chemically enhanced sedimentation, also known as clarification by coagulation and flocculation (to enhance the removal of suspended solids), chemical precipitation, ion exchange or adsorption (to remove specific contaminants), and disinfection (to inactivate bacteria and viruses), are quite

TABLE 9.1
Overview of the Chemical Processes Used in Water and Wastewater Treatment

Chemical-Based Processes	Description	Target Contaminants
Coagulation	Destabilisation of particles to facilitate aggregation during flocculation	Colloidal particles (particle size 0.01–1 μm)
Flocculation	Promoting the aggregation of small particles into larger ones to enhance removal by sedimentation	Total suspended solids
Neutralisation	Control of pH	Not applicable
Chemical precipitation	Enhancement of the removal of suspended solids and target contaminants by the addition of chemicals	Phosphorus, heavy metals, total suspended solids, and BOD
Disinfection/advanced oxidation	Use of strong oxidants (e.g. UV light, ozone, chlorine compounds, and hydrogen peroxide) to inactivate bacteria and viruses (disinfection)	Bacteria and viruses (disinfection)
Ion exchange	Process in which ions of a given species are displaced from an insoluble exchange material by ions of a different species in solution (i.e. contaminants)	Wastewater: Ammonia-contaminated drinking water: hardness due to calcium and magnesium
Adsorption	Process of accumulating contaminants that are in solution on an adsorbent	Organic compounds

often used for the treatment, despite the fact that different chemicals or dosages are used (as described in the following sections) (Yargeau, 2012).

9.3 SCOPE OF VARIOUS CONVENTIONAL TECHNIQUES IN WASTEWATER TREATMENT

The prominent reason behind the higher level of wastewater containing chemically heterogeneous compounds is the increasing population and the socio-economic buildout (Zinicovscaia & Cepoi, 2016). The quality factor of natural water is used to show its dependency on the degree of wastewater treatment (Zinicovscaia & Cepoi, 2016). The social change began to produce more polluted water, because of which the developing countries became victims, as people are facing sanitation problems besides consumption of polluted water (Rajasulochana & Preethy, 2016). The wastewater treatment process starts with effluent collection from industries through underground drainage systems and subjecting it to treatment plants, wherein, through primary, secondary, and tertiary levels, purified water can be collected (Sathya et al., 2022). Industrial wastewater declares notable threat to health and the environment so proper treatment and advanced practices are necessary (Sathya et al., 2022). Chemical precipitation, carbon adsorption, ion exchange, evaporation, and membrane processes are the conventional techniques used for removing heavy metals (Rajasulochana & Preethy, 2016). The wastewater treatment aims to overcome water pollution and provide protection to public health through the consumption of useful water resources against the spread of diseases (Rajasulochana & Preethy, 2016). To purify industrial wastewater, the highly efficient and recommended conventional treatment method, among many other methods, includes the implementation of an integrated water reuse design that allows the retrieved wastewater to be used. In the case of water supply, this method ensures adequate workability, through which fidelity can be assured (Sathya et al., 2022). For resource recycling and purification, the RRR technique is used in wastewater treatment plants in a more efficient way. The major objective of wastewater treatment is to eliminate contamination so that the required amount of potable water can be obtained. Thus, during the design of the treatment plant, there should be surety that the treatment plant must be composed of all efficient parameters. The wastewater treatment process shows effects on saving water and preventing water scarcity, which at the end shows its impact on the treatment plants. Wastewater treatment involves various remarkable effects on saving water and preventing water scarcity, which show both direct and indirect impacts on the treatment plant economically (Sathya et al., 2022).

9.4 ADVANTAGES AND DISADVANTAGES OF TECHNIQUES USED FOR WASTEWATER TREATMENT

9.4.1 Techniques Available for Pollutant Removal

Domestic sewage, urban run-off, industrial effluents, agriculture wastewater STPs, and the food processing industry are the main causes of organic pollution comprising organic compounds (Airport & Free, 2016). There is significant interest in techniques that are efficiently used for removing highly toxic organic compounds from water, which are limited. High-cost methods include coagulation, filtration with coagulation, precipitation, ozonation, adsorption, ion exchange, reverse osmosis, and advanced oxidation processes.

Some methods ensure the recovery of pollutant values and their removal. Unlikely, these are not easily feasible methods because of their high cost; ion exchange and reverse osmosis are the most recommended methods (Ahmed et al., 2021). Among many methods, simple design-based adsorption method by solid adsorbent is a common technique that is also advantageous because of its low land and initial cost involving the elimination of effluent from wastewater (Alasadi, 2019). This technique is also famous among researchers due to its wide usage in treating industrial wastewater from pollutants.

9.4.2 Biological Process

Biological wastewater treatment is a common and impactful operation for treating wastewater. It relies on the natural processes of biodegradation to break down the organic matter present in the wastewater. This process involves the use of microorganisms such as bacteria, fungi, yeasts, and algae. These microorganisms consume the organic matter in the wastewater, converting it into simpler substances that are less harmful to the environment (Samer, 2015). The process is both cheap and easy to implement, making it a popular choice for wastewater treatment. Wastewater typically contains a variety of organic matter, including garbage, waste, and partially digested food. It may also contain harmful substances like pathogenic organisms, heavy metals, and toxins. The goal of biological wastewater treatment is to break down this organic matter and remove any harmful substances present in the wastewater (Rajasulochana & Preethy, 2016). By doing so, the resulting wastewater can be safely disposed of without causing harm to the environment. Biological wastewater treatment is commonly used worldwide because of its effectiveness and economic benefits. It is often used as a secondary treatment process after primary treatment with methods like dissolved air flotation (DAF). In the primary treatment process, sediments and substances like oil are removed from the wastewater. The secondary treatment process, using biological methods, then removes any remaining organic matter and harmful substances. In addition to its effectiveness and economic benefits, biological wastewater treatment has several other advantages (Roesler & Jenny, 2021). For example, it produces very little sludge compared to other treatment methods. This means that disposal costs are lower and there is less environmental impact. The process is also relatively simple and can be easily integrated into existing wastewater treatment plants.

9.5 DIVERSITY IN WASTEWATER TREATMENT: MICROBIAL

9.5.1 Bacteria

Industrial and wastewater treatment plants often use bacteria and other microorganisms to clean sewage, but choosing the right type of bacteria can be challenging. Anaerobic bacteria are commonly found in wastewater treatment plants and are involved in fermentation processes that produce hydrogen sulphide, methane, and volatile organic compounds. The sewer environment creates ideal conditions for anaerobic bacteria to grow (Cyprowski et al., 2018). These bacteria play a crucial role in reducing the volume of sludge and producing methane gas, which can be used as an alternative energy source if properly handled. Phosphorous removal from wastewater is another benefit of using anaerobic microbes in sewage treatment.

The effluent generated by industrial activities can cause water pollution, posing a significant threat to water quality management. Non-biodegradables and pollutants can persist in natural ecosystems for extended periods and accumulate in the biological food chain (Sathya et al., 2022). Therefore, treating sewage with anaerobic bacteria is essential for maintaining a healthy environment. It not only helps reduce pollution but also generates alternative energy, leading to more sustainable

TABLE 9.2
Temperature Classification for Working of Microbes

Type	Temp. Range (°C)	Optimum Temp Range (°C)	Temp. Range (°F)	Optimum Temp Range (°F)
Psychrophilic	10–30	12–18	50–86	53.6–64.4
Mesophilic	30–50	25–40	68–122	77–104
Thermophilic	35–75	55–65	95–167	131–149

wastewater treatment. Phosphorous removal from wastewater is another benefit of anaerobic microbes used in sewage treatment, as shown in Table 9.2.

9.5.2 Protozoa

Protozoa are single-celled microbes found in wastewater treatment systems, making up 4% of the microbial ecosystem. They are larger and more complex than bacteria, and majorly the known types of protozoa found in wastewater include amoeba, flagellates, and ciliates. Protozoa feed on free bacteria and small floc, which helps to improve the cleansing parameters of the final effluent. Protozoa play an important role in the microbial loop and the classical food web by connecting the highly productive and nutrient-retaining microbial loop with the metazoans (Pauli et al., 2005).

They are mainly found in aerobic processes and biofilms, with only a few specialised species found in anaerobic processes (Pauli et al., 2005). In current-era scenarios with low loads and high sludge retention times, protozoa such as ciliates, flagellates, and amoebae are common and more likely to feed on suspended bacteria or some other particulate matter. While the immensity of nutrient removal in wastewater treatment is efficiently performed by bacteria, protozoa and metazoa help balance these populations and offer prominent insight into wastewater conditions (Madoni, 2011). Ciliates, which belong to the phylum *Ciliophora*, are the most commonly found protozoa in percolating filters, ranging from 500 to 10,000 individuals per millilitre of wastewater (Madoni, 2011). Evidence and some scientific reports suggest that the complexity and species structure of the protozoa population in activated sludge (AS) are associated with the quality of effluent delivered, with certain species bound to be found in plants delivering effluents within a particular range of biochemical oxygen demand (BOD) than in plants delivering effluents outside this range (Curds & Cockburn, 1970).

9.5.3 Filamentous Bacteria

Filamentous bacteria are a naturally occurring component of the AS wastewater treatment process. These long, thin bacteria grow in strands that more dominantly resemble threads and form a mesh that ultimately aids in the formation of flocs, or clusters of bacteria and other particles. While some filamentous bacterial species are necessary for good floc production and settling, others could result in issues including bulking and foaming (Nierychlo et al., 2020).

The two different types of filamentous bacteria, *Microthrix* and *Trichococcus*, are dominant in many wastewater treatment plants, chiefly in colder seasons. In opposition to this, the recently identified genus *Ca. Amarolinea*, which is a member of the phylum *Chloroflexi*, has been a persistent as well as predominant filament (Nierychlo et al., 2020). In the context of settling and eliminating sediment from wastewater, filamentous bacteria generate floc, which acts as support structure machinery for other bacteria to cling onto as they are part of floc generation. In the AS process, there are several types of filamentous bacteria that can proliferate, and some can be problematic too. In a more factual way and in accordance with a survey of bulking AS plants in the United States, 20–30 different types of filamentous microorganisms are known to be part of AS, whereas bulking is caused by around 15 major species of filamentous bacteria. On average, around 15 major species of filamentous bacteria cause bulking phenomena, with *Nocardia* being one of the most predominant and responsible for foaming patterning. There have been more than 30 distinct filament morphotypes identified in AS that predominantly treats municipal waste (Noble et al., 2015).

9.5.4 Algae

By providing oxygen that enables aerobic bacteria to break down organic pollutants in the water and excess nitrogen and phosphorus while carrying out the process, algae benefits wastewater treatment methodologies. It is also a justifiable and inexpensive substitute for current wastewater treatment practices. The chronicle use of commercial algal cultures spans about 75 years, with implementation to wastewater treatment and mass production of different strains such as Chlorella and *Dunaliella* (Abdel-Raouf et al., 2012). Algae can be collectively used in wastewater treatment for many different purposes, some of which are used to remove coliform bacteria, reduce both chemical and biochemical oxygen demand, remove N and/or P, and also remove heavy metals (Abdel-Raouf et al., 2012). Efficiency in reducing toxic components found in industrial, home, or municipal wastewater is the second key goal of any wastewater treatment plant, and algae offers an utterly new way to accomplish both of these goals (Pandey & Gupta, 2022). The algae accomplish the wastewater treatment by eliminating nitrogen, phosphorus, and dissolved organic carbon. These growth conditions reduce dangerous bacteria while using more biomass with a higher-energy content than existing wastewater treatment methodologies. Algae can be utilised in wastewater treatment for a variety of purposes, such as BOD reduction, N and/or P removal, coliform inhibition, and heavy metal removal. The excessive concentration of N and P in most wastewaters also indicates that these wastewaters could be used as inexpensive nutrient sources for algal biomass production (Pandey & Gupta, 2022). Resolution of Algae Growth Potential is based on the relation of a maximum biomass yield regarding the biologically used nutrients for microalgae growth (Airport & Free, 2016). Algae play a significant part in the self-purification of organic contaminants in natural waters, as is widely known. In the past 50 years, biological wastewater treatment systems with microalgae have gained particular attention (Figure 9.2). *tb 2: Utilisation of Algal Cultivation.* It is now

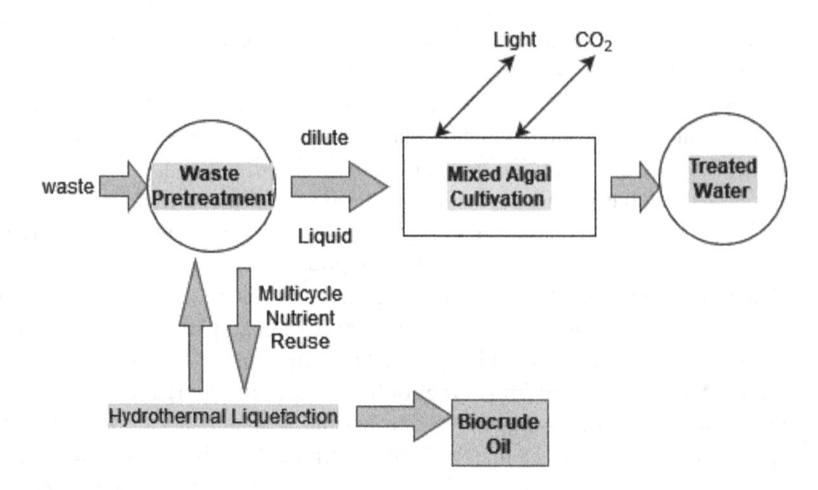

FIGURE 9.2 Utilisation of algal cultivation.

widely acknowledged that algal wastewater treatment systems are just as effective as traditional treatment systems. These specific and distinguishing features have made algal wastewater treatment systems a notable low-cost alternative to tedious, expensive treatment systems, particularly for the purification of municipal wastewaters (Pandey & Gupta, 2022). Heavy metals can be effectively absorbed by microalgae.

Metals could be biomagnified by algae to produce a workable solution for the mannered treatment of wastewater that has been contaminated with the metals (Abdel-Raouf et al., 2012).

9.5.5 Fungi

Fungal groups are essential components of AS and act as crucial decomposers in wastewater treatment structures due to the enormity of fungal biomass and diversity, specifically having a dominant position in organic matter biodegradation and nutrient cycling techniques (Evans & Seviour, 2012). Merging wastewater remediation with the recovery of precious assets might also, in all likelihood, result in an economically feasible solution for sustainable waste management. In the context of this, the fungal wastewater treatment process may be an appealing substitute that makes use of a low-priced organic substrate as a feed to generate high-value fungal by-products with collateral wastewater remediation (Sankaran et al., 2010). Mycoremediation is any other type of bioremediation wherein some species of fungi play a vital role in lowering the impact of pollution in the surrounding area (Kadhim et al., 2021). This kind of bioremediation is considered a herbal and effective approach that doesn't produce any toxic by-products or everlasting hazardous compounds, while the environmental pollutants are eliminated completely because of mineralisation tactics. Out of the six identified fungi species, only three (*Aspergillus terreus*, *A. niger*, and *Penicillium digitatum*) have been identified after the physical and chemical stages inside the wastewater treatment plant (Kadhim et al., 2021). Numerous studies illuminate the role of some fungi species (*A. niger*, *A. terreus*, and *P. digitatum*) in treating a few pollutants in sewage water. It verified the ability of the use of these species to lower the concentration of chlorides (22–43%), nitrites (97%), and phosphates (22.8–32.1%). Comparative sequence analysis of their 18S rRNA genes was used to determine the fungal diversity of groups in numerous AS plants that process various influent wastes (Evans & Seviour, 2012). Our most current knowledge of the diversity of fungi in plants used for wastewater treatment has mostly come from the use of culture-based approaches. It can be worthwhile to look at fungal cultures for biological treatment if your waste stream contains high concentrations of lignin or other materials that are resistant to bacterial repair.

Fungi have the ecological and biochemical potential to reduce organic and inorganic chemicals and the risk posed by contaminants, either by chemical alterations or by modifying chemical bioavailability in wastewater on a global scale.

9.6 WASTEWATER TREATMENT TECHNIQUES USING MICROBES

Microbes may remediate wastewater in an environmentally beneficial manner. Bioremediation employs techniques that make use of the microbial metabolism's innate capacity to break down poisonous and complex xenobiotic substances that are harmful to the majority of living forms and the environment. Complex hazardous macromolecules are broken down into smaller ones, such as carbon dioxide, water, certain sugars, acids, and microbial biomass, which are not harmful to the environment or living things (White & Rosenblatt, 1963).

9.6.1 Factors Affecting Bioremediation

Bioremediation is the use of biological organisms or their derivatives to degrade, remove, or transform environmental contaminants into less harmful substances. In order to remediate and maintain a dynamic equilibrium in the concentration of a variety of chemicals existing in soil and water, the microbial community is crucial. A variety of physical and chemical variables, including the following, are necessary for the bacterial bioremediation of effluents with various toxicant loads:

- The temperature, pH levels, and nutrition availability.
- Waste produced through stereochemistry and toxicity.
- The potency of the biological therapy and strain employed.
- Duration of waste retention in the purification procedure.
- The occurrence and concentration of other wastewater components.

The type, concentration, and complexity of the contaminant can influence bioremediation efficiency and also affect the process. Some contaminants may be highly toxic to microorganisms, and some may require special enzymes or conditions for degradation.

9.6.2 Processes Involved in Treatment

Primary treatment involves the elimination of grit, oil, and suspended particles. The use of microbes in anaerobic or aerobic conditions to reduce biological oxygen demand (BOD) and remove colour, oil, and phenol are examples of secondary treatment. Tertiary treatment involves the use of reverse osmosis, electrodialysis, and ion exchange for the effluent's ultimate purification and elimination. Wastewater is treated biologically using techniques that have been around for a while to handle industrial effluents. Accordingly, these processes can be classified as either aerobic or anaerobic and include membrane batch reactors, AS, anaerobic sludge reactors, sequence batch reactors, anaerobic filters, and anaerobic film reactors, among others (Srivastava et al., 2021).

9.7 ROLE OF MICROBES IN WASTEWATER TREATMENT IN DIFFERENT INDUSTRIES

The increasing rates of industrial wastewater pollution need to be controlled, predominantly to reduce overall water pollution. The AS processes are the most extensively used biological water treatment processes in various wastewater treatment plants all over the world (Yang et al., 2020). The performance and functional stability of such wastewater treatment plants depend on the structure and diversity of their microbial community. The metabolic functions of microbes, especially their enzymatic activities, treat industrial or municipal wastewater. This is implied in various industrial sectors for treating the wastewater coming out of them.

9.7.1 Food Industries

The wastewater being discharged from the food industry these days includes plenty of organic and inorganic matter. This is capable of depleting the oxygen in receiving streams and is poisonous to a variety of living forms in an ecosystem (Rani et al., 2019). The food industry has a high amount of lofty water intake, resulting in the generation of wastewater rich in organic compounds. The inappropriate dumping of this wastewater is a major trouble and a cause of concern to the government and industry as well. The pretreatment of industrial wastewater is also not an effective way of treating water, as it leaves behind toxic pollutants in the industrial effluent. Consequently, this leads to fatal health hazards for humans and animals due to the great degree of water pollutants being dumped into the water bodies (Emmanuel et al., 2013).

The quality parameters of wastewater produced in the food industry are temperature, BOD, pH, total suspended solids, total dissolved solids (TDS), chemical oxygen demand (COD), and odour. The treatment of wastewater from food industries involves physical, chemical, and biological methods, or a combination of all three processes. Microbes play a very crucial role in biological wastewater treatment procedures. These majorly include bacteria, algae, and fungi. These microbes use the organic matter in industrial waste as food because nutrients like phosphorus, potassium, nitrogen, and several other elements are required in microquantities for their metabolic activities. Most of the bacteria that are involved in wastewater treatment plants belong to various genera, such as *Candidatus*, *Microthrix*, *Tetrasphaera*, *Rhodoferax*, *Hyphomicrobium*, *Rhodobacter*, and *Trichococcus* (Rani et al., 2019).

9.7.2 Pharmaceutical Industry

The pharmaceutical industry is India's first science-based industry and is ranked first in terms of its production. The pharmaceutical domain is of immense importance for modern civilisation as a source of lifesaving drugs and medicines. The pharma industry, working on various drugs, releases major pharmaceutical pollutants and metabolite components that affect both terrestrial and aquatic ecosystems. The major components of such pollutants are drugs and chemicals like plant steroids, hormones, antibiotics, animal steroids, lipid derivatives, etc. (Rani et al., 2019).

A number of physiochemical techniques have been used so far for the removal and biodegradability of the pharmaceutical waste generated by the industry. These involve ultrasonic irradiations, advanced oxidation processes, and Fenton reactions. However, these methods could only remove the suspended solids and colloidal impurities but were unable to eliminate the refractory compounds. The stable and recalcitrant organic compounds in pharmaceutical wastes can be mineralised through microbial processes that only involve microbial enzymes. The biological treatment of this wastewater involving microbes is more effective in biodegradation as well as mineralisation. The fungal strains such as *Bjerkandera adusta MUT 2295*, *Aspergillus* sp., and *Penicillium* sp. are involved in the reduction of COD. Bacterial strains such as *Comamonas*, *Arthrobacter*, *Rhodococcus*, and *Enterobacte*r help in the biodegradation of phenol and many other complex organic compounds, such as resorcinol and catechol. These microbes are promising for the future as a zero-waste production technology, especially when combined with advanced oxidation processes for pharmaceutical waste removal (Kavitha and Beebi, 2003).

9.7.3 Textile Industry

The textile industries require large quantities of water for the application of chemicals and to rinse manufactured textiles. This, however, produces a lot of waste effluent, which pollutes environmental resources all around us. This includes surface soils and water ecosystems as the waste is rich in dyes and other chemicals (Rani et al., 2019). The treatment methods involve physical, chemical, and biological techniques, wherein biological treatment involving microbes has been the most effective and economical alternative. Various microbes, including bacteria, algae, yeasts, and fungi, have been found to be capable of the accumulation and degradation of different pollutants in the textile industry. These microbes break down complex compounds into less hazardous compounds with fewer complexes, thereby decreasing the BOD and COD. Biological treatment of wastewater with an AS process is commonly used for biological wastewater management of industrial effluent. The most commonly used fungi species are *Penicillium geastrivous*, *Umbelopsis isabellina*, *Aspergillus foetidus*, *Rhizopus oryzae*, etc. Also, bacterial species like *Brevibacterium casei*, *Bacillus* sp., *Acinetobacter* sp., *Pseudomonas aeruginosa*, *E. coli*, and *Arthrobacter* sp. are also efficient microbes in the textile industry wastewater treatment, especially for chromium and phenol biodegradation (Rani et al., 2019).

9.7.4 Distillery Industry

The distillery industry is an agriculture-based industry that uses agro-products like wheat, barley, rice, sugar cane juice, molasses, cassava, and corn. Around 10–15 L of wastewater is generated per litre of alcohol that is produced. Huge quantities of wastewater, known as spent wash, are generated. This water, when disposed of, affects the flora and fauna, reduces soil alkalinity, and increases the concentration of heavy metals. However, it also suppresses the activity of microorganisms.

Scientists have come up with environmentally friendly products that have the potential to degrade and decolourise the effluents. These microorganisms utilise techniques such as enzymatic degradation, adsorption, and absorption mechanisms. Cyanobacteria like *Oscillatoria boryana*, *Lyngbya*, *Synechocystis*, *Chlorella* sp., *Nostoc muscorum*; bacteria including *Lactobacillus hilgardii* WNS, *Bacillus* sp., and *Alcaligenes* sp.; and fungi like *Coriolus hirsutus*, *C. versicolor*, *Geotrichum candidum*, *P. chrysosporium*, *Flavodon candidum*, *P. chrysosporium*, and *Flavodon flavus* are capable of decolourising spent wash by degrading melanoidin. Further, microalgae are now being cultured and harvested for their valuable products, i.e., livestock feed, ethanol, methane, and organic fertiliser.

9.8 TREATMENT OF INDUSTRIAL WASTEWATER USING MICROBIAL FUEL CELL

The industries produce a large amount of wastewater that needs a lot of energy and freshwater for its processing and discharge. Discharging this wastewater directly into the surrounding environment will affect its ecosystem. Therefore, investigations have been done for energy generation by utilising the different industrial effluents. It mainly targets the dairy industry, leather industry, sugar industry, and brewery industry wastewater as a feed for the fuel cells. This is an eco-friendly energy source and a feasible bioenergy generation method. Bioelectrogenesis through microbial fuel cell (MFC) technology is a promising biological process for bioenergy generation (Sahu, 2019).

A microbial fuel cell (MFC) is defined as a biologically catalysed electrochemical system that is efficient in converting the chemical energy in organic waste to electrical energy. In this cell, the bacteria present can oxidise reduced substrates, and a coenzyme, which is carried down the respiratory chain, acts as an electron carrier. This in turn translocates protons and, subsequently, a chemical energy potential called a proton gradient. It is this chemical energy that can be converted to electrical energy in the MFCs (Sahu, 2019). The final electron acceptor, such as oxygen, completes the oxidising reaction. Microbial fuel setups are such that the bacterial cells are employed as catalysts in oxidising both organic and inorganic substrates to generate an electrical flow or simply current production.

The major significance was to check the open-voltage production by various industries, such as the sugar industry, brewery industry, leather industry, and dairy industry. It is found that the voltage increases with an increase in the number of experiment days at a fixed internal resistance. The anode plays an important role in fuel to bring good oxidation that thus allows for defined reaction pathways on the

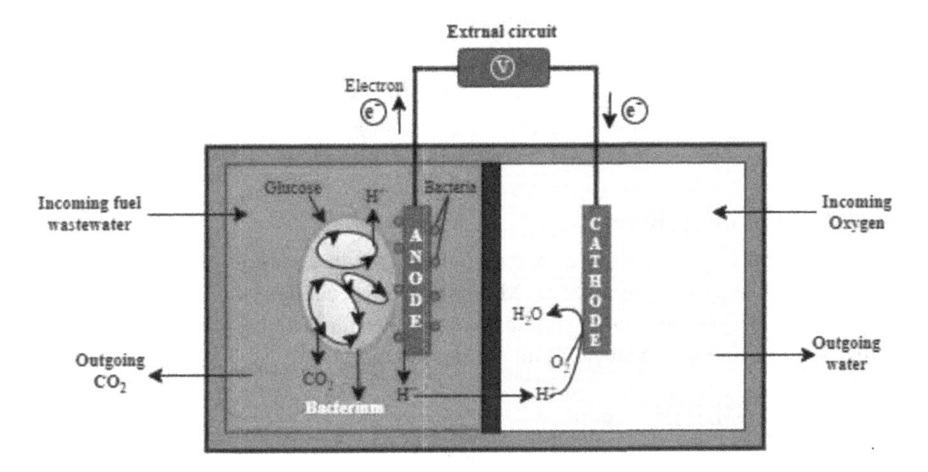

FIGURE 9.3 Microbial fuel cell.

electrode surface. The cathode is predominantly magnesium oxide, which is favourable for electrochemical conductivity, ion diffusivity, and solid-state properties better than any other cathode material (Sahu, 2019). The MFCs, as shown in Figure 9.3, can be tested with pure microbial culture in the near future. That would actually be a single cell converting into a stack for more power production. Therefore, it can be practically done by treating industrial wastewater to generate clean green energy and make the surrounding area clean as well.

9.9 MICROBIAL ENZYME APPLICATION IN INDUSTRIAL WASTEWATER TREATMENT

The microbial enzymes are found to specifically react on the recalcitrant pollutants by precipitation and structural modification to other products. They also make the waste susceptible to treatment or aid in converting waste material into value-added products by modifying its characteristics (Katiyar & Katiyar, 2021). These enzymes are highly substrate-specific in their action. They act as biocatalysts to cause specific biochemical reactions that result in the formation of the various metabolic processes of the cells. The use of enzymes in wastewater treatment in place of toxic chemical reactants is extremely important to meet the increasing demands for cleaner and greener technologies, especially when the world is looking for eco-friendly techniques.

9.9.1 Microbial Enzyme Type

9.9.1.1 Microbial Oxidoreductases Enzyme

Oxidoreductases are enzymes that are involved in the oxidative coupling process used by a variety of bacteria, fungi, and higher plants to detoxify hazardous chemical molecules.

9.9.1.2 Microbial Oxygenases

They are oxidoreductases, which use molecular oxygen to transport oxygen during the oxidation of reduced substrates. They primarily affect the metabolism of organic molecules by enhancing their reactivity, making them more soluble in water, or causing the cleavage of aromatic rings.

9.9.1.3 Monooxygenase

Enzymes such as these catalyse oxidative processes in a variety of substrates, such as fatty acids, steroids, and alkanes. Monooxygenases catalyse the biodegradation of several aliphatic and aromatic chemicals in wastewater (Pandey et al., 2017). It is an important enzyme in industrial pollutant degradation.

9.9.1.4 Microbial Dioxygenase

Dioxygenases particularly oxygenate a variety of substrates and typically oxidise aromatic molecules. They are essentially soil microorganisms that play a role in converting benzene precursors into aliphatic compounds.

9.9.1.5 Microbial Proteases

Peptide bonds are hydrolysed by proteases in an aqueous environment. They create them in nonaqueous surroundings and use them in a variety of industries, including food, detergent, leather, and pharmaceuticals.

9.9.1.6 Microbial Lipases

Among the many processes that lipases catalyse are hydrolysis, aminolysis, esterification, and alcohol lysis. They are used in biological remediation in a particular diagnostic capacity. It has several potential uses in the culinary, chemical, detergent, cosmetic, and paper sectors.

9.9.1.7 Microbial Cellulases

Enzymatic hydrolysis of cellulose uses cellulases to break it down into reducing sugars. They induce the cellulose microfibrils that develop when cotton-based textiles are washed and used to be removed.

9.9.2 Technologies Used for the Enzymatic Treatment of Wastewater

There are a number of technologies being used for the treatment of wastewater, including harsh conditions that might change the conformations of the enzymes being used. These harsh physical and chemical conditions of temperature, pH, and ionic strength thus result in an alteration of the function of enzymes, which is often encountered in effluent systems. The immobilisation method is used to enhance their reusability and also minimise the chances of loss of enzyme activity under severe unfavourable conditions. Thus, it helps reduce the loss of enzymes. The use of immobilised enzymes also has additional advantages like increased stability, easy handling, and reusability (Pandey et al., 2017).

One of the simplest methods of enzyme administration to target effluent is the direct introduction of cells or tissues that are capable of producing an enzyme into the effluent. The suitably adapted strains of microorganisms are used to co-metabolise target pollutants from industrial wastewater. The latest area of research for this purpose is nanotechnology. It involves the use of nanoparticles and has gained much interest due to its capability to completely degrade contaminants into harmless forms such as carbon dioxide and water. This comes under the integration of enzyme technology and nanotechnology, known as single-enzyme nanoparticles. These enzymes can easily degrade a variety of recalcitrant compounds, such as phenols, pesticides, and polyaromatic dyes.

Another technique used, possibly for wastewater treatment, is membrane bioreactors. It has also been combined with enzyme technology to come up with a better and more promising technique of biodegradation by enzymes. In these systems, the use of hollow-fibre bioreactors or capillary membrane reactors successfully increases the surface area volume ratio. This further increases the wastewater treatment capacity of the system.

9.9.3 Advantages of Enzymatic Techniques Over Other Techniques

- There is a constant demand for enzymatic technology because of its biocatalyst property, which is gaining importance in wastewater treatment. The microbial enzymes can remove recalcitrant pollutants by precipitation or transformation into other products by acting on them.
- These techniques provide ease of operation and are tie saving as compared to the existing traditional methods. Also, these enzymatic processes, being biological, are a more eco-friendly technique and a more sustainable alternative.
- The various physicochemical treatments like chemical precipitation, coagulation membrane filtration flocculation, and floatation that have been used so far have become out-dated due to the chemicals used, high energy consumption, and handling costs for sludge disposal, along with the high operational cost.
- These enzymes can selectively degrade a target pollutant. They can operate under mild reaction conditions, especially in terms of temperature and pH (Pandey et al., 2017).
- Also, these enzymes are appropriate on the grounds of being biodegradable, as they can selectively degrade a target pollutant without affecting the other components in the effluent coming from industrial wastewater.

9.10 USE OF NANOTECHNOLOGY FOR BIODEGRADATION OF WASTEWATER

The pollutants coming from industries are a serious challenge for discharge and removal, and also a major concern for the environment. It is possible only with the use of advanced technologies like nanotechnology to reduce these contaminants by creating environmentally friendly nanostructures. Nanomaterials are acquiring importance due to their improved mechanical, chemical, and physical characteristics. They are also a sustainable production and cost-reduction method under green biotechnology. This technology, in combination with enzyme technology, has increased the reusability and activity of enzymes (Mandeep & Shukla, 2020). The membrane-associated nanomaterials have special properties of permeability, resistance to odour, mechanical strength, and temperature resistance that make them effective for pollutant degradation.

9.10.1 Nano-Adsorbents and Nanofiltration Methods

Nano-absorbents have efficiently been utilised for the degradation of inorganic and organic pollutants. Carbon-based nanomaterials, such as toxic compounds from manufacturing industries or pharmaceutical effluent, are adsorbents in carbon nanotubes. The removal of fluoride ions from industrial effluent has also been proven to be beneficial when using activated carbon-modified nano-magnets. The recent modifications in MFCs to include bio-nanocatalysts are used for the generation of bioelectricity. Other nano-absorbents that are involved in the removal of such wastewater pollutants are meta- and metal-oxide nanomaterials, magnetic polymers of Co_3O_4 and SiO_2, and magnetic nanoparticles coated with nylon 6.

Nanofiltration, on the other hand, is the method used for nutrient recovery from waste from industry (Mandeep & Shukla, 2020). Nanofiltration membranes have the ability to dismiss almost all metal ions, including nickel, copper, lead, and cadmium, from industrial pollutants that are potentially harmful for the environment. With the use of microorganism-assisted nanotechnological applications, scientists have been able to overcome the commercialisation issues with the generation of green nanostructures. This has sparked a revolution in green nanotechnology.

9.10.2 Nano-technology and Enzyme Technology

The combination of enzyme technology and nanotechnology is of utmost importance, predominantly as they reduce enzyme- cell contact caused by steric obstructions and a drop in energy from the surface when present with nanomaterials. Due to their resistance to unfolding, immobilised enzymes on nanomaterials are very stable and have enhanced kinetic characteristics (Mandeep & Shukla, 2020). Enzyme immobilisation on solid substrates leads to changes in structures; such modifications are not observed when nanomaterials are used for enzyme immobilisation. The research makes it evident that this combination gives an environment that is steady and effective for the degradation of industrial wastewater discharges.

9.11 ECO-FRIENDLY, ECONOMIC, AND SUSTAINABILITY INFERENCE OF REMEDIATION METHODS FOR WASTEWATER CONTAMINATED WITH HEAVY METALS

There are several environmentally friendly techniques for cleaning up wastewater contaminated with heavy metals, including:

- Phytoremediation: This technique uses plants to absorb and accumulate heavy metals from water and soil. Phytoremediation is a viable and affordable approach for recycling heavy metal remediation in wastewater.

- Bioremediation: This technique uses microorganisms like fungi and bacteria to break down and remove heavy metals from the wastewater. Bioremediation can be applied to large-scale contaminated sites and has a low impact on the environment.
- Electro-remediation: This technique uses electrical current to remove heavy metals from wastewater. Electro-remediation is a highly efficient and eco-friendly method for heavy metal remediation in wastewater.
- Adsorption: This technique uses adsorbent materials, such as activated carbon or clay, for heavy metal removal from wastewater. Adsorption is a cost-effective and sustainable technique for wastewater heavy metal treatment.
- Nanoremediation: This technique uses nanotechnology to remove heavy metals from wastewater. Nanoremediation is a promising wastewater heavy metal treatment method using an environmentally acceptable technique.

It is essential to remember that the effectiveness of these methods may change based on the precise kind and quantity of heavy metals in the wastewater, as well as the site-specific conditions. Therefore, a combination of techniques may be needed for the complete and effective remediation of heavy metal-contaminated wastewater (Elbasiouny et al., 2021).

9.11.1 Through Water Hyacinth (WH)

Eichhornia crassipes, often known as water hyacinth (WH), is a member of the Pontederiaceae family and one of the very dangerous water plants in the world (Elbasiouny et al., 2021). The plant has the ability to accumulate large amounts of heavy metals like lead, cadmium, copper, chromium, and zinc in its tissues, making it a promising eco-friendly remediation technique for effluent water mixed with heavy metals. Many photosynthetic species need sunlight penetration, which can be reduced by the production of WH on the water's surface. Hence, WH reduces the development rate of photosynthetic organisms, upsetting the ecological balance. Moreover, its extensive surface covering on bodies of water restricts oxygen transport into the water. The long roots of the WH are suspended in the water, and this root structure might offer the right substrate for aerobic bacteria that function in a sewage system (Elbasiouny et al., 2021).

These microbes convert the nutrients and organic elements in the effluent water into inorganic compounds that the plants can ingest. WP typically expands quickly because non-native nations lack natural adversaries or rivals. More than two million WHs, or 270–400 tonnes of plant biomass, may be detected in one acre of water. It can spread unchecked as an invasive species in areas like wetlands or irrigation systems. Moreover, it hinders river transportation. Thus, many nations have concentrated on eliminating WHs because of these factors as well as the level of eutrophication that results from the microbial and fungal populations dying off (Joseph et al., 2019). WH is a positive, eco-friendly remediation method for wastewater containing high metal contamination. The plant could be used in constructed wetlands, phytofiltration, bioreactors, and floating treatment wetlands to eliminate heavy metals from waste water. The techniques are effective, sustainable, and can be used in small- and large-scale applications.

9.11.2 Through Black Tea Wastes

Among the most consumed beverages worldwide is tea. Every year, over 3.4 million tonnes of tea are drunk nation-wide (Mohammed, 2012). Tea is an infusion produced from powdered black tea (BT) or dehydrated green tea leaves. BT is often produced using the CTC (crush, tear, curl) technique, which results in a dense, fibrous residue. Every day, tea businesses, restaurants, and homes produce and discard thousands of tonnes of BT trash. This problem extends beyond the waste produced during manufacture and after use, since BT wastes are frequently dumped in nearby bodies of water. It is interesting to utilise BT leaves as a cheap sorbent. The structural elements of its disposal or sorption capabilities have received a lot of attention in some studies. Others were concerned about kinetics, an essential physical and chemical factor in the evaluation of the essential sorbent quality, and the application of the adsorption mechanism.

Wastewater contaminated with heavy metals can be harmful to the environment and human health. However, there are eco-friendly remediation techniques that can be used to treat this type of wastewater, including the use of BT waste. Here are some potential steps for using BT waste as a remediation technique for heavy metal-contaminated wastewater:

- Collect BT waste: The first step is to collect BT waste, which can be obtained from tea production factories, tea shops, or households that consume tea on a regular basis.
- Prepare the BT waste: The collected BT waste should be dried and ground into small particles or powder.
- Prepare the contaminated wastewater: The contaminated wastewater should be collected and tested to determine the concentration and types of heavy metals present.
- Treat the contaminated wastewater with BT waste: Once the concentration and types of heavy metals are determined, the BT waste can be added to the wastewater. The tannins and other organic compounds in the BT waste can bind with the heavy metals, removing them from the wastewater. The ratio of BT waste to wastewater will depend on the concentration of heavy metals in the wastewater.
- Monitor and filter the treated wastewater: The treated wastewater should be monitored to ensure that the heavy metal concentration is reduced to safe levels. The treated wastewater can then be filtered through a sand or activated carbon filter to remove any remaining organic compounds and particles.
- Reuse or dispose of the treated wastewater: The treated wastewater can be reused for non-potable purposes, such as irrigation, or disposed of in an environmentally safe manner.

Overall, the use of BT waste as a remediation technique for heavy metal-contaminated wastewater is an eco-friendly approach that can be effective and affordable. However, it is important to properly test and monitor the treated wastewater to ensure that it is safe for the environment and human health.

9.12 CONCLUSION

In conclusion, the feasibility of producing biofuel from effluent will rise with the adoption of various approaches to address the operational constraints of mixed bacterial metabolisms and other related issues (Sahu, 2019). Various papers that are released each year claim that there is presently ongoing research work being done on the development of more affordable, effective, and distinctive decontamination methods. Everyone, including members of the public, companies, researchers, scientists, and judgement on a national, European, or worldwide level – is now highly concerned about the environment, and particularly the issue of water contamination. Disinfection is a serious problem as a result of the public's demand for pollutant-free wastewater disposal in receiving waterways (Crini & Lichtfouse, 2019). Future developments might result in the development of more effective systems that require less energy, water, and chemical input to work at their best. The utilisation of contemporary biological technology methods would result in enzymatic systems with enhanced effects at various optimum pH and temperature parameters, resulting in greater effectiveness with less energy used and cheaper prices (Pandey et al., 2017).

REFERENCES

Abdel-Raouf, N., Al-Homaidan, A. A., & Ibraheem, I. B. M. (2012). Microalgae and wastewater treatment. *Saudi Journal of Biological Sciences, 19*(3), 257–275. https://doi.org/10.1016/j.sjbs.2012.04.005

Ahmed, S. F., Mofijur, M., Nuzhat, S., Chowdhury, A. T., Rafa, N., Uddin, M. A., Inayat, A., Mahlia, T. M. I., Ong, H. C., Chia, W. Y., & Show, P. L. (2021). Recent developments in physical, biological, chemical, and hybrid treatment techniques for removing emerging contaminants from wastewater. *Journal of Hazardous Materials, 416*(March), 125912. https://doi.org/10.1016/j.jhazmat.2021.125912

Airport, S., & Free, I. (2016). *Waste Water Treatment Using MBR. 9*(2), 226–228.

Alasadi, T. (2019). *A New Simple Method for the Treatment of Waste Water Containing Cu (II) and Zn (II) Ions Using Adsorption on Dried Conocarpus erectus Leaves.* February.

Crini, G., & Lichtfouse, E. (2019). Advantages and disadvantages of techniques used for wastewater treatment. *Environmental Chemistry Letters, 17*(1), 145–155. https://doi.org/10.1007/s10311-018-0785-9

Curds, C. R., & Cockburn, A. (1970). Protozoa in biological sewage-treatment processes-II. Protozoa as indicators in the activated-sludge process. *Water Research, 4*(3), 237–249. https://doi.org/10.1016/0043-1354(70)90070-9

Cyprowski, M., Stobnicka-Kupiec, A., Ławniczek-Wałczyk, A., Bakal-Kijek, A., Gołofit-Szymczak, M., & Górny, R. L. (2018). Anaerobic bacteria in wastewater treatment plant. *International Archives of Occupational and Environmental Health, 91*(5), 571–579. https://doi.org/10.1007/s00420-018-1307-6

David, M. K. (2017). *A Review Paper on Industrial Waste Water Treatment Processes.* January.

Elbasiouny, H., Darwesh, M., Elbeltagy, H., Abo-alhamd, F. G., Amer, A. A., Elsegaiy, M. A., Khattab, I. A., Elsharawy, E. A., Ebehiry, F., El-Ramady, H., & Brevik, E. C. (2021). Ecofriendly remediation technologies for wastewater contaminated with heavy metals with special focus on using water hyacinth and black tea wastes: A review. *Environmental Monitoring and Assessment, 193*(7). https://doi.org/10.1007/s10661-021-09236-2

Evans, T. N., & Seviour, R. J. (2012). Estimating biodiversity of fungi in activated sludge communities using culture-independent methods. *Microbial Ecology, 63*(4), 773–786. https://doi.org/10.1007/s00248-011-9984-7

Joseph, L., Jun, B. M., Flora, J. R. V., Park, C. M., & Yoon, Y. (2019). Removal of heavy metals from water sources in the developing world using low-cost materials: A review. *Chemosphere, 229*, 142–159. https://doi.org/10.1016/j.chemosphere.2019.04.198

Kadhim, N. F., Mohammed, W. J., Al Hussaini, I. M., Al-Saily, H. M. N., & Ali, R. N. (2021). The efficiency of some fungi species in wastewater treatment. *Journal of Water and Land Development, 50*, 248–254. https://doi.org/10.24425/jwld.2021.138180

Katiyar, K. (2022). AI-Based Predictive Analytics for Patients' Psychological Disorder. In *Predictive Analytics of Psychological Disorders in Healthcare: Data Analytics on Psychological Disorders* (pp. 37–53). Springer.

Katiyar, K., Kumari, P., & Srivastava, A. (2022). Interpretation of Biosignals and Application in Healthcare. In *Information and Communication Technology (ICT) Frameworks in Telehealth* (pp. 209–229). Springer.

Katiyar, S., & Katiyar, K. (2021). Recent Trends Towards Cognitive Science: from Robots to Humanoids. In *Cognitive Computing for Human-Robot Interaction* (pp. 19–49). Elsevier.

Madoni, P. (2011). Protozoa in wastewater treatment processes: A minireview. *Italian Journal of Zoology, 78*(1), 3–11. https://doi.org/10.1080/11250000903373797

Mandeep, & Shukla, P. (2020). Microbial nanotechnology for bioremediation of industrial wastewater. *Frontiers in Microbiology, 11*(November). https://doi.org/10.3389/fmicb.2020.590631

Mohammed, R. R. (2012). Removal of heavy metals from waste water using black teawaste. *Arabian Journal for Science and Engineering, 37*(6), 1505–1520. https://doi.org/10.1007/s13369-012-0264-8

Nierychlo, M., McIlroy, S. J., Kucheryavskiy, S., Jiang, C., Ziegler, A. S., Kondrotaite, Z., Stokholm-Bjerregaard, M., & Nielsen, P. H. (2020). *Candidatus amarolinea* and *Candidatus microthrix* are mainly responsible for filamentous bulking in Danish municipal wastewater treatment plants. *Frontiers in Microbiology, 11*(June), 1–17. https://doi.org/10.3389/fmicb.2020.01214

Noble, H., Smith, J., Fuchs, T. A., Abed, U., Goosmann, C., Hurwitz, R., Schulze, I., Wahn, V., Weinrauch, Y., Brinkmann, V., & Zychlinsky, A. (2015). Untitled _ enhanced reader. pdf. *Clinical Infectious Diseases, 176*(2), 231.

Ochando-Pulido, J. M., Pimentel-Moral, S., Verardo, V., & Martinez-Ferez, A. (2017). A focus on advanced physico-chemical processes for olive mill wastewater treatment. *Separation and Purification Technology, 179*, 161–174. https://doi.org/10.1016/j.seppur.2017.02.004

Pandey, B. C., & Gupta, D. S. (2022). Effectiveness of algae in wastewater treatment systems. *International Journal for Research in Applied Science and Engineering Technology, 10*(8), 520–526. https://doi.org/10.22214/ijraset.2022.46241

Pandey, K., Singh, B., Pandey, A. K., Badruddin, I. J., Pandey, S., Mishra, V. K., & Jain, P. A. (2017). Application of microbial enzymes in industrial waste water treatment. *International Journal of Current Microbiology and Applied Sciences, 6*(8), 1243–1254. https://doi.org/10.20546/ijcmas.2017.608.151

Pauli, W., Jax, K., & Berger, S. (2005). Protozoa in Wastewater Treatment: Function and Importance. *Biodegradation and Persistance*, *2*, 203–252. https://doi.org/10.1007/10508767_3

Piaskowski, K., Świderska-Dąbrowska, R., & Zarzycki, P. K. (2018). Dye removal from water and wastewater using various physical, chemical, and biological processes. *Journal of AOAC International*, *101*(5), 1371–1384. https://doi.org/10.5740/jaoacint.18-0051

Rajasulochana, P., & Preethy, V. (2016). Comparison on efficiency of various techniques in treatment of waste and sewage water – A comprehensive review. *Resource-Efficient Technologies*, *2*(4), 175–184. https://doi.org/10.1016/j.reffit.2016.09.004

Rani, N., Sangwan, P., Joshi, M., Sagar, A., & Bala, K. (2019). Microbes: A Key Player in Industrial Wastewater Treatment. In *Microbial Wastewater Treatment*. Elsevier Inc. https://doi.org/10.1016/B978-0-12-816809-7.00005-1

Roesler, T. A., & Jenny, C. (2021). Introduction to treatment. *Medical Child Abuse*, 155–164. https://doi.org/10.1542/9781581105131-ch08

Sahu, O. (2019). Sustainable and clean treatment of industrial wastewater with microbial fuel cell. *Results in Engineering*, *4*, 100053. https://doi.org/10.1016/j.rineng.2019.100053

Samer, M. (2015). Biological and chemical wastewater treatment processes. *Wastewater Treatment Engineering*, *October*. https://doi.org/10.5772/61250

Sankaran, S., Khanal, S. K., Jasti, N., Jin, B., Pometto, A. L., & Van Leeuwen, J. H. (2010). Use of filamentous fungi for wastewater treatment and production of high value fungal byproducts: A review. *Critical Reviews in Environmental Science and Technology*, *40*(5), 400–449. https://doi.org/10.1080/10643380802278943

Sathya, K., Nagarajan, K., Carlin Geor Malar, G., Rajalakshmi, S., & Raja Lakshmi, P. (2022). A comprehensive review on comparison among effluent treatment methods and modern methods of treatment of industrial wastewater effluent from different sources. *Applied Water Science*, *12*(4), 1–27. https://doi.org/10.1007/s13201-022-01594-7

Sharma, S., Bahuguna, A., Singh, S. K., Bahuguna, A., & Dadarwal, B. K. (2021). Physical method of wastewater treatment-a review. *Quest Journals Journal of Research in Environmental and Earth Sciences*, *7*(6), 2348–2532.

Singh Asiwal, R., Kumar Sar, S., Singh, S., & Sahu, M. (2016). Wastewater treatment by effluent treatment plants. *International Journal of Civil Engineering*, *3*(12), 19–24. https://doi.org/10.14445/23488352/ijce-v3i12p105

Srivastava, A., Seth, A., & Katiyar, K. (2021). Microrobots and nanorobots in the refinement of modern healthcare practices. In *Robotic Technologies in Biomedical and Healthcare Engineering* (pp. 13–37). CRC Press.

White, B. W., & Rosenblatt, F. (1963). Principles of neurodynamics: Perceptrons and the theory of brain mechanisms. *The American Journal of Psychology*, *76*(4), 705. https://doi.org/10.2307/1419730

Yargeau, V. (2012). Water and wastewater treatment: Chemical processes. *Metropolitan Sustainability: Understanding and Improving the Urban Environment*, *1854*, 390–405. https://doi.org/10.1533/9780857096463.3.390

Zinicovscaia, I., & Cepoi, L. (2016). Cyanobacteria for bioremediation of wastewaters. *Cyanobacteria for Bioremediation of Wastewaters*, 1–124. https://doi.org/10.1007/978-3-319-26751-7

10 Anaerobic Biological Reactors for Wastewater Treatment

Khalida Bloch and Sougata Ghosh

10.1 INTRODUCTION

The continuous growth of population and industrialization has led to an increase in the level of refractory compounds in the environment, may it be soil, water, or air. The water ecosystem is polluted due to the discharge of untreated or insufficiently treated municipal and industrial effluents. Hence, it becomes necessary to treat the wastewater before it is discharged into the water bodies (Zhang et al., 2018). The wastewater contains numerous inorganic and organic pollutants such as aliphatic and aromatic hydrocarbons, insecticides, and polychlorinated biphenyls that are recalcitrant in nature, which in turn increases the levels of biological oxygen demand (BOD), chemical oxygen demand (COD), and total suspended solids (TSS) of the water that are detrimental to the water ecosystem (Siddique et al., 2014). Among both aerobic and anaerobic digestions that are employed for the treatment of wastewater, anaerobic digestion (AD) acts as an outstanding process for the reduction of pollutants with simultaneous energy generation (Tansel and Surita, 2014). In the AD process, the conversion of complex organic molecules takes place in the absence of oxygen by a diverse group of microorganisms (Yu et al., 2014). It removes highly complex organic matter from the wastewater. The anaerobic treatment generates less amount of biomass sludge, requires low energy, and is more effective on wet wastes. Hence, this process has received considerable attention as it is eco-friendly and can treat complex industrial wastewater contaminated with toxic pollutants (Abdelgadir et al., 2014). Several biological reactors, such as anaerobic granular sludge blankets, continuous stirred tanks, expanded granular sludge beds, plug flow, and anaerobic membrane bioreactors (AnMBRs), are used for the treatment of acids, palladium, tellurite, pentachlorophenol (PCP), and acrylic-containing wastewater. AD can treat the wastewater generated from wineries, distilleries, slaughterhouses, and textile industries. AD includes hydrolysis, acetogenesis, and methanogenesis. The microorganisms hydrolyze complex molecules into monomers using hydrolytic enzymes, followed by their conversion into volatile fatty acids (VFAs) by acidogens. Eventually, the methanogenic organisms convert acids into methane using acetotrophic and methanogenic pathways (Khanal et al., 2017). This chapter emphasizes the use of different types of anaerobic digesters and their mechanisms to treat various municipal, domestic, and industrial wastewaters listed in Table 10.1.

 DOI: 10.1201/9781003381327-10

TABLE 10.1
Various Anaerobic Bioreactors and Their Wastewater Treatment Efficiencies

Sr. No.	Reactor	Type of Wastewater	COD Removal (%)	Metal Removal Efficiency (%)	References
1	UASB	Acidic wastewater	90	–	Keyser et al. (2003)
2	UASB	Tellurite (TeO_3^{2-})	>98	92–95	Mal et al. (2017)
3	UASB	Palladium (Pd(II))	57.4 ± 15.1, 83.9 ± 5.0	98.9 ± 0.7, 97.7 ± 1.8	Pat-Espadas et al. (2016)
4	UASB	Pentachlorophenol	90–100	97.4–100	Shen et al. (2005)
5	UASB	Acrylic acid	95	–	Show et al. (2020)
6	UASB	Slaughterhouse wastewater	93.4	–	Borja et al. (1998)
7	CSTR	Sulfur black dye	75.24	84.53	Andleeb, et al. (2010)
8	CSTR	Distillery effluent	72	–	Mohite and Salimath (2020)
9	CSTR	Synthetic wastewater	–	94–100	Kieu et al. (2011)
10	CSTR coupled with MFU	Artificial wastewater, sauerkraut brine wastewater, chicken slaughterhouse wastewater	90	–	Fuchs et al. (2003)
11	CSTR	Petrochemical wastewater	88	–	Siddique et al. (2014)
12	EGSB	Domestic wastewater	76–81	–	Chu et al. (2005)
13	EGSB	Petrochemical wastewater	85.6 ± 2.5	81.5 ± 4.8	Liang et al. (2019)
14	EGSB	Domestic sewage	71.5 ± 2.3	–	Xu et al. (2018)
15	EGSB	High-nitrate wastewater	–	99.2	Liao et al. (2013)
16	EGSB	Synthetic wastewater	82.3	–	Yang et al. (2018)
17	PFR	Agro-industrial waste	70–80	–	Eftaxias et al. (2021)
18	PFR	Municipal solid waste	–	–	Rossi et al. (2022)
19	PFR	Piggery wastewater	–	–	Gorecki et al. (1993)
20	AnMBR	Swine wastewater	–	–	Jiang et al. (2020)

10.2 ANAEROBIC GRANULAR SLUDGE BLANKET

Among the various modes of AD-mediated wastewater treatment, an anaerobic granular sludge blanket is employed for the treatment of the effluents of various food and agricultural industries that are discussed in this section. The successful treatment of acidic wastewater produced from wineries using an upflow anaerobic sludge blanket (UASB) was reported by Keyser et al. (2003). The waste generated from wineries is rich in organic compounds, including sugars, acids, proteins, and alcohol. Microbial isolation and characterization were also carried out on winery effluent. The isolated organisms metabolized the raw waste and produced VFAs. *Enterobacter sakazakii* was added to the granular biosolid combination as it showed maximum production of VFAs. The bioreactor containing a solid separator on top was used in combination with the UASB system. The upflow velocity was 2 m/h at 35°C. The raw anaerobic sludge (1000 g) was seeded into the control bioreactor along with wastewater collected from wineries. The winery effluent was supplemented for 5 d with sodium lactate (5 g/L), dipotassium phosphate (K_2HPO_4, 100 mg/L), urea (100 mg/L), and trace elements (1 mL). The hydraulic retention time (HRT) was 2.2 d at pH 6. The anaerobic granules of 700 g were added to the normal start-up bioreactors. After the acclimatization of the bacteria, the wastewater was added to the bioreactor. The pH of the reaction was adjusted to 6, and HRT was set at 2.2 d. The accelerated bioreactor was seeded with 700 g of the granular sludge. The organisms identified from winery effluents were *E. sakazakii*, *Bacillus licheniformis*, *Bacillus megaterium*, *Brevibacillus laterosporus*, and *Staphylococcus* species. The production of VFAs was associated with the biodegradation and production of metabolites by acidogenic microbes. The concentration of VFAs was analyzed using gas chromatography (GC), which showed significant degradation of granules. Under agitation (150 rpm) and stationary conditions, *E. sakazakii* produced 455 mg and 477 mg VFAs/L, respectively. The UASB efficiency of the controlled bioreactor containing only sewage sludge showed COD removal after 90 d. The normal startup bioreactor containing granules from UASB showed 80–86% COD removal by 50 d. The COD removal efficiency did not reach above 90% with an HRT of 30 h and an organic loading rate (OLR) of 5.1 kg COD m^3/d. The accelerated bioreactor containing anaerobic granular sludge synthesized in an anaerobic batch reactor with *E. sakazakii* showed 90% COD removal with a pH of 7.3. The biogas of 2.3 L/d and the OLR of 6.3 kg COD m^3/d were produced. The reaction was carried out for a further 100 d. The increase in OLR to 10.12 kg COD m^3/d was achieved. Hence, granule seeding showed >90% removal of COD with a reduced start-up time of 17 d making UASB an effective wastewater treatment.

The removal and recovery of tellurite (TeO_3^{2-}) were carried out using an UASB in the study conducted by Mal et al. (2017). The bioreactor was loaded with lactate, organic compounds, and anaerobic granular sludge at a rate of 0.6 g COD/L/d. The reduction of tellurite and biogenic retention of Te(0) were monitored. The anaerobic sludge (200 g wet weight) was inoculated in the UASB reactor with an HRT of 12 h at 30°C for 70 d. It was observed that the removal of COD was 75–91% within 7 d of the start-up period. The COD removal within 2–3 weeks was >98% with 10 mg/L tellurite. When the concentration of tellurite was increased to 20 mg/L, the removal of COD also increased. Lactate utilization remains unaffected at a concentration of 10 mg/L of tellurite. The removal of tellurite was significantly improved in 3 weeks. During the end of period III (50–70 days), the removal efficiency of tellurite and Te increased to 92–95%. The characterization of Te associated with granular sludge was determined using scanning electron microscope (SEM) analysis, which revealed the presence of crystalline structure in tellurite-reducing granular sludge. The energy dispersive spectroscopy (EDX) showed the presence of elemental Te(0) along with carbon, oxygen, nitrogen, and sulfur that could be attributed to the microbial cells, the exopolysaccharide (EPS) matrix, and inorganic minerals in the granules. The Raman spectra of granular sludge showed vibrational peaks at 121.9 and 140.5/cm. The X-ray diffraction (XRD) spectrum exhibited that the granules consisted of hexagonal tellurium. Centrifugation of the granular sludge at 10,000 rpm for 20 min led to the recovery of the loosely bound (LB)-EPS and the associated

Te(0) nanoparticles. Up to 74–78% of Te was recovered. The transmission electron microscope (TEM) analysis of the LB-EPS showed the nanospheres, nanorods, and shrads of Te(0) with a size of 20–25 nm. Hence, tellurite was continuously removed by anaerobic granules from the synthetic wastewater using UASB reactors.

In the study conducted by Pat-Espadas et al. (2016), the removal and recovery of palladium (Pd) from aqueous streams were accomplished using an UASB reactor. The methanogenic anaerobic granular sludge was collected from the UASB bioreactor used for the treatment of brewery wastewater. The laboratory-scale UASB reactors were seeded with 10 g VSS/L volatile suspended solids of anaerobic granular sludge. The reactors were fed with mineral water containing macroelements, yeast extract, and trace elements. Resazurin was used as a redox indicator for anaerobic conditions. Ethanol was added in reactor one (R1) and acetate in reactor two (R2). The reaction was carried out at room temperature. The reactors were operated without Pd(II) for an initial 19 d, while from days 20 to 40, they were supplemented with 5 mg/L of Pd(II). The concentration of Pd(II) was increased to 15 mg/L in period 3, which was carried out for 41–54 d. The HRT was maintained at 7.8 ± 0.1 h. The COD removal efficiency was 99% ± 1.2% in the initial phase. In the second phase, it was observed that the presence of palladium in the reactor containing acetate reduced the COD removal efficiency from 99% to 59.9% ± 12.2%, which was still higher than the reactor containing ethanol (99% to 86.7% ± 5.1%). The COD removal efficiency in phase 3 was 57.4% ± 15.1% and 83.9% ± 5.0% in R2 and R1, respectively. The Pd(II) removal efficiency of R1 was 98.9% ± 0.7%, while it was 97.7% ± 1.8% for R2. It can be further concluded that the presence of ethanol in R1 showed Pd(II) removal by biogenic H_2 stimulation from the fermentation of ethanol. The H_2 served as an excellent electron donor and a reducing agent, which helped in the microbial reduction of the Pd. The acetate acted as a precursor for methanogenesis but served as a poor electron donor. The removal of Pd(II) was attributed to biosorption by the microbial biomass. The SEM images showed the presence of aggregates on the surface of the microbes, while EDS analysis revealed the presence of Pd in the product obtained from R1, as seen in Figure 10.1. The XRD analysis further confirmed the presence of Pd(0) in the samples.

The degradation of PCP using a UASB reactor was conducted by Shen et al. (2005). The activated sludge was obtained from the wastewater treatment plant. The synthetic wastewater containing PCP (10 g/L) prepared in methanol was used in the experiment. The synthetic water and PCP were stored at 4°C. Several nutrients, such as sucrose, peptone, meat extract, and other trace elements, were added. Additionally, sodium carbonate was added to maintain the buffering capacity. The PCP was added to the bottom of the UASB reactor using a peristaltic pump. The anaerobic sludge was obtained from the wastewater treatment plant. Reactor 1 (#1) was anaerobic sludge from a citric acid plant containing only PCP, while reactor 2 (#2) was seeded with a mixture of anaerobic sludge from a citric acid plant containing 2-CP, 3-CP, 4-CP, and PCP for 6 months with a ratio of 1:1:1:1. The dichlorination and degradation of the PCP were noted in the anaerobic sludge containing chlorophenol. The reactor containing sludge without chlorophenol did not show any PCP dechlorination and degradation. The UASB reactor containing PCP degrading anaerobic sludge could treat PCP at a concentration of 170–181 mg/L with a loading rate of

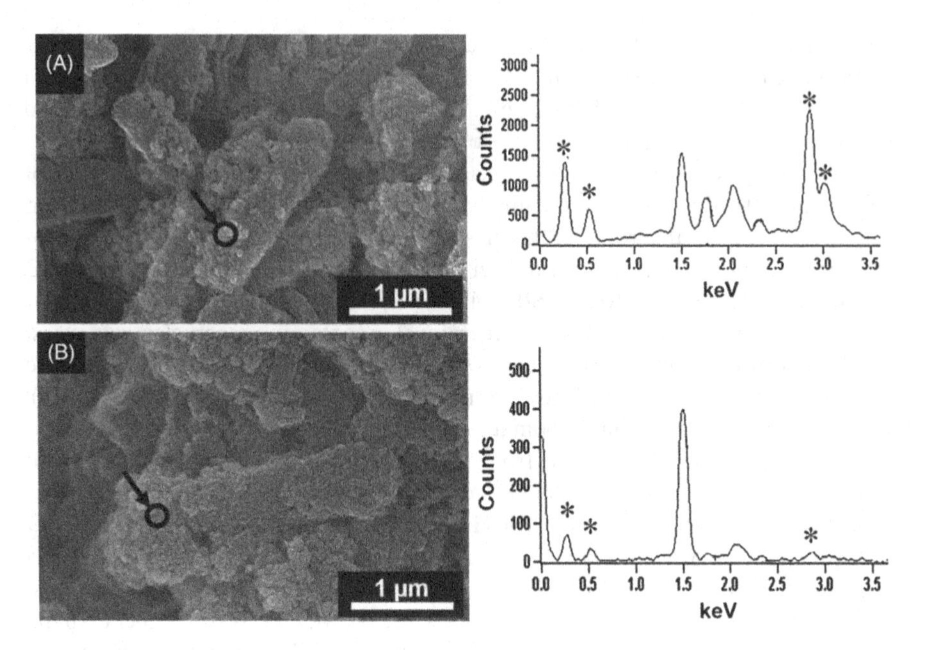

FIGURE 10.1 On the left, SEM images of microbial samples obtained from UASB reactor R1 fed with ethanol (A) and UASB reactor R2 fed with acetate (B) after 55 days of operation. On the right, energy-dispersive X-ray spectroscopy diagrams correspond to the samples shown in the SEM images. The circle and arrow indicate the point of analysis. The asterisks (*) indicated the peaks corresponding to palladium as detected in the EDS analysis [images were collected on Hitachi S-4800 Type II/Thermo NORAN NSS EDS at 30 kV, magnification (A) 30,000× and (B) 50,000×]. (Reprinted with permission from Pat-Espadas et al. (2016). Continuous removal and recovery of palladium in an upflow anaerobic granular sludge bed (UASB) reactor. J. Chem. Technol. Biotechnol. 91(4), 1183–1189. Copyright © 2015 Society of Chemical Industry.)

200–220 mg/L/d. The percent COD removal rate, percent PCP removal, and HRT of reactor #1 were 86.72%, 89.2–100%, and 5.8–1.2 d, respectively, with an operation period of 10–30 d, while it was 97.56%, 99.6–100%, and 1.1–0.9 d when operated for 35–75 d. The PCP removal efficiency, COD removal efficiency, and HRT of reactor #2, when operated for 10–30 d, were 90–100%, 85.75%, and 4.5–1.1 d, respectively. The percent COD removal was 96.35%, and the percent PCP removal was 97.4–100% with an HRT of 1.0–0.85 d when operated for 35–75 d.

In another study carried out by Show et al. (2020), the anaerobic granular sludge blanket (GSB) reactor was used to treat acrylic acid polluted wastewater. The acrylic acid wastewater was obtained from the polymer production plant. The COD generated was 110,000 mg/L. The GSB reactor was seeded with anaerobic sludge containing 2000 mg/L of VSS obtained from a chemical wastewater treatment plant. The feed was added from the bottom of the GSBs. The flow was adjusted across the microbial blanket formed by anaerobic bacteria. After the microbial acclimatization, the loading rate was maintained at around 10%. Thereafter, it was increased to 50% within 6 months to yield biogas and reduce COD. A significant increase in the COD of the effluent equivalent to 1900 mg/L was recorded on day 103. Overall, 95% of COD loading up to 9800 mg/L and 3074 kg/d was removed by the GSBs. The final effluent, ranging from 173 to 278 mg COD/L, conformed to the public sewer limits of 500 mg/L. The granules of the sludge, ranging from 0.2 to 1.0 mm, were formed during the startup. It was then distributed as tiny granules of 1.2–3.1 mm over a few years. This resulted in better stability. The granules aid in the growth of various bacterial populations, which are responsible for the degradation of complex pollutants.

The anaerobic reactor with a sludge blanket at the bottom and submerged cubes of polyurethane foam on the upper side was used for the treatment of wastewater collected from the slaughterhouse. Borja et al. (1998) used a reactor of 9.7 cm diameter and 69 cm height for the study. 1000 L of wastewater was stored at –20°C while the reactor was allowed to operate at 35°C. The biomass concentration of VSSs was 10.1 g/L. The OLR of 2 g COD/L/d with an HRT of 3 d was fed into the reactor. After 20 d of the acclimatization period, the experiments were carried out. In the initial five phases, the COD was increased to 10.41 g/L from 3.74 g/L at an HRT of 1.5 d. Various parameters like COD, BOD, total organic carbon (TOC), TS, VS, non-volatile solids (NVS), phosphorous, and pH were evaluated. The COD removal of 96.2% was achieved in phase 5, while it was 275 mg/L in phase 3. At an OLR concentration of 20.82 g COD/L/d, 93.4% of COD removal was obtained at an HRT of 0.5 d. The electron micrographs showed an increase in the size of granules up to 5.1 ± 1.4 mm. The concentration of VSS increased from 10.1 to 15.5 g/L. The production of methane was 0.345 L CH_4 STP per gram of COD removed, achieved at an HRT of 0.5 d.

10.3 CONTINUOUSLY STIRRED TANK REACTOR

A laboratory-scale stirred tank bioreactor was used to remove the sulfur black dye from the wastewater generated from the textile industry. Andleeb et al. (2010) used *Aspergillus terreus* SA3 in this study for treating the wastewater. The fungus was able to decolorize and degrade the dye. The fungal inoculum was prepared using Sabouraud dextrose agar (SDA). A spore suspension of 20 mL along with 0.05% Tween20 was used. The pH was adjusted to 5. The lab scale stirred tank bioreactor was set up with a working volume capacity of 2000 cm^3. The reactor was operated at pH 5 in a continuous mode. The bioreactor was seeded with a 5% spore suspension along with the simulated textile effluent (STE) at a 50 mg/L concentration. The retention time was 24–72 h when it was fed with various concentrations (100, 150, 200, and 300 mg/L) of dye every time. It was observed that a maximum of 84.53% of color removal was obtained within 1 d with HRT of 24 h and 50 ppm of sulfur black dye. The dye removal decreased to 80.06%, 71.55%, 54.55%, 41.89%, and 30.09% with an increase in the concentration of dye from 100, 150, 200, 300, and 500 ppm of dye, respectively, when operated for 72 h. The highest COD removal of 75.24% was obtained at a low concentration of dye (50 ppm). Low COD removal of 44.23% and 44.15% was noted with 300 and 500 ppm concentrations of dye, respectively. The reduction in the BOD up to 66.50% was obtained with a 200-ppm concentration of dye at 24-h HRT.

In the study conducted by Mohite and Salimath (2020), the wastewater collected from the distilleries was treated using a continuously stirred tank bioreactor (CSTR) of 20 L capacity along with 12 L liquid and 8 L gas volume. Activated seed culture with a dilution factor of 3 was used in the process. The wastewater was continuously added into the reactor at a COD rate of 0.01 kg/d. The water was rich in total dissolved solids (TDS) and TSS, while the COD and BOD were 126,000 and 57,000 mg/L, respectively. The AD was carried out using various loading rates equivalent to 0.01 to 0.16 kg COD/d. The temperature played a very crucial role in the efficiency of the anaerobic treatment. The concentration of mixed liquor-suspended solids (MLSS) was affected by the temperature. The highest COD removal recorded was up to 72% at 37°C ± 1°C with 36,000–44,000 mg/L MLSS. The OLR of 9.166 kg COD/m^3/d showed a maximum COD removal of 72.66%. The production of biogas ranging from 29 to 32 L was accompanied by COD removal up to 72–73%. The increase in temperature from 32°C to 37°C showed a COD removal of 72% with an HRT of 14 d.

The heavy metal removal efficiency was investigated using a sulfate-reducing bacteria (SRB) consortium in semi-continuous stirred tank reactors (CSTR) by Kieu et al. (2011). The inoculum comprised of Cr (82 mg/L), Ni (64 mg/L), Fe (76 mg/L), Zn^{2+} (18.5 mg/L), Cu^{2+} (38 mg/L), Mn^{2+} (76 mg/L), and SO_4^{2-} (202 mg/L) was used for the treatment in five reactors. The temperature of the reactors was maintained at 30°C in a hot water bath and agitated at 400 rpm. The reactors were soaked for 72 h in HNO_3 (3 M), followed the addition of synthetic wastewater using a peristaltic pump with a flow rate of 100 mL/d. The anaerobic condition was maintained using sterile nitrogen gas. The synthetic wastewater with KH_2PO_4 (0.5 g/L), NH_4Cl (1.0 g/L), Na_2SO_4 (3.7 g/L), sodium lactate (4.42 g/L), and trisodium citrate (0.3 g/L) had a pH 6.0 ± 0.2. The bioreactors were seeded with 10% v/v of the SRB consortium and incubated for 9 d. The reactors were continuously seeded with wastewater with various elements such as Cu^{2+}, Zn^{2+}, Ni^{2+}, and Cr^{6+} at concentrations of 30, 60, 90, 120, and 150 mg/L, respectively. The loading rate of each metal in reactors R1, R2, R3, R4, and R5 was increased up to 1.5, 3, 4.5, 6, and 7.5 mg/L/d, respectively and operated at an HRT of 12 weeks. The heavy metal removal efficiency for Cu^{2+} was 96–100%, while it was 94–100% for Zn^{2+} and Ni^{2+} and 96–100% for Cr^{6+} in reactors R1, R2, and R3 with loading rates of 1.5, 3, and 4.5 mg/L/d, respectively. The removal efficiency of Cu^{2+}, Zn^{2+}, Ni^{2+}, and Cr^{6+} in reactor R4 was 98–100% with a loading rate of 6 mg/L/d. The metal removal efficiency of reactor R5 was reduced to 78–91% from 98–100% while no precipitation was noted after 4 weeks of operation.

The treatment of wastewater containing high organic content was reported in the study carried out by Fuchs et al. (2003) using a stirred tank reactor. The reactor was attached to an external filtration device that was used for the treatment of wastewater from different sources, such as the vegetable processing industry, slaughterhouses, and artificial wastewater. The artificial wastewater with peptone, glucose, yeast extract, and NaCl was used and prepared freshly twice a week, while another wastewater was stored at 4°C. The sludge was collected from an anaerobic digester and maintained at 30°C. Several parameters, such as time of operation, degradation, pH, production of gas, and amount of methane, were monitored along with VFA generation. In the treatment of artificial wastewater, a concentration of sludge of 20–5 g/L with a loading rate of 20 g COD/L/d was achieved. The yield of methane was obtained in the range of 0.20–0.30 L_n of methane/g of COD. In the wastewater of sauerkraut brine, the concentration of solids was maintained from 22 to 38 g/L. The concentration of VFAs was 1700 mg/L after 49 d. In the case of the slaughterhouse wastewater, the loading rate of 7.4 g of COD/L/d was noted, while the VFAs of 3000 mg/L were obtained after 60 d. The complete removal of COD was achieved at a loading rate of 4.3 g/L/d.

The performance of CSTR in the treatment of petrochemical wastewater (PWW) with and without biogas recirculation was evaluated in the study conducted by Siddique et al. (2014), as illustrated in Figures 10.2 and 10.3. The recirculation of biogas was examined at a different rate of 10.15, 15.81, 21.14, and 36.25 L/d for 100 d. It was observed that the efficiency of COD removal and generation of bio-methane improved after the increase in the recirculation of biogas. The COD removal of 98.5% and the removal of VFA of 94% were achieved in the CSTR along with the biogas recirculation with an HRT of 9 d. The generation of biogas and methane of about 9.2 ± 0.5 and 6.08 ± 0.5 m^3/m/d, respectively was obtained. The methane and biogas generation efficiency increased by 26% and 55%, respectively, compared to CSTR without biogas recirculation. The retention efficiency of biomass of CSTR showed an increase of 16.78%, 20%, and 25%, with an increase in the rate of recirculation of biogas of 10.15, 15.81, and 24.14 L/d, respectively. After 154 d, a biomass generation of 82% was achieved with 88% COD and 95% BOD removal.

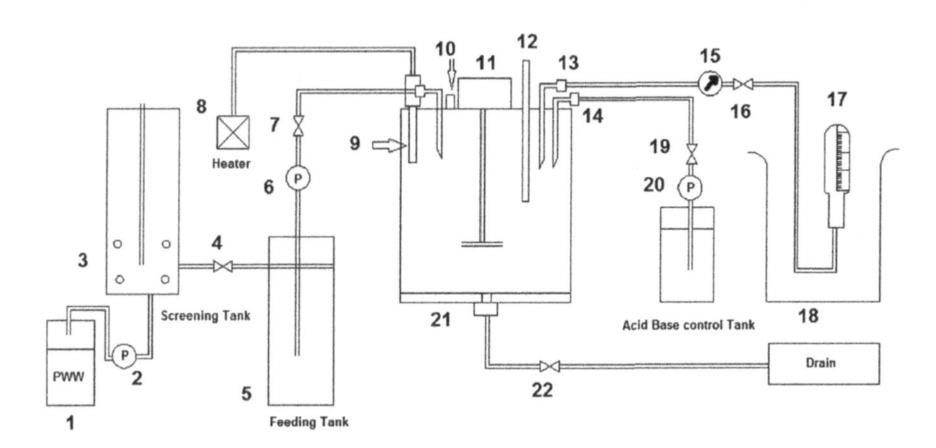

FIGURE 10.2 Experimental setup of CSTR. 1 – raw PWW dosing zone; 4, 7, 16, 19, 22 – control valve; 3 – screening tank, 5 – feeding tank; 2, 6, 20 – peristaltic pump; 8 – heater; 9 – temperature sensor; 10 – pressure controller; 11 – stirrer motor with stirrer; 12 – pH sensor; 13 – gas collection wire; 14 – acid base control wire; 15 – biogas flow meter; 17 – biogas collection tank; 21 – process reactor (capacity 4.5 l, working volume 2.7 L); 18 – water displacement system. (Reprinted with permission from Siddique, N.I., Munaim, M.S.A, Wahid, Z.A., 2014. Role of biogas recirculation in enhancing petrochemical wastewater treatment efficiency of continuous stirred tank reactor. J. Clean. Prod., 91, 229–234. Copyright © 2014 Elsevier Ltd.)

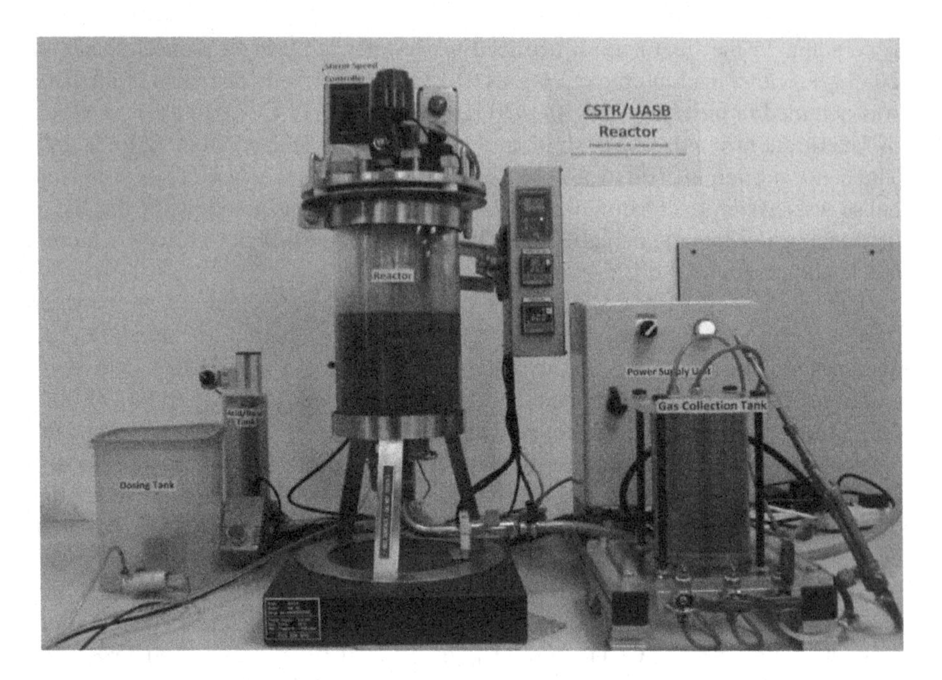

FIGURE 10.3 Photograph of experimental setup. (Reprinted with permission from Siddique, N.I., Munaim, M.S.A, Wahid, Z.A., 2014. Role of biogas recirculation in enhancing petrochemical wastewater treatment efficiency of continuous stirred tank reactor. J. Clean. Prod., 91, 229–234. Copyright © 2014 Elsevier Ltd.)

10.4 EXPANDED GRANULAR SLUDGE BED DIGESTION

The hollow fiber membrane filtration unit was attached along with the expanded granular sludge bed (EGSB) for treating domestic wastewater in the study carried out by Chu et al. (2005). The reactor was maintained for 7 months at 11–25°C with an HRT of 3.5–5.7 h. The reactor was seeded with granular sludge with a 15.8 g VSS/L granule concentration. The COD was in the range of 383–849 mg/L, while the ratio of the BOD to COD was 81–84%. The TOC concentration was in the range of 147–264 mg/L. The COD and TOC removal was 73–88% and 70–80%, respectively. The COD removal efficiency was independent of the HRT at 15°C for 3.5–5.7 h, while it was influenced by membrane filtration. The temperature also affected the rate of the anaerobic process. The methanogenic activity of sludge was low at low temperatures. The size of the granules was checked during the 7-month operation period. The diameter of the granules was about 1.15 mm, with 38 wt% of the total volume. At 120th d, the size of the granules was around 9% at the bottom and 25% in the middle. The SEM images showed that the granules have some cavities, which aid in gas production.

The treatment of petrochemical wastewater was reported using an EGSB bioreactor in the study conducted by Liang et al. (2019). The petrochemical wastewater had 4649 ± 651 mg/L of COD and more than 50 mg/L of petrochemicals. The reactor was divided into four units, and the treatment was done in an EGSB of 2200 m^3

with an HRT of 62.8 h. The reaction was allowed to run for more than 450 d. The COD removal was 85.6% ± 2.5%, while the removal of petrochemicals was about 81.5% ± 4.8%. The petrochemical wastewater was analyzed using GC/MS. The presence of various organic compounds such as 2,5-hexanedione, 2,5-bishydroxymethyl tetrahydrofuran, 3-methyl-2-cyclopenten-1-one, 3,5-diacetyl-2-pentone, [1-cyclohexyl-2,2-dimethyl] propyl acetate, 4,4-dimethyl-2-pentanone, 2-methyl-1-propanol, 5-methyl-2-hexanone, and 2,8-dimethyl-5-nonanone were obtained, which are non-biodegradable in nature. The microbial diversity present in the EGSB after 350 d of operation was evaluated, which revealed the predominance of *Proteobacteria* and *Firmicutes*. The dominant species was *Desulfomicrobium baculatum*, a sulfate-reducing bacteria. These bacteria made the excess amount of sulfate available in EGSB and added sulfate to the reactor. The bacteria oxidized pyruvate and lactate in the presence of sulfate.

The low-strength domestic wastewater was treated using an EGSB reactor with a 28.6 L capacity in the study conducted by Xu et al. (2018). The synthetic wastewater containing potato starch, peptone, sucrose, broth of beef, carbamide, and several other components was treated. The reaction was kept for 135 d, which was divided into periods I, II, III, and IV. The anaerobic sludge of 12.0 g/L, along with mixed liquor volatile suspended solids (MLVSS) and liquor with suspended solids (MLSS), was used with an HRT of 4–8 h. The OLR increased to 2.7 kg COD/m^3/d from 1.35 kg COD/m^3/d. The COD removal in the first 24 d of period I was exponentially increased with HRT. The stability was achieved within 3 d in period II with 71.5% ± 2.3% COD removal. At the end of period III, 72.0% ± 0.8% COD removal was obtained due to the formation of the granular sludge. The COD removal was decreased in period IV as HRT was lowered from 5 to 4 h. The amount of acetic acid obtained in period I was 136.6 mg/L, while it was 39.0–102.2 mg/L in period III. The yield of propionic acid was 32.2 mg/L. The change in the characteristics of the sludge was determined along with MLSS and MLVS. The decrease in the MLSS was noted during the start-up phase, but the amount of OLR and MLSS increased in period II. The granules of <200 μm was formed in the reactor. The granulation was formed with an OLR of 1.35–1.80 kg COD/m^3/d in periods I and II. At HRT of 5–6 h, the number of granules (>2.0 mm) was increased to 29% ± 2% from 7% ± 1%, and the diameter was increased to 780 ± 150 μm from 140 ± 10 μm. The extracellular polymeric substances (EPS) released by the microbes played an important role in the granulation of sludge. The composition and distribution of tightly bound (TB) EPS and LB EPS were determined. A significant decrease in the LB-EPS from 11.3 mg/g VSS was obtained in period I, while it increased in period IV to 23.3 mg/g/VSS. The TB-EPS was increased to 53.3 mg/g/VSS from 21 mg/g/VSS. The composition of the microbial community was investigated, which revealed the presence of several bacterial and archaeal communities such as *Bacteroidetes*, *Proteobacteria*, *Firmicutes*, *Chloroflexi*, *Nitrospirae*, and *Spirochaetae*.

Liao et al. (2012) investigated the use of an EGSB reactor in the treatment of high-nitrate wastewater and the presence of microbial diversity in it. The synthetic wastewater composed of several components was treated in the EGSB reactor of 60 mm diameter, 910 cm height, and 3.02 L volume at 35°C ± 1°C in mesophilic conditions.

The inoculum was added to the EGSB reactor. The batch and continuous experiments were conducted, followed by a chemical analysis of the wastewater. After 5 weeks, the biomass concentration was 25.81 g/L MLVSS. The pH of the reactor was around 9.0 to 9.6. The nitrate nitrogen (NO_3-N) was removed within 24 h, while its total removal was 99.2% within 8 h and the accumulation was 216.63 mg/L. The effect of various parameters was checked. The optimum carbon/nitrogen ratio, liquid upflow velocity (V_{up}), and pH were 2.0, 3.0 m/h, and 6.2–8.2, respectively. The microbial diversity showed the dominance of *Proteobacteria*, *Firmicutes*, *Actinobacteria*, *Bacteroidetes*, and *Chloroflexi*.

In another study conducted by Yang et al. (2018), low-strength wastewater was treated using an EGSB reactor. The EGSB reactor was seeded with granular activated carbon (GAC), and the reactor was operated with an HRT of 8, 6, and 5, and a duration of 4 h with V_{up} ranging from 1.09 to 2.44 m/h. The reactor with a capacity of 28.6 L was inoculated with 9 L of sludge and operated with wastewater. The size of the granular sludge was in the range of 0.2–0.6 mm. At 69 d, the granular size was increased to 19.3% with a 0.6–1.0 mm size. At 135 d, the granules were >1 mm. The COD removal was 45.9–52.1%, which increased to 82.3% from 62.0% with an increase in the duration of the treatment from 13 to 30 d. Several acidogenic and hydrolytic microbial species release the VFAs, while the amount of acetic acid was 84%. Microbial species such as *Proteobacteria*, *Firmicutes*, *Spirochaetae*, *Nitrospirae*, *Actinobacteria*, *Synergistetes*, *Bacteroidetes*, *Chloroflexi*, *Planctomycetes*, and *Gemmatimonadetes* were most dominant.

10.5 PLUG FLOW ANAEROBIC DIGESTION

The AD of complex agro-industrial waste and cooking oil/animal fat was reported using a plug flow reactor (PFR) made up of plexiglass in a study conducted by Eftaxias et al. (2021). The PFR was operated using diluted poultry manure (DPM), used cooking oil (UCO), dairy manure, and animal fat. The reactor was operated with a high OLR of 21 g/L/d. The COD removal efficiency and VFA production along with biogas and methane were evaluated. The experiment was conducted in a set of three periods, where period I was substrate with UCO, and periods II and III had dairy manure with animal fat. The concentration of COD was adjusted using water. The PFR was 130 cm long, with a liquid level of 12 cm and a volume of 20 L. The temperature was maintained at 36.7°C ± 1.6°C, while the HRT decreased to 5.5 d from 20 d in period I and 30 d from 10 d in periods II and III. The amount of lipids in period I was 27% of the total COD, while it was 49% and 29% in the case of periods II and III, respectively. The COD concentration of the influent supernatant was 55% for period I and 38% for periods II and III. The HRTs for periods I, II, and III were 7.5 ± 1.5, 6.1 ± 1.6, and 6.0 ± 1.8 d, respectively. The biogas production rate (BPR) was 3.4 ± 0.8 L/d when the methane (CH_4) production varied from 63% to 78%. The CH_4 was <70% in period I. The amount of acetic and propionic acid decreased from 2 to 0.2 g/L. The COD removal was between 70% and 80%.

Rossi et al. (2022) investigated the AD of municipal solid waste using a PFR. The organic fraction of municipal solid waste (OFMSW) was digested to recover the VFAs and biogas through thermophilic AD. The pilot-scale PFR of 37 L volume was operated at 55°C ± 2°C with an OFMSW of 260 kg for 150 d. The amounts of proteins, carbohydrates, lipids, lignin, and cellulose were 14.69% ± 2.95%, 35.12% ± 7.53%, 1.63% ± 0.46%, 17.84% ± 3.62%, and 17.19% ± 5.43%, respectively. The concentration of VFAs increased with a decrease in HRT from 22 to 16 d. The VFA concentration was 4421.8 mg/L at HRT of 16 d, while the biogas production was 145 NL biogas/d. The bacterial community of *Defluviitoga* sp. was the most abundant, accounting for 72.7% of the total bacterial population, followed by 12.73% of unclassified bacterial species, 5.52% of unidentified *Lentimicrobiaceae*, and 4.36% of *Proteiniphilum*, as evident from Figure 10.4. These microbial communities were speculated to be responsible for the hydrolysis of the complex polysaccharides.

In a study conducted by Gorecki et al. (1993), the piggery wastewater was treated using an inclined PFR with a 5 L volume anaerobically under steady-state conditions at 35°C. The reactor was fed with anaerobic sludge collected from the piggery waste

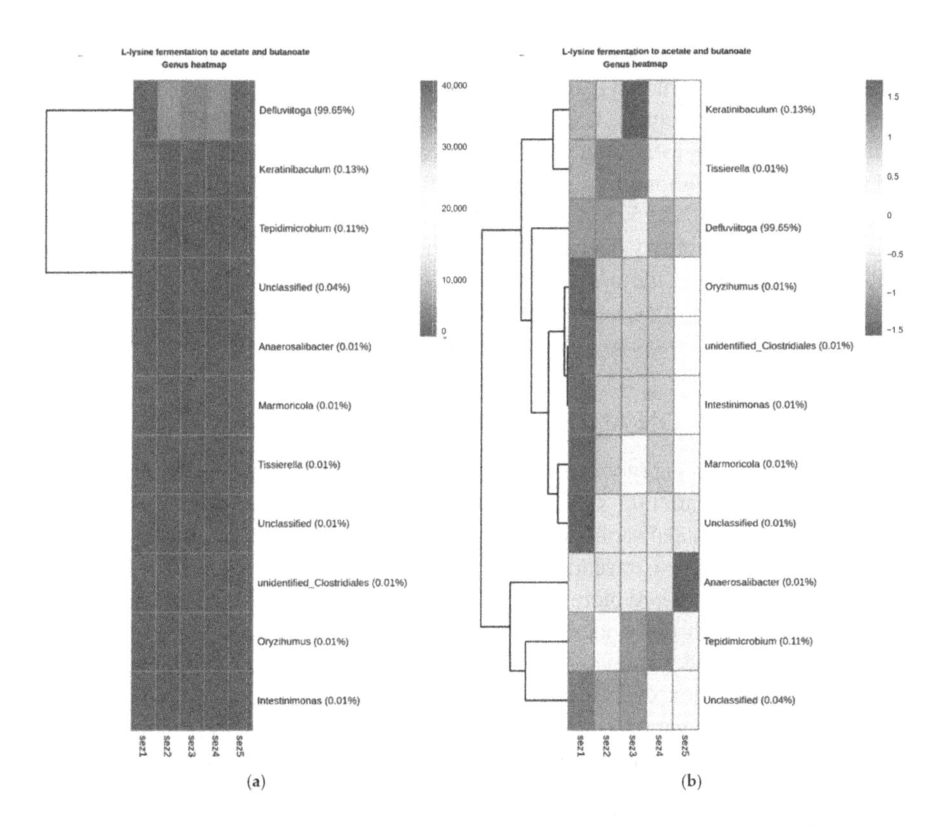

FIGURE 10.4 Functional heatmaps of L-lysine fermentation to acetate and butanoate: (a) absolute counts and (b) autoscaled counts, aggregated at the genus level. (Reprinted from Rossi, E., Becarelli, S., Pecorini, I., Gregorio, S., Iannelli, R., 2022. Anaerobic digestion of the organic fraction of municipal solid waste in plug-flow reactors: Focus on bacterial community metabolic pathways. Water, 14 (2), 195. (Open Access)).

storage tank. The TS was 7.5% at the start-up phase. The duration of the adaptation phase was 1 month, with an HRT of 60–30 d. The OLR of 5 g COD/L/d increased, followed by reduction, and again increased to 10 g COD/L/d. The concentration of VFAs was around 4 g/L during the first phase. The VFAs of the effluent were lower than 1 g/L during the first phase, while they were less than 200 mg/L at the end. The yield of CH_4 at 30°C was 0.35 L/g COD removed. The removal efficiency of organic matter was 60%, while the level of ammonia nitrogen was 2 g/L at pH 8.5. The granulation was observed within one month, with sizes ranging from 1 to 3 mm diameter at 46% VS and 156 g/kg of TS.

10.6 ANAEROBIC MEMBRANE BIOREACTOR

The swine wastewater was treated using a flat sheet AnMBR with an 8-L working volume in the study conducted by Jiang et al. (2020). The wastewater was collected, filtered using a 40-mm mesh to remove the unwanted materials, and stored at 4°C. The inoculum was collected from the mesophilic anaerobic digester. The digester was operated for 25 d HRT. The microfiltration membrane, made up of chlorinated polyethylene with a size of 0.2 μm, was put inside the reactor. The AnMBR was run under 5, 3, 2, and 1 d of HRT. The TS, biogas, and CH_4 production increased with the change in HRT. The CH_4 increased from 0.75 to 0.81, 0.96, and 1.63 L/(L.d) at an HRT equivalent to 5, 3, 2, and 1 d, respectively. The TCOD was 14.9 g/L, while the TCOD/VS ratio was 3.5. The CH_4 was 76–77% with 20.5–22.6% CO_2 at pH 8.1, while the ammonium nitrogen was 0.5–1.2 g/L throughout the process. The removal of VS at HRT of 5, 3, 2, and 1 d was 76%, 64%, 57%, and 40%, respectively. The methanogenic activity decreased under short HRT. With the shortening of SRT/HRT, the presence of the *Methanocorpusculum* became dominant in the sludge, while *Methanosaeta* dominated in the biofilm. After 4 months, membrane fouling was observed, which increased the transmembrane pressure.

10.7 CONCLUSION AND FUTURE PERSPECTIVES

The anaerobic treatment of wastewater is a promising approach, as both energy consumption and generation of sludge are reduced in this process. It does not require expensive chemicals and sophisticated equipment. It also shows significant potential for energy generation. Conventional anaerobic systems have a long HRT, but the newly developed advanced bioreactors show the removal of organic matter in a short HRT and yield high amounts of biogas and biofuels. The treatment efficiency is influenced by various parameters such as temperature, pH, OLR, SRT, the velocity of upflow, distribution of size, and gas-liquid-solid phase, which should be thoroughly optimized to get maximum efficiency. It also depends on the microbial community growing inside the bioreactor. Hence, identification of microbial diversity and their associated metabolomics will help establish the cellular events during the process (Bloch and Ghosh, 2022). Further, investigation of the molecular events and genetic study will help in finding the genes involved in the process that can be used for developing genetically engineered microbes with high anaerobic treatment efficiency. The microbial biomass generated during this treatment process can be used for the production of biochar, which has high potential for removing hazardous dyes, heavy metals, and toxic organic compounds (Ghosh and Sarkar, 2022). The components of the reactor play a crucial role in the treatment process, including retention sludge capacity, fluidization quality, and the formation of granules. However, the bioreactors require a scale-up along with an evaluation of their performance.

In view of the background, it can be concluded that the anaerobic treatment of wastewater can be a powerful strategy for treating industrial as well as domestic wastewater in order to address environmental pollution.

REFERENCES

Abdelgadir, A., Chen, X., Liu, J., Xie, X., Zhang, J., Zhang, K., Wang, H., Liu, N., 2014. Characteristics, process parameters, and inner components of anaerobic bioreactors. BioMed. Res. Int. 2014, 841573.

Andleeb, S., Atiq, N., Ali, M.I., Razi-Ul-Hussnain, R., Shafique, M., Ahmad, B., Ghumro, P.B., Hussain, M., Hameed, A., Ahmad, S., 2010. Biological treatment of textile effluent in stirred tank bioreactor. Int. J. Agric. Biol. 12(2), 256–260.

Bloch, K., Ghosh, S., 2022. Cyanobacteria mediated toxic metal removal as complementary and alternative wastewater treatment strategy. In: Kumar, V., Kumar, M. (Eds.) Integrated Environmental Technologies for Wastewater Treatment and Sustainable Development, Elsevier. pp. 533–547.

Borja, R., Banks, C.J., Wang, Z., Mancha, A., 1998. Anaerobic digestion of slaughterhouse wastewater using a combination sludge blanket and filter arrangement in a single reactor. Bioresour. Technol. 65(1–2), 125–133.

Chu, L.-B., Yang, F.-L., Zhang, X.-W., 2005. Anaerobic treatment of domestic wastewater in a membrane-coupled expended granular sludge bed (EGSB) reactor under moderate to low temperature. Process Biochem. 40(3–4), 1063–1070.

Eftaxias, A., Georgiou, D., Diamantis, V., Aivasidis, A., 2021. Performance of an anaerobic plug-flow reactor treating agro-industrial wastes supplemented with lipids at high organic loading rate. Waste Manag. Res. 39(3), 508–515.

Fuchs, W., Binder, H., Mavrias, G., Braun, R., 2003. Anaerobic treatment of wastewater with high organic content using a stirred tank reactor coupled with a membrane filtration unit. Water Res. 37(4), 902–908.

Ghosh, S., Sarkar, B., 2022. Nanotechnology for advanced oxidation based water treatment processes. In: Shah, M.P., Bera, S.P., Töre, G.Y. (Eds.) Advanced Oxidation Processes for Wastewater Treatment: An Innovative Approach, CRC Press. pp. 79–91.

Gorecki, J., Bortone, G., Tilche, A., 1993. Anaerobic treatment of the centrifuged solid fraction of piggery wastewater in an inclined plug flow reactor. Wat. Sci. Tech. 28(2), 107–114.

Jiang, M., Westerholm, M., Qiao, W., Wandera, S.M., Dong, R., 2020. High rate anaerobic digestion of swine wastewater in an anaerobic membrane bioreactor. Energy. 193, 116783.

Keyser, M., Witthuhm, R.C., Ronquest, L.C., Britz, T.L., 2003. Treatment of winery effluent with upflow anaerobic sludge blanket (UASB) – Granular sludges enriched with *Enterobacter sakazakii*. Biotechnol. Lett. 25, 1893–1898.

Khanal, S.K., Giri, B., Nitayavardhana, S., Gadhamshetty, V., 2017. Anaerobic bioreactors/digesters: Design and development. In: Lee, D.-J., Jegatheesan, V., Ngo, H.H., Hellenbeck, P.C., Pandey, A. (Eds.) Current Developments in Biotechnology and Bioengineering: Biological Treatment of Industrial Effluents, Elsevier. pp. 261–279.

Kieu, H.T.Q., Muller, E., Horn, H., 2011. Heavy metal removal in anaerobic semi-continuous stirred tank reactors by a consortium of sulfate-reducing bacteria. Water Res. 45(13), 3863–3870.

Liang, J., Mai, W., Tang, J., Wei, Y., 2019. Highly effective treatment of petrochemical wastewater by a super-sized industrial scale plant with expanded granular sludge bed bioreactor and aerobic activated sludge. Chem. Eng. J. 360, 15–23.

Liao, R., Shen, K., Li, A.-M., Shi, P., Li, Y., Shi, Q., Wang, Z., 2013. High-nitrate wastewater treatment in an expanded granular sludge bed reactor and microbial diversity using 454 pyrosequencing analysis. Bioresour. Technol. 134, 190–197.

Mal, J., Nancharaiah, Y.V., Maheshwari, N., van Hullebusch, E.D., Lens, P.N.L., 2017. Continuous removal and recovery of tellurium in an upflow anaerobic granular sludge bed reactor. J. Hazard. Mater. 327, 79–88.

Mohite, D.D., Salimath, S.S., 2020. Anaerobic biological treatment of distillery wastewater-study on continuous stirred tank reactor. IOP Conf. Ser.:Mater. Sci. Eng. 814, 012030.

Pat-Espadas, A.M., Field, J.A., Razo-Flores, E., Cervantes, F.J., Sierra-Alvarez, R., 2016. Continuous removal and recovery of palladium in an upflow anaerobic granular sludge bed (UASB) reactor. J. Chem. Technol. Biotechnol. 91(4), 1183–1189.

Rossi, E., Becarelli, S., Pecorini, I., Gregorio, S.D., Iannelli, R., 2022. Anaerobic digestion of the organic fraction of municipal solid waste in plug-flow reactors: Focus on bacterial community metabolic pathways. Water. 14(2), 195.

Shen, D.-S., Liu, X.-W., He, Y.-H., 2005. Studies on adsorption, desorption, and biodegradation of pentachlorophenol by the anaerobic granular sludge in an upflow anaerobic sludge blanket (UASB) reactor. J. Hazard. Mater. 125(1–3), 231–236.

Show, K.-Y., Yan, Y.-G., Zhao, J., Shen, J., Han, Z.-X., Yao, H.-Y., Lee, D.-J., 2020. Startup and performance of full-scale anaerobic granular sludge blanket reactor treating high strength inhibitory acrylic acid wastewater. Bioresour. Technol. 317, 123975.

Siddique, N.I., Munaim, M.S.A., Wahid, Z.A., 2014. Role of biogas recirculation in enhancing petrochemical wastewater treatment efficiency of continuous stirred tank reactor. J. Clean. Prod. 91, 229–234.

Tansel, B., Surita, S.C., 2014. Differences in volatile methyl siloxane (VMS) profiles in biogas from landfills and anaerobic digesters and energetics of VMS transformations. Waste Manage. 34(11), 2271–2277.

Xu, H., Liu, Y., Gao, Y., Li, F., Yang, B., Wang, M., Ma, C., Tian, Q., Song, X., Sand, W., 2018. Granulation process in an expanded granular sludge blanket (EGSB) reactor for domestic sewage treatment: Impact of extracellular polymeric substances compositions and evolution of microbial population. Bioresour. Technol. 269, 153–161.

Yang, B., Wang, M., Wang, J., Song, X., Wang, Y., Xu, H., Bai, J., 2018. Mechanism of high contaminant removal performance in the expanded granular sludge blanket (EGSB) reactor involved with granular activated carbon for low strength wastewater treatment. Chem. Eng. J. 334, 1176–1185.

Yu, B., Xu, J., Yuan, H., Lou, Z., Lin, J., Zhu, N., 2014. Enhancement of anaerobic digestion of waste activated sludge by electrochemical pretreatment. Fuel. 130, 279–285.

Zhang, W., Liu, F., Wang, D., Jin, Y., 2018. Impact of reactor configuration on treatment performance and microbial diversity in treating high-strength dyeing wastewater: Anaerobic flat-sheet ceramic membrane bioreactor versus upflow anaerobic sludge blanket reactor. Bioresour. Technol. 269, 269–275.

Index

Note: Locators in *italics* represent figures and **bold** indicate tables in the text.

B

C

D

K

L

M

U

W

X

Y

Z

Printed in the United States
by Baker & Taylor Publisher Services